INTERNATIONAL CENTRE FOR MECHANICAL SCIENCES

COURSES AND LECTURES - No. 327

# BIFURCATION AND STABILITY OF DISSIPATIVE SYSTEMS

EDITED BY

Q.S. NGUYEN

ECOLE POLYTECHNIQUE, PALAISEAU

SPRINGER - VERLAG WIEN - NEW YORK

Le spese di stampa di questo volume sono in parte coperte da
contributi del Consiglio Nazionale delle Ricerche.

This volume contains 65 illustrations.


Printed in Italy

In order to make this volume available as economically and as rapidly as possible the authors' typescripts have been reproduced in their original forms. This method unfortunately has its typographical limitations but it is hoped that they in no way distract the reader.

ISBN 3-211-82437-5 Springer Verlag Wien-New York
ISBN 0-387-82437-5 Springer Verlag New York-Wien

## *PREFACE*

*This book contains six lectures delivered at the International Centre for Mechanical Sciences, Udine, Italy in the session "Bifurcation and Stability of Dissipative Systems", June 1991.*

*The first theme concerns the plastic buckling of structures in the spirit of Hill's classical approach. Non-bifurcation and stability criteria are introduced and post-bifurcation analysis performed by asymptotic development method in relation with Hutchinson's work. Links with Koiter's elastic analysis are underlined. Some additional mathematical problems of plastic bifurcation are discussed, in particular the possibility of smooth bifurcation is considered. Some recent results on the generalized standard model are also given and their connection to Hill's general formulation is also be presented. Instability phenomena of inelastic flow processes such as strain localization and necking are discussed in the same spirit.*

*The second theme concerns stability and bifurcation problems in internally damaged or cracked solids. In brittle fracture or brittle damage, the evolution law of crack lenghts or damage parameters is time-independent like in plasticity and leads to a similar mathematical description of the quasi-static evolution. Stability and non-bifurcation criteria in the sense of Hill can be again obtained from the discussion of the rate response.*

*The book is intended for post-graduate students, researchers and engineers who are interested in bifurcation and stability analysis in anelasticity and represents a self-consistent treatise on plastic buckling with an unified presentation covering other standard time-independent processes such as instability and bifurcation problems in damage and fracture.*

*Chapter 1, by A. Benallal, R. Billardon and J. Geymonat, gives some general considerations on bifurcation and localization in time-independent materials. A full and complete analysis of the rate problem for incremental linear solid is carried out. The rate problem is formulated and discussed in the framework of the modern theory of linear elliptic boundary value problems.*

*Chapter 2, by Q.S. Nguyen, discusses bifurcation and stability of irreversible systems obeying to maximum dissipation principle in solid mechanics (friction, plasticity, fracture, and damage). General results on the quasi-static response such as uniqueness, bifurcation and stability are discussed. Illustrations in fracture or damage mechanics are considered: interacting linear cracks, plane crack of arbitrary shape, criteria of bifurcation stability in relation with the second variation of energy with respect to a geometric domain.*

*Chapter 3, by H. Petryk, is devoted to the theory of bifurcation and instability in time-independent plasticity. Hill's general theory of bifurcation and stability in time-independent inelastic solids is presented. The effect of formation of a vertex on the yield surface and linearization of the bifurcation problem in velocities is discussed. Several recent extensions of Hill's theory are given: energy aspects in the bifurcation theory, energy criterion of instability of a quasi-static deformation process.*

*Chapter 4, by A. Léger and M. Potier-Ferry, presents a general discussion on bifurcation and post bifurcation analysis of elastic-plastic beams and plates in the light of Hutchinson's results. The bifurcated branch is explored and some general generic bifurcation modes are proposed.*

*Chapter 5, by Triantafyllidis, addresses the possibility of smooth bifurcation and its illustration by simple examples. Bifurcation and post-bifurcation analysis is given through the description of first, second, and higher order problems.*

*Chapter 6, by V. Tvergaard, is devoted to the problem of tensile instabilities at large straints. Basic formulations for elastic-plastic and elastic-visco-plastic behaviour at finite strain are considered. Specific analyses of tensile instabilies at large strains are given: diffuse bifurcation modes, effect of visco-plastic behaviour, localization of plastic flow in shear bands, cavitation instabilities.*

*It is true that nowadays, the theory of plastic buckling is well understood and has reached a certain degree of mathematical perfection as its counterpart in elasticity. Yet, if elastic buckling has been the subject of several textbooks with strong mathematical ramifications such as the catastrophe theory ..., very few treatises have been devoted to plastic buckling in the literature! This book has been proposed to fill the gap in this spirit and to underline the fact that general methods of investigation in plastic buckling can also be successfully applied to the study of all time-independent behaviours such as brittle damage and brittle fracture.*

*The editor is indebted to the CISM staff for all organization facilities, in particular to Prof. S. Kaliszky who represents the Scientific Comity and to Prof. C. Tasso the CISM Edition.*

*Q.S. Nguyen*

# CONTENTS

# BIFURCATION AND LOCALIZATION IN RATE-INDEPENDENT MATERIALS. SOME GENERAL CONSIDERATIONS

**A. Benallal, R. Billardon and G. Geymonat**
**University of Paris 6, Cachan Cedex, France**

## ABSTRACT

This work deals with some aspects of bifurcation and localization phenomena for solids made of rate-independent materials. Only the theoretical developments are presented. Physical non-linearities (plasticity, damage, ...) and geometrical non-linearities are taken into account. The analysis is limited to quasi-static loadings. A full and complete analysis of the rate problem for incrementally linear solids is carried out. The first order rate problem is formulated and analysed in the framework of modern theory of linear elliptic boundary value problems. Three conditions are necessary and in the same time sufficient for this problem to be well-posed. These conditions are local in nature and are used to describe localization phenomena.

## INTRODUCTION

Bifurcations, instabilities and localization are very common phenomena in various fields of physics. They are very important in fluid and solid mechanics. This work deals with some aspects of these bifurcation and localization phenomena for solids made of rate-independent materials. In this paper, only the theoretical developments are presented. Physical non-linearities such as plasticity or damage and geometrical non-linearities are included for completeness. The analysis is further limited to quasi-static loadings.

These bifurcation and localization phenomena are studied by analysing the boundary value problems arising in a given solid mechanics context. More precisely, localization phenomena are treated as particular bifurcation modes and the bifurcation modes are obtained by the analysis of uniqueness of the first order rate boundary value problem and eventually by that of higher rate boundary value problems.

Although the general and complete results to come are given for the rate problem corresponding to incrementally linear solids, their implications for non-linear and particularly bi-linear type rate constitutive relations investigated here are underlined.

In chapter I, the constitutive equations that we have in mind are presented and some examples in different fields of solid mechanics are given. These will serve as pilot examples in the following chapters.

In chapter II, the general nonlinear initial boundary value problem is formulated. Then, a fundamental result of existence and uniqueness (valid under some particular circumstances) is recalled. Moreover various examples of loss of uniqueness for this problem are presented in order to give a first sight on bifurcation phenomena.

Chapter III is devoted to the presentation of the nonlinear first order rate boundary value problem as it arises in general situations. There again some general results of existence and uniqueness are provided.

In chapter IV, a full and complete analysis of the rate problem for incrementally linear solids is carried out. The first order rate problem is presented and formulated following the general guidelines developed by HILL [30, 32, 33]. Systematic use of modern theory of linear elliptic boundary value problems is made and one of the objectives of this paper is to make this theory more available in the mechanician community. Application of these results to the analysis of this first order rate problem reveals *three conditions that are necessary and in the same time sufficient* for this problem to be well-posed in a meaning to be given later. These conditions are local in nature, which allows their numerical implementation if desired (in finite element discretization for example); moreover, they have simple mechanical significations :

i) the first one is *the ellipticity condition*, by now very classical in many fields of continuum mechanics and is linked to bifurcation modes involving *jumps of the velocity gradient* and to *stationary acceleration waves* (e.g. HADAMARD [29], HILL [31], MANDEL [45] and RICE [56]).

ii) the second one, called *the boundary complementing condition* in the following, is a relation between the coefficients appearing in the boundary conditions and the constitutive properties at a given point of the boundary of a solid and is related to surface modes, interfa. bifurcation *modes decaying exponentially beneath the boundary of the solid*. It can also be given a dynamical interpretation by analogy to *stationary surface waves* such as RAYLEIGH waves [1] for instance in elasticity;

iii) the third one, present only when *the solid is heteregeneous* (i.e. made of different materials) and named hereafter *the interfacial complementing condition*, is concerned with interfaces between these different materials. Here again, a physical interpretation can be done in terms of interfacial modes of deformation, i.e. bifurcation *modes decaying exponentially beneath the interface and inside the different materials*. The dynamical counterpart here is the occurence of *stationary interfacial waves* by analogy to STONELY or LOVE waves[1] in elastic media.

The three conditions summarized above are given explicit forms for general constitutive properties and boundary conditions and can be used in many practical situtations. These conditions are also related to some bifurcation analyses carried out in the literature for simple boundary value problems but studied in their own merit. For sake of clearness, mathematical developments are avoided whenever possible; however, validity of the results is underlined as far as possible in order to keep rigor.

When these three conditions are met, the first order rate boundary value problem admits *at most a finite number of linearly independent solutions* which represent in classical words *diffuse modes of deformation*..

The ellipticity condition is frequently used to describe localization phenomena as proposed by RICE [55,56] but it is also now suggested to consider the two complementing conditions (at the boundaries and interfaces) as localization criteria at boundaries or interfaces of solids (see BENALLAL, BILLARDON, GEYMONAT [6,9] and NEEDLEMAN, ORTIZ [50] ).

The paper is closed by some remarks on the consequences of the ill-posedness of the rate-boundary value-problem associated to the loss of the above conditions. The current trends to overcome the corresponding shortcomings are also briefly underlined.

# I. CONSTITUTIVE EQUATIONS AND EXAMPLES

## I.1 CONSTITUTIVE EQUATIONS

In these notes, we will consider a general class of materials obeying the following rate constitutive equations :

$$\overset{\circ}{\sigma} = \mathbb{L} : \mathbb{D} = \begin{cases} \mathbb{E} : \mathbb{D} \\ \mathbb{H} : \mathbb{D} = \{ \mathbb{E} - \dfrac{\mathbb{E} : \mathbf{a} \otimes \mathbf{b} : \mathbb{E}}{h + \mathbf{b} : \mathbb{E} : \mathbf{a}} \} : \mathbb{D} \end{cases} \tag{I.1.1}$$

where $\overset{\circ}{\sigma}$ denotes any objective derivative of the Cauchy stress tensor, $\mathbb{D}$ the deformation rate and $\mathbb{L}$ the tangent modulus. This form of constitutive equations arises in many situations.

As an example, in the framework of thermodynamics of irreversible processes and under the assumption of small strain and isothermal conditions, the behaviour of the material is described by means of the free energy potential $\Psi(\varepsilon, \alpha)$ which depends on the strain tensor $\varepsilon$ and a set of internal variables $\alpha$ . These internal variables may be scalars, vectors or tensors. The Cauchy stress tensor $\sigma$ and the thermodynamical forces **A** associated to the internal variables are derived from the following state laws :

$$\sigma = \rho \frac{\partial \Psi}{\partial \varepsilon} \qquad\qquad \mathbf{A} = - \rho \frac{\partial \Psi}{\partial \alpha} \tag{I.1.2}$$

where $\rho$ denotes the mass density.

The second law of thermodynamics essentially requires that for every admissible process, the following mechanical dissipation inequality holds :

$$d = \mathbf{A} \bullet \dot{\alpha} \geq 0 \tag{I.1.3}$$

where symbol "•" denotes the adequate scalar product (e.g multiplication for scalars, scalar product for vectors and full contraction for tensors).
The irreversible behaviour of the material is given by the evolution of the internal variables.

This is determined by a yield function f(**A**; α) via the domains of reversibility $\mathcal{C}(\alpha) = \{\mathbf{A}; f(\mathbf{A}; \alpha) \leq 0\}$ and an inelastic potential F(**A**; α) via the normality rule :

$$\dot{\alpha} = \lambda \frac{\partial F}{\partial \mathbf{A}} \tag{I.1.4}$$

and the classical Kuhn-Tucker relations :

$$\lambda \geq 0 \ , \ f(\mathbf{A}; \alpha) \leq 0 \quad \text{and} \quad \lambda \dot{f}(\mathbf{A}; \alpha) = 0 \tag{I.1.5}$$

The functions f and F are chosen so that (I.1.3) is satisfied. When λ is strictly positive, it is computed by the consistency condition $\dot{f} = 0$ as

$$\lambda = < \frac{\mathbb{b} : \mathbb{E} : \dot{\varepsilon}}{h + \mathbb{b} : \mathbb{E} : \mathbb{a}} > \tag{I.1.6}$$

with the following notations :

$<x> = \mathrm{Max}(x,0)$

$$\mathbb{E} = \rho \frac{\partial^2 \Psi}{\partial \varepsilon \partial \varepsilon} \ , \quad \mathbb{b} : \mathbb{E} = -\frac{\partial f}{\partial \mathbf{A}} \bullet \rho \frac{\partial^2 \Psi}{\partial \alpha \partial \varepsilon} \ , \quad \mathbb{E} : \mathbb{a} = \rho \frac{\partial^2 \Psi}{\partial \varepsilon \partial \alpha} \bullet \frac{\partial F}{\partial \mathbf{A}} \tag{I.1.7}$$

$$h = -\frac{\partial f}{\partial \alpha} \bullet \frac{\partial F}{\partial \mathbf{A}} + \frac{\partial f}{\partial \mathbf{A}} \bullet \{ \rho \frac{\partial^2 \Psi}{\partial \alpha \partial \alpha} + \rho \frac{\partial^2 \Psi}{\partial \alpha \partial \varepsilon} : \mathbb{E}^{-1} : \rho \frac{\partial^2 \Psi}{\partial \varepsilon \partial \alpha} \} \bullet \frac{\partial F}{\partial \mathbf{A}} \tag{I.1.8}$$

The hardening modulus h is such that $h > h_{th} = - \mathbb{b} : \mathbb{E} : \mathbb{a}$ which denotes the so-called snap-back threshold. Moreover, the elastic moduli are such that $\mathbb{E}$ is positive definite. In this case, the tangent modulus $\mathbb{L}$ (see (I.1.1)) takes the following forms :

$$\mathbb{L} = \mathbb{E} \qquad \text{whenever } f(\mathbf{A}; \alpha) < 0 \tag{I.1.9}$$

$$\mathbb{L} = \begin{cases} \mathbb{E} & \text{whenever } f(\mathbf{A}; \alpha) = 0 \text{ and } \mathbb{b} : \mathbb{E} : \dot{\varepsilon} < 0 \\ \mathbb{H} = \mathbb{E} - \dfrac{\mathbb{E} : \mathbb{a} \otimes \mathbb{b} : \mathbb{E}}{h + \mathbb{b} : \mathbb{E} : \mathbb{a}} & \text{whenever } f(\mathbf{A}; \alpha) = 0 \text{ and } \mathbb{b} : \mathbb{E} : \dot{\varepsilon} \geq 0 \end{cases} \tag{I.1.10}$$

Modulus $\mathbb{H}$ is also called the tangent modulus in loading.

Let us remark that $\mathbb{L}$ obviously satisfies the minor symmetries $\mathbb{L}_{ijhk} = \mathbb{L}_{jihk} = \mathbb{L}_{ijkh}$ and in the case of associated behaviour (i.e. when f=F) $\mathbb{L}$ is also *symmetric*, i.e. it also has the major symmetries $\mathbb{L}_{ijhk} = \mathbb{L}_{hkij}$.

Other type of evolution laws for the internal variables can also be considered. For instance, in the case of non-smooth associated behaviour, it is classical to write :

$$\dot{\alpha} \in \partial\Phi_{\mathcal{C}(\alpha)}(\mathbf{A}) \qquad \text{(I.1.11)}$$

where $\partial\Phi_{\mathcal{C}(\alpha)}(\mathbf{A})$ is the subdifferential of the indicator function $\Phi_{\mathcal{C}(\alpha)}$ at point $\mathbf{A}$. The indicator function $\Phi_{\mathcal{C}(\alpha)}$ of the convex set $\mathcal{C}(\alpha)$ is defined by :

$$\Phi_{\mathcal{C}(\alpha)}(\mathbf{A}) = \begin{cases} 0 & \text{if } \mathbf{A} \in \mathcal{C}(\alpha) \\ +\infty & \text{if } \mathbf{A} \notin \mathcal{C}(\alpha) \end{cases} \qquad \text{(I.1.12)}$$

In this case, the following inequality holds :

$$\dot{\alpha} \bullet (\mathbf{A} - \mathbf{A}^*) \geq 0 \text{ for all } \mathbf{A}^* \in \mathcal{C}(\alpha) \qquad \text{(I.1.13)}$$

Moreover, if the convex set $\mathcal{C}(\alpha)$ is independent of $\alpha$, the maximum dissipation principle which generalizes HILL's maximum work principle is recovered. This corresponds to the generalized standard materials framework.

## I.2 EXAMPLES

### I.2.1 Elastic plastic materials with isotropic and kinematic hardenings

In this case, three internal variables are used : the plastic strain $\varepsilon^p$, the cumulated plastic strain p associated to isotropic hardening and the tensorial variable $\boldsymbol{\alpha}$ related to kinematic hardening. A classical model corresponds to the following expressions :

$$\rho\,\Psi(\varepsilon, \varepsilon^p, p, \boldsymbol{\alpha}) = \frac{1}{2}(\varepsilon - \varepsilon^p) : \mathbb{E} : (\varepsilon - \varepsilon^p) + \frac{1}{2}\boldsymbol{\alpha} : \mathbb{K} : \boldsymbol{\alpha} + w(p) \qquad \text{(I.2.1)}$$

$$f(\boldsymbol{\sigma}, R, \mathbb{X}) = J(\boldsymbol{\sigma} + \mathbb{X}) + \mu\,\mathrm{tr}(\boldsymbol{\sigma}) + R - \sigma_y \qquad \text{(I.2.2)}$$

$$F(\boldsymbol{\sigma}, R, \mathbb{X}) = J(\boldsymbol{\sigma} + \mathbb{X}) + \beta\,\mathrm{tr}(\boldsymbol{\sigma}) + R - \sigma_y \qquad \text{(I.2.3)}$$

where $R = -\rho \frac{\partial \Psi}{\partial p} = -w'(p)$ and $\mathbb{X} = -\rho \frac{\partial \Psi}{\partial \alpha} = -\mathbb{K} : \alpha$ describe the isotropic and kinematic hardenings respectively and where $J(a) = \sqrt{\frac{3}{2}\{a : a - \frac{1}{3}[tr(a)]^2\}}$ . Besides, $\sigma_y$ denotes the initial yield stress, $\mathbb{E}$ the matrix of elastic constants, $\mathbb{K}$ the matrix of hardening moduli, w(p) a given function, $\mu$ and $\beta$ material parameters. In this case, the tangent modulus in loading is given by :

$$\mathbb{H} = \mathbb{E} - \frac{\mathbb{E} : (\frac{3}{2}\frac{s + X}{J(\sigma + X)} + \beta\, 1) \otimes (\frac{3}{2}\frac{s + X}{J(\sigma + X)} + \mu\, 1) : \mathbb{E}}{w''(p) + \frac{9}{4}\frac{(s + X) : \mathbb{K} : (s + X)}{J^2(\sigma + X)} + (\frac{3}{2}\frac{s + X}{J(\sigma + X)} + \mu 1) : \mathbb{E} : (\frac{3}{2}\frac{s + X}{J(\sigma + X)} + \beta 1)} \tag{I.2.4}$$

where s denotes the stress deviator and 1 the second order unit tensor.
These constitutive equations are apt to describe the behaviour of some geomaterials (with $\mu$ and $\beta$ standing for the friction coefficient and the dilatancy factor respectively) and of many metals (with $\mu = \beta = 0$).

### I.2.2 Elastic perfectly plastic materials

In this case, the only internal variable used is the plastic strain $\varepsilon^p$. This model is formally recovered from the previous one by setting $\mathbb{K} = 0$ , $w(p) = 0$ and $\mu = \beta = 0$.
This situation schematizes what may occur with some ductile materials. Moreover this model is tractable for limit analysis.

### I.2.3 Isotropic elastic fully damageable materials

In this case, a scalar internal variable $D \in [0, D_c]$ is also introduced to describe the degradation of elastic materials. Classical models correspond to the following potentials :

$$\rho\, \Psi(\varepsilon, D) = \frac{1}{2}\, g(D)\ \varepsilon : \mathbb{E} : \varepsilon \tag{I.2.5}$$

$$F(Y, D) = f(Y, D) = Y - a(D) \tag{I.2.6}$$

where $Y = -\rho \frac{\partial \Psi}{\partial D} = -\frac{1}{2}\, g'(D)\ \varepsilon : \mathbb{E} : \varepsilon$. Function g(D) is decreasing and satisfies $0 \le g(D) \le 1$, $g'(0) = 1$, $g(0) = 1$ and $g(D) \to 0$ when $D \to D_c$ ; moreover, a(D) is an increasing function which satisfies $a(D) \ge 0$ . Usual examples are :

$$g(D) = 1 - D \qquad \text{with } D_c = 1 \tag{I.2.7}$$

$$g(D) = \exp(-D) \qquad \text{with } D_c = +\infty \tag{I.2.8}$$

$$g(D) = \frac{1}{1 + D} \qquad \text{with } D_c = +\infty \tag{I.2.9}$$

The corresponding tangent modulus in loading is the following :

$$\mathbb{H} = g(D)\,\mathbb{E} - \frac{g(D)\,\mathbb{E} : \varepsilon \otimes \varepsilon : g(D)\,\mathbb{E}}{\frac{g(D)}{(-g'(D))^3}[a(D)\{g(D)g''(D)-2(g'(D))^2\}-a'(D)g(D)g'(D)]+g(D)\ \varepsilon:\mathbb{E}:\varepsilon} \tag{I.2.10}$$

### I.2.4 Brittle elastic partially damageable materials

In this case, one scalar internal variable $D \in [0, 1]$ is also introduced to describe the partial degradation of elastic materials. Such a model corresponds to the following potential :

$$\rho\,\Psi(\varepsilon , D) = \frac{1}{2}\ \ \varepsilon : ( (1\text{-}D)\,\mathbb{E}_0 + D\ \mathbb{E}_1 ) : \varepsilon \tag{I.2.11}$$

and to the following reversibility domains :

$$\begin{cases} \mathcal{C}_{(D)} = \{\ Y \ /\ Y - k \leq 0\ \} & \text{if } 0 \leq D < 1 \\ \mathcal{C}_{(1)} = \mathbf{R} \end{cases} \tag{I.2.12}$$

where $Y = -\rho\,\frac{\partial\Psi}{\partial D} = \frac{1}{2}\ \varepsilon : (\,\mathbb{E}_0 - \mathbb{E}_1\,) : \varepsilon$ and $k > 0$.

The evolution of damage is given by relation (I.1.11) i.e. :

$$\dot{D} \in \partial\Phi_{\mathcal{C}(D)}(Y) \tag{I.2.13}$$

This implies in particular that $D = 1$ whenever $\underset{0 \leq \tau \leq t}{\text{Sup}}\ Y(\tau)\ > k$ .

This type of model can be applied to some reinforced brittle composites in some simple situations.

## II. THE GENERAL NONLINEAR INITIAL AND BOUNDARY VALUE PROBLEM

### II.1 FORMULATION

Let us now consider, under the assumption of small strain and isothermal conditions, a body occupying the bounded region $\Omega$ in $\mathbf{R}^n$ ( n = 1, 2, 3 for applications) and subjected during the time interval [0, T], where T > 0, to body forces $\mathbf{f}^d(x,t)$ in $\Omega$, surface tractions $\mathbf{F}^d(x,t)$ on a part $\partial_F\Omega$ and displacements $\mathbf{u}^d(x,t)$ on the complementary part $\partial_u\Omega$ of its boundary $\partial\Omega$ (whose unit outward normal is denoted by $\mathbf{n}$).

In order to find the behaviour of the body during the time interval [0, T], one has to solve the following nonlinear quasi-static initial and boundary value problem ($\mathcal{P}$) :

($\mathcal{P}$) *Find* $\mathbf{u}(x,t)$ , $\sigma(x,t)$ *and* $\alpha(x,t)$ *defined for* $x \in \Omega$ *and* $0 \le t \le T$ *satisfying* :

(a) the equilibrium equations :

$$\operatorname{div}\sigma + \mathbf{f}^d = 0 \qquad \text{in } \Omega \times [0, T] \qquad \text{(II.1.1)}$$

(b) the compatibility conditions :

$$\varepsilon(\mathbf{u}) = \frac{1}{2}[\nabla\mathbf{u} + (\nabla\mathbf{u})^T] \qquad \text{in } \Omega \times [0, T] \qquad \text{(II.1.2)}$$

(c) the constitutive equations given in section I.1.

(d) the boundary conditions

$$\begin{cases} \sigma . \mathbf{n} = \mathbf{F}^d & \text{on } \partial_F\Omega \times [0, T] \\ \mathbf{u} = \mathbf{u}^d & \text{on } \partial_u\Omega \times [0, T] \end{cases} \qquad \text{(II.1.3)}$$

(e) the initial conditions :

$$\begin{cases} \mathbf{u}(x,0) = \mathbf{u}_0(x) & \text{in } \Omega \\ \alpha(x,0) = \alpha_0(x) & \text{in } \Omega \end{cases} \qquad \text{(II.1.4)}$$

This nonlinear initial and boundary value problem is so general that there is little hope to derive an existence and/or uniqueness theorem for it. Conversely a numerical solution for this kind of problem can always be obtained by finite element codes with suitable local numerical integration of the constitutive equations. Howewer, this numerical solution may in some situations be meaningless and in particular completely mesh dependent. One of the objectives of these notes consists in deriving criteria for the limit of validity of such computations.

## II.2 A RESULT OF C. JOHNSON

Let us consider the case of Von Mises plasticity with kinematic hardening: in this case, two internal variables are used, viz. the plastic strain $\varepsilon^p$ and the tensorial variable $\alpha$. This model is recovered by setting $w(p) = 0$ and $\mu = \beta = 0$ in relation (I.2.1-3).

In this case, the inequality (I.1.13) reads

$$\dot{\sigma} : \mathbb{E}^{-1} : (\sigma^* - \sigma) + \dot{\alpha} : \mathbb{K} : (\alpha^* - \alpha) - \dot{\varepsilon} : (\sigma^* - \sigma) \geq 0$$
$$\text{for all } \; (\sigma, \alpha^*) \in \mathcal{D} = \{ (\sigma, \alpha^*) \; ; \; f(\sigma, -\mathbb{K} : \alpha^*) \leq 0 \} \qquad \text{(II.1.5)}$$

Let us remark that $\mathcal{D}$ is a closed, convex, non-empty set and that 0 belongs to its interior. Let us summarize without giving details the results first established by JOHNSON [38] and their improvements by SUQUET [65] and BLANCHARD, LE TALLEC [20] :

*Under a natural compatibility condition on the loads, and under the fundamental assumption that the tensors* $\mathbb{E}$ *and* $\mathbb{K}$ *are strictly definite positive, the nonlinear quasi-static initial and boundary value problem* ($\mathcal{P}$) *has a unique solution.*

## II.3 SOME EXAMPLES OF NON-UNIQUENESS

The fundamental assumption in Johnson's theorem is that $\mathbb{K}$ is strictly positive definite. (It is recalled that $\mathbb{E}$ is always strictly positive definite.) This assumption means that the material is hardening since this model is an associative plasticity model.
In order to have a flavour of the difficulties that may occur for associative non-hardening materials, we give below three elementary examples corresponding to the same uniaxial problem (see ERICKSEN [24] for nonlinear elasticity).

Let us consider a bar whose reference configuration is the interval $\Omega = [0, L]$ of the real line with $L > 0$. The bar is fixed at $x = 0$ and the displacement at $x = L$ is prescribed to

$u^d = At$ with $A > 0$. No body forces are considered.
Let us denote by $u(x,t)$, $\sigma(x,t)$ and $\varepsilon(x,t)$ the displacement, stress and strain fields respectively. These fields must satisfy the following set of equations :

$$\begin{cases} \varepsilon(x, t) = \dfrac{\partial u}{\partial x}(x, t) & 0 \le x \le L, \quad t \ge 0 \\ \dfrac{\partial \sigma}{\partial x}(x, t) = 0 & 0 \le x \le L, \quad t \ge 0 \\ u(0, t) = 0 \text{ and } u(L, t) = A\,t & t \ge 0 \end{cases} \tag{II.3.1}$$

These equations must be complemented with the constitutive equations and the initial conditions. We shall consider three examples for which the initial boundary value problem presents, after a uniqueness regime, an *infinite number* of solutions.

### II.3.1 Elastic-perfectly plastic material (SUQUET [65])

In this special case, the constitutive equations are the following :

$$\begin{cases} \sigma = E\,(\varepsilon - \varepsilon^p) \\ \dot{\varepsilon}^p \in \partial\Phi_{\mathcal{C}(\sigma)} \end{cases} \tag{II.3.2}$$

where $\mathcal{C}(\sigma) = \{\, \sigma \,/\, |\sigma| \le \sigma_y \,\}$, and where $E > 0$ and $\sigma_y$ denote the Young modulus and the yield stress respectively.
With the initial conditions $u(x, 0) = 0$, $\varepsilon^p(x, 0) = 0$, the following fields are solutions to problem $(\mathcal{P})$ :

(i) for $0 \le t < \dfrac{L\,\sigma_y}{E\,A}$

$$u(x, t) = \frac{A\,x\,t}{L} \tag{II.3.3}$$

$$\varepsilon^p(x, t) = 0 \tag{II.3.4}$$

$$\sigma(x, t) = \frac{E\,A\,t}{L} \tag{II.3.5}$$

(ii) for $t \ge \dfrac{L\,\sigma_y}{E\,A}$

$$u(x, t) = \int_{\frac{L\sigma_y}{EA}}^{t} \int_0^x \lambda(\tau,s)\, d\tau\, ds + \frac{x\,\sigma_y}{E} \qquad \text{(II.3.6)}$$

$$\varepsilon^p(x, t) = \int_{\frac{L\sigma_y}{EA}}^{t} \lambda(\tau,x)\, d\tau \qquad \text{(II.3.7)}$$

$$\sigma(x, t) = \sigma_y \qquad \text{(II.3.8)}$$

where $\lambda(\tau,s)$ is any summable function such that $\lambda(\tau, s) \geq 0$ and

$$\int_{\frac{L\sigma_y}{EA}}^{t} \int_0^L \lambda(\tau,s)\, d\tau\, ds = A\,t - \frac{L\,\sigma_y}{E} \qquad \text{(II.3.9)}$$

Among this infinity of solutions the choice $\lambda(\tau, s) = \frac{A}{L}$ gives $u(x, t) = \frac{A\,x\,t}{L}$.

### II.3.2 Brittle elastic partially damageable material (FRANCFORT, MARIGO [25])

Here, the uniaxial version of the model described in section I.2.4 is adopted. It reads

$$\sigma = \{(1-D)\,E_0 + D\,E_1\}\,\varepsilon \qquad \text{(II.3.10)}$$

With $Y(t) = \frac{1}{2}\,(E_0 - E_1)\,\varepsilon^2$ damage is governed by

$$D(t) = \begin{cases} 0 & \text{whenever } \sup\{Y(\tau),\ 0 \leq \tau \leq t\} \leq k \\ 1 & \text{whenever } \sup\{Y(\tau),\ 0 \leq \tau \leq t\} > k \end{cases} \qquad \text{(II.3.11)}$$

Here again, it is possible to construct an infinite number of solutions. To this end, let $\Sigma$ be arbitrarily chosen such that

$$\sqrt{\frac{2k}{E_0 - E_1}}\,E_1 \leq \Sigma \leq \sqrt{\frac{2k}{E_0 - E_1}}\,E_0 \qquad \text{(II.3.12)}$$

For $\frac{\Sigma L}{A E_0} \leq t \leq \frac{\Sigma L}{A E_0}$, let $\Omega(t)$ be an arbitrarily increasing family of measurable subsets of [0,L], such that its measure is Measure($\Omega(t)$) = a(t) L and $\Omega(t_0) \supset \Omega(t_1)$ for $t_0 \geq t_1$ with

$$a(t) = \frac{E_1 E_0}{E_0 - E_1} \left\{ \frac{A\, t}{\Sigma\, L} - \frac{1}{E_0} \right\} \tag{II.3.13}$$

We denote by $\Omega^c(t)$ its complement in [0,L] and by $\chi(x,t)$ the characteristic function of $\Omega(t)$. With the initial conditions u(x, 0) = 0, D(x, 0) = 0, the following fields are solutions to problem ($\mathcal{P}$) :

(i) for $0 \leq t \leq \frac{L\, \Sigma}{A\, E_0}$

$$u(x,t) = \frac{A\, x\, t}{L} \tag{II.3.14}$$

$$D(x,t) = 0 \tag{II.3.15}$$

$$\sigma(x,t) = \frac{A\, E_0\, t}{L} \tag{II.3.16}$$

(ii) for $\frac{L\, \Sigma}{A\, E_0} \leq t < \frac{L\, \Sigma}{A\, E_1}$

$$u(x,t) = \int_0^x \varepsilon(s,t)\, ds \text{ with } \varepsilon(s,t) = \frac{\Sigma}{E_1}\, \chi(s,t) + \frac{\Sigma}{E_0}(1 - \chi(s,t)) \tag{II.3.17}$$

$$D(t) = \begin{cases} 0 & \text{on } \Omega^c(t) \\ 1 & \text{on } \Omega(t) \end{cases} \tag{II.3.18}$$

$$\sigma(x,t) = \Sigma \tag{II.3.19}$$

(iii) for $t \geq \frac{L\, \Sigma}{A\, E_0}$

$$u(x,t) = \frac{A\, x\, t}{L} \tag{II.3.20}$$

$$D(x,t) = 1 \tag{II.3.21}$$

$$\sigma(x,t) = \frac{A\, E_1\, t}{L} \tag{II.3.22}$$

Let us remark that for each solution, the bar is not damaged in the phase (i), is partially damaged on an *arbitrary* subset of measure a(t)L during the phase (ii) and is completely damaged in the phase (iii). However the global response of the bar is always the same.

### II.3.3 Elastic fully damageable material (BILLARDON [16])

The behaviour of the material has been described in I.2.3. We make the following additional assumption concerning functions a(D), g(D) :

$$a(D)\, g''(D) - a'(D)\, g'(D) > 0 \quad \text{for } 0 < D < D_c \tag{II.3.23}$$

Under this assumption the irreversible behaviour of D(t) is completely described by :

$$\text{(i) } D(t) = 0 \quad \text{if } Z(t) \leq a(0) \tag{II.3.24}$$

$$\text{(ii) } - g'(D(t))\, Z(t) = a(D(t)) \quad \text{if } a(0) \leq Z(t) \leq \frac{a(D_c)}{- g'(D_c)} \tag{II.3.25}$$

$$\text{(iii) } D(t) = D_c \quad \text{if } Z(t) \geq \frac{a(D_c)}{- g'(D_c)} \tag{II.3.26}$$

where:

$$Z(t) = \sup_{0 \leq \tau \leq t} \frac{1}{2}\, \varepsilon(\tau) : \mathbb{E} : \varepsilon(\tau). \tag{II.3.27}$$

To obtain a first solution of the problem we define

$$t_0 = \frac{L}{A} \sqrt{\frac{2\, a(0)}{E}} \tag{II.3.28}$$

$$t_c = \frac{L}{A} \sqrt{\frac{2\, a(D_c)}{E\, (- g'(D_c))}} \tag{II.3.29}$$

Then a solution is given for $0 \leq t \leq t_c$ by $u(x,t) = \frac{A\, x\, t}{L}$ and $D(x,t) = D(t)$ obtained through (ii) with $Z(t) = \frac{1}{2} \frac{A^2 E\, t^2}{L^2}$. This homogeneous solution is unique if and only if :

$$[a(D)\, \{g(D)\, g''(D) - 2\, (g'(D))^2\} - a'(D)\, g(D)\, g'(D)] \geq 0 \quad \text{for } 0 < D < D_c \tag{II.3.30}$$

If a(D) and g(D) are such that :

$$[a(D)\ \{g(D)\ g''(D) - 2\ (g'(D))^2\} - a'(D)\ g(D)\ g'(D)] \geq 0 \quad \text{for } 0 \leq D \leq D_* \qquad \text{(II.3.31)}$$

$$[a(D)\ \{g(D)\ g''(D) - 2\ (g'(D))^2\} - a'(D)\ g(D)\ g'(D)] < 0 \quad \text{for } D_* < D < D_c \qquad \text{(II.3.32)}$$

then, the problem has an infinite number of solutions. Some of them can be constructed in the following way. Let $\hat{D}$ fixed such that $D_* \leq \hat{D} < D_c$ and let :

$$\hat{t} = \frac{L}{A} \sqrt{\frac{2\ a(\hat{D})}{E\ (-\ g'(\hat{D}))}} \qquad \text{(II.3.33)}$$

$$t_* = \frac{L}{A} \sqrt{\frac{2\ a(D_*)}{E\ (-\ g'(D_*))}} \qquad \text{(II.3.34)}$$

The following solution fields can then be exhibited.

(i) for $0 \leq t \leq t_*$ :

$$u(x,t) = \frac{A\ x\ t}{L} \qquad \text{(II.3.35)}$$

$$D(x,t) = D(t) \qquad \text{(II.3.36)}$$

(ii) for $t_* \leq \hat{t} < t < t_c$, let $\alpha$ such that $0 < \alpha < 1 - \frac{t}{t_c}$ and let $D^{\#}(t)$ the solution of the equation

$$\{1 - \alpha\ (1 - \frac{g(D)}{g(\hat{D})})\}^2\ \frac{a(D)}{-\ g'(D)} = \frac{1}{2}\ \frac{A^2\ E\ t^2}{L^2}$$

satisfying $\hat{D} < D^{\#}(t) < D_c$. Then we set :

$$D(x,t) = \begin{cases} \hat{D} & \text{for } 0 < x < \alpha\ L \\ D^{\#}(t) & \text{for } \alpha\ L < x < L \end{cases} \qquad \text{(II.3.37)}$$

$$\varepsilon(x,t) = \begin{cases} \dfrac{g(D^{\#}(t))}{g(\hat{D})} \dfrac{A\,t}{L[1-\alpha\,(1-\dfrac{g(D^{\#}(t))}{g(\hat{D})})]} & \text{for } 0 < x < \alpha\,L \\[2ex] \dfrac{A\,t}{L[1-\alpha\,(1-\dfrac{g(D^{\#}(t))}{g(\hat{D})})]} & \text{for } \alpha\,L < x < L \end{cases} \tag{II.3.38}$$

$$u(x,t) = \int_0^x \varepsilon(s,t)\,ds \tag{II.3.39}$$

Let us remark that for $\alpha\,L < x < L$ , the bar is still damaging, and that for $0 < x < \alpha\,L$ the unloading condition is satisfied. The constant tension in the bar is given by

$$\sigma(x,t) = g(D(t))\,\frac{E\,A\,t}{L} \qquad \text{for } 0 \leq t \leq t^* \tag{II.3.40}$$

$$\sigma(x,t) = \frac{g(D^{\#})}{1-\alpha\,(1-\dfrac{g(D^{\#})}{g(\hat{D})})}\,\frac{E\,A\,t}{L} \qquad \text{for } \hat{t} \leq t \leq t_{cr} \tag{II.3.41}$$

## II.4 CONCLUDING REMARKS

In section II.3 we have displayed some examples of non-uniqueness in the case of non-hardening associative models. One must emphasize here that similar difficulties may also arise in the hardening regime for non-associated models.

Starting from the analysis of these elementary problems, it appears that the evolution problem may be not well posed since it can possess in general too many solutions. Many questions can then be raised :

*Are there some safety regions where the solution is unique (or where the number of solutions is finite) ?*

*When there are too many solutions, are all of them equally satisfactory or can we make a choice on the basis of some physical reasons ?*

In the sequel, we shall try to give some partial answers to these questions.

# III. THE NON-LINEAR RATE BOUNDARY-VALUE PROBLEM

## III.1 GENERAL SETTING

Since the general problem ($\mathcal{P}$) can be not well posed, a first simplification is to study, for every fixed time instant t, the so-called first order rate boundary value problem ($\mathcal{RP}$; t) stated below. Indeed, if problem ($\mathcal{P}$) is well-posed for the time interval [0, T], then the rate boundary value problem ($\mathcal{RP}$; t) has a unique solution for every $t \in [0, T]$.

*Let us suppose that at the generic instant* t *during the deformation of the body, its current shape* $\Omega$, *the distributions of stress and the complete state of the material are determined.* Attention is then focused on the response of the body to an imposed infinitesimal change of external loading which is represented by nominal traction rates $\dot{\mathbf{F}}$ acting on the part $\partial_F\Omega$ of the boundary $\partial\Omega$ of the body (whose unit outward normal is denoted by $\mathbf{n}$) and nominal body forces $\dot{\mathbf{f}}$ per unit volume acting in its interior. On the remainder part $\partial_u\Omega$ of the boundary, the velocity $\mathbf{v}$ is prescribed as $\mathbf{v}^d$.

We will be concerned in this section by the analysis of the first order rate problem which consists in the determination of the velocity field $\mathbf{v}$ throughout the body. When this field is known, the stress field and the internal state rate fields are indeed readily obtained by the rate constitutive equations (I.1.1), (I.1.4) and (I.1.6) so that the overall incremental process is determined.

Following HILL [32], and in order to encompass a wide class of configuration dependent loadings, it is stipulated here that the nominal tractions and the body forces rates are linear functions of the particle velocity, whereas nominal traction rates, in addition, can depend linearly on the velocity gradient of the particle so that :

$$\dot{\mathbf{f}} = \dot{\mathbf{f}}_0 + \mathbb{h} \, . \mathbf{v} \qquad \text{(III.1.1)}$$

$$\dot{\mathbf{F}} = \dot{\mathbf{t}} + \mathbb{k} \, . \, \mathbf{v} + \mathbf{n} \, . \, \Theta : (\nabla \mathbf{v})^T \qquad \text{(III.1.2)}$$

where the second order tensors $\mathbb{h}$, $\mathbb{k}$ and the fourth order tensor $\Theta$ may be functions of the state and the positions of the particle, and where $\mathbf{n}$ denotes the unit outward normal to the boundary.

For the formulation and analysis of the first order rate problem, it is convenient (see HILL [30, 33]) to consider the unsymmetrical nominal stress tensor $\mathbb{S}$. As there is no reason for preferring a particular choice of objective stress rate to another and in order to retain some generality, this rate will be taken in the form (see for instance RANIECKI and BRUHNS [54]):

$$\overset{\circ}{\sigma} = \dot{\mathbb{S}} - \mathfrak{T}(\sigma) : (\nabla \mathbf{v})^T \tag{III.1.3}$$

where $\mathfrak{T}$ is a fourth order symmetric tensor which has to be specified for any particular choice. Then we consider the following rate constitutive equations in terms of the nominal stress tensor

$$\dot{\mathbb{S}} = \mathfrak{C} : (\nabla \mathbf{v}) \tag{III.1.4 }$$

with

$$\mathfrak{C} = \mathbb{L} + \mathfrak{T} \tag{III.1.5}$$

The nonlinear first order rate boundary-value problem ($\mathcal{RP}$; t) can then be formulated in the following way :

($\mathcal{RP}$; t) *Find* **v**(x) *defined for* $x \in \Omega$ *satisfying* :

(a) <u>the balance equations for continued quasi-static motions inside the body</u> :

$$\begin{cases} \operatorname{div} \dot{\mathbb{S}}^T + [\dot{\mathbf{f}}_0 + \mathbb{h} \, . \mathbf{v}] = 0 \\ \dot{\mathbb{S}} - \dot{\mathbb{S}}^T = \sigma \, . \, (\nabla \mathbf{v})^T - \nabla \mathbf{v} \, . \, \sigma \end{cases} \quad \text{in } \Omega \tag{III.1.6}$$

(b) <u>the compatibility conditions</u> :

$$\mathbb{D} = \varepsilon(\mathbf{v}) = \frac{1}{2}[\nabla \mathbf{v} + (\nabla \mathbf{v})^T] \quad \text{in } \Omega \tag{III.1.7}$$

(c) <u>the nonlinear rate constitutive equations</u> (III.1.4), (III.1.5) and (I.1.1).

(d) <u>the boundary conditions</u> :

$$\begin{cases} \dot{\mathbb{S}}^T . \mathbf{n} = \dot{\mathbf{F}} = \dot{\mathbf{t}} + \mathbf{k} \, . \, \mathbf{v} + \mathbf{n} \, . \, \Theta : (\nabla \mathbf{v})^T & \text{on } \partial_F \Omega \\ \mathbf{v} = \mathbf{v}^d & \text{on } \partial_u \Omega \end{cases} \tag{III.1.8}$$

This problem ($\mathcal{RP}$; t) is non-linear because of the dependence of the modulus $\mathbb{L}$ on the velocity field as can be seen from relations (I.1.1), (I.1.10) and (I.1.11). We will assume in the next of the chapter that the boundary $\partial\Omega$ as well as the coefficients of $\mathbb{L}$ are sufficiently regular and that $\mathbb{E}$ is uniformly positive definite in $\Omega$.

## III.2 EXISTENCE AND UNIQUENESS RESULTS

### III.2.1 Variational formulation

Consider the following functional spaces :

$$\mathcal{V} = [\,\mathbb{H}^1(\Omega)\,]^3, \quad \mathcal{V}_0 = \{\mathbf{w} \in \mathcal{V} \,/\mathbf{w} = \mathbf{0} \text{ on } \partial_2\Omega\} \tag{III.2.1}$$

Next, let $\hat{\mathbf{v}}^d \in \mathcal{V}$ be such that $\hat{\mathbf{v}}^d = \mathbf{v}^d$ on $\partial_2\Omega$ ; then, we define

$$\mathcal{V}_{ad} = \{\mathbf{w} \in \mathcal{V} \,/\, \mathbf{w} - \hat{\mathbf{v}}^d \in \mathcal{V}_0\} \tag{III.2.2}$$

Last, let $\mathsf{A} : \mathcal{V} \rightarrow \mathcal{V}'$ be the nonlinear operator defined by

$$(\mathsf{A}(\mathbf{v}), \mathbf{w}) = \int_\Omega \nabla\mathbf{w} : \mathbb{C} : \nabla\mathbf{v} \; d\omega \; - \; \int_\Omega \mathbf{w} : \mathbb{h} : \mathbf{v} \; d\omega$$
$$- \int_{\partial_1\Omega} \mathbf{w} : \mathbb{k} : \mathbf{v} \; ds \; - \; \int_{\partial_1\Omega} \mathbf{n} \,.\, \Theta : (\nabla\mathbf{v})^T \,.\, \mathbf{w} \; ds \tag{III.2.3}$$

Let us remark that the first three integrals appearing in this expression are well defined for any $\mathbf{v}, \mathbf{w} \in \mathcal{V}$. For the operator $\mathsf{A}$ to be well defined the fourth integral should *involve only tangential derivatives of* $\mathbf{v}$. Therefore, we assume throughout that this condition holds. It is then easily verified that $\mathsf{A}$ is *bounded* (i.e. transforms bounded sets into bounded sets) and *hemicontinuous* (that is to say the application $\mathbf{l} : \mathbf{R} \rightarrow \mathbf{R}$ defined by $\mathbf{l}(\lambda) = (\mathsf{A}\,(\mathbf{u}+\lambda\mathbf{v}), \mathbf{w})$ is continuous for every $\mathbf{u}, \mathbf{v}, \mathbf{w}$ in $\mathcal{V}$). Let us also notice that

$$\mathsf{L}(\mathbf{w}) = \int_\Omega \dot{\mathbf{f}}_0 . \mathbf{w} \; d\omega + \int_{\partial_1\Omega} \dot{\mathbf{t}} \,.\, \mathbf{w} \; ds \tag{III.2.4}$$

is a linear continuous functional on $\mathcal{V}_0$.

The principle of virtual powers shows that the solutions $\mathbf{v}$ of problem ($\mathcal{RP}$; t) satisfy the following problem :

($\mathcal{Q}$) *Find* $\mathbf{v} \in \mathcal{V}_{ad}$ *such that for every* $\mathbf{w} \in \mathcal{V}_0$

$$(\mathsf{A}(\mathbf{v}), \mathbf{w}) = \mathsf{L}(\mathbf{w}) \qquad \text{(III.2.5)}$$

### III.2.2 Existence and uniqueness results

i) An obvious *necessary* condition for the existence of solutions to problem ($\mathcal{Q}$) is that for all fixed $\mathbf{w} \in \mathcal{V}_0$ one has :

$$\ell(\mathbf{w}) = \sup_{\mathbf{v} \in \mathcal{V}_{ad}} (\mathsf{A}(\mathbf{v}), \mathbf{w}) \geq \mathsf{L}(\mathbf{w}) \qquad \text{(III.2.6)}$$

Let us remark that $\ell$ is a *proper, convex function, positively homogeneous of degree one.* When $\mathsf{A}$ is a *linear* operator, then $\ell$ is the indicator function of the kernel of the transposed operator $\mathsf{A}^T$; one should emphasize here that in finite dimensional spaces (III.2.6) is nothing but the usual necessary and sufficient condition for solvability of linear equations also recalled in the Appendix. Howewer this condition for nonlinear operators is in general only necessary.

ii) A sufficient condition for problem ($\mathcal{Q}$) *to have solutions* is that operator $\mathsf{A}$ be monotonic from $\mathcal{V}_{ad}$ to $\mathcal{V}'$ and coercive on $\mathcal{V}_{ad}$ (see LIONS [40]).
Operator $\mathsf{A}$ is said to be *monotone* from $\mathcal{V}_{ad}$ to $\mathcal{V}'$ if $(\mathsf{A}(\mathbf{v}) - \mathsf{A}(\mathbf{w}), \mathbf{v} - \mathbf{w}) \geq 0$ for every pair of fields $\mathbf{v}$, $\mathbf{w}$ belonging to $\mathcal{V}_{ad}$. Moreover, it is *coercive* if for every $\mathbf{z} \in \mathcal{V}_0$, $\| \mathbf{z} \| = 1$, it is coercive in the direction $\mathbf{z}$, i.e. satisfies

$$\lim (\mathsf{A}(\hat{\mathbf{v}}^d + \lambda \mathbf{z}), \mathbf{z}) = \infty \text{ when } \lambda \to \infty \qquad \text{(III.2.7)}$$

When $\mathsf{A}$ is monotone let us remark that $\ell(\mathbf{w}) \geq (\mathsf{A}(\hat{\mathbf{v}}^d), \mathbf{w})$. Moreover, we see that

$$\ell(\mathbf{w}) = +\infty \qquad \text{(III.2.8)}$$

for every $\mathbf{w} \neq \mathbf{0}$ such that $\mathsf{A}$ is coercive in the direction $\mathbf{z} = \dfrac{\mathbf{w}}{\| \mathbf{w} \|}$.

Let us now define the closed convex cone :

$$\Phi = \{ \mathbf{w} \in \mathcal{V}_0 \,/\, \ell(\mathbf{w}) = \mathsf{L}(\mathbf{w}) \} \qquad \text{(III.2.9)}$$

obviously $\mathbf{0} \in \Phi$ ; moreover if $\mathsf{A}$ is coercive then $\Phi = \{\mathbf{0}\}$. $\Phi$ is the coincidence set of the convex function $\ell$ and the function $\mathsf{L}$.

iii) When A is monotone but no more coercive, for the existence of a solution it is sufficient that $\Phi$ be a vector space and that a compactness condition be satisfied (without entering in all the details let us only remark that when A is linear this condition means that the image of A is a closed subspace of $\mathcal{V}'$) (see GASTALDI and TOMARELLI [27]. Then the solution set is convex and closed.

iv) A sufficient condition for *existence and uniqueness* of solution to problem ($\mathbb{Q}$) is that A be *strictly* monotone from $\mathcal{V}$ to $\mathcal{V}'$ and coercive [40]. The definition of strict monotony follows from that of monotony by replacing the inequality with a strict inequality.

### III.2.3 An example

In the case of small strains ( $\mathcal{T} = \mathbb{O}$) and when $\mathbb{k} = \mathbb{h} = \Theta = \mathbb{O}$, the previous formulae simplify greatly and the following conclusions hold :

($\alpha$) Operator A is monotonic from $\mathcal{V}$ to $\mathcal{V}'$ if the condition

$$(\mathbf{a} : \mathbb{E} : \mathbf{b}) + (\mathbf{a} : \mathbb{E} : \mathbf{a})^{1/2} \ (\mathbf{b} : \mathbb{E} : \mathbf{b})^{1/2} \leq 2\,h \qquad \text{(III.2.10)}$$

holds almost everywhere in $\Omega_p = \{\mathbf{x} \in \Omega \,/\, f = 0\}$ where f denotes the yield function (see (I.1.5) and (I.1.10)).

($\beta$) Under the condition

$$(\mathbf{a} : \mathbb{E} : \mathbf{b}) + (\mathbf{a} : \mathbb{E} : \mathbf{a})^{1/2} \ (\mathbf{b} : \mathbb{E} : \mathbf{b})^{1/2} < 2\,h \qquad \text{(III.2.11)}$$

holding almost everywhere in $\Omega_p$ the operator A is strictly monotone and coercive. Therefore problem (Q) has a unique solution.

($\gamma$) If the fourth order tensor $\mathbb{L}$ is moreover *symmetric* (i.e. $\mathbf{a} = \beta\,\mathbf{b}$ for some real $\beta$), the solutions of ($\mathbb{Q}$) render stationary the functional J defined over $\mathcal{V}_{ad}$ by

$$J(\mathbf{w}) = \frac{1}{2}\int_{\Omega} \mathbb{E} : \varepsilon(\mathbf{w}) : \varepsilon(\mathbf{w})\, d\omega - \int_{\Omega_p} \frac{h}{2\beta} < \beta\, \frac{\mathbb{E} : \mathbf{a} : \varepsilon(\mathbf{w})}{h} >^2 d\omega$$

$$- \int_{\Omega} \mathbf{f}^d . \mathbf{w}\, d\omega - \int_{\partial_1\Omega} \mathbf{F}^d . \mathbf{w}\, ds \qquad \text{(III.2.12)}$$

### III.2.4 Remarks

1) The strict monotony condition is RANIECKI and BRUHNS' condition [54] which generalizes HILL's uniqueness condition [30]. Condition (II.2.10) has been proposed in [54] for uniqueness under homogeneous states. In both cases, the existence question was not considered. The inequality (III.2.11) is the statement that the operator $\mathbb{L}^*$ defined by

$$\mathbb{L}^* = \frac{1}{2}\ (\mathbb{L} + \mathbb{L}^T) = \mathbb{E} - \frac{(\mathbb{E} : \mathbf{a}) \otimes (\mathbf{b} : \mathbb{E}) + (\mathbf{b} : \mathbb{E}) \otimes (\mathbb{E} : \mathbf{a})}{2h} \qquad \text{(III.2.13)}$$

be positive definite.

2) Specialization of (III.2.12) to a scalar situation permits to understand at least qualitatively what may happen when the above uniqueness condition fails. Indeed, let us consider the scalar function (very similar to functional (III.2.12)) defined by

$$I(x) = E\, x^2 - \langle g\, x \rangle^2 - f\, x \qquad \text{(III.2.14)}$$

for which we are seeking the stationary (critical) points in case they exist. It is clear that if $E > g^2$, then I is strictly convex and has a unique minimum for every value of f. This is the case where the inequality (III.2.11) holds. Now if $E = g^2$, I is still convex and the possible situations are sketched in figure 1 according to the sign of f.

It can be seen from figure 1, that when $f < 0$, I has only one minimum. When $f = 0$, I has an infinite number of minima whereas when $f > 0$, no critical point exists at all for I.

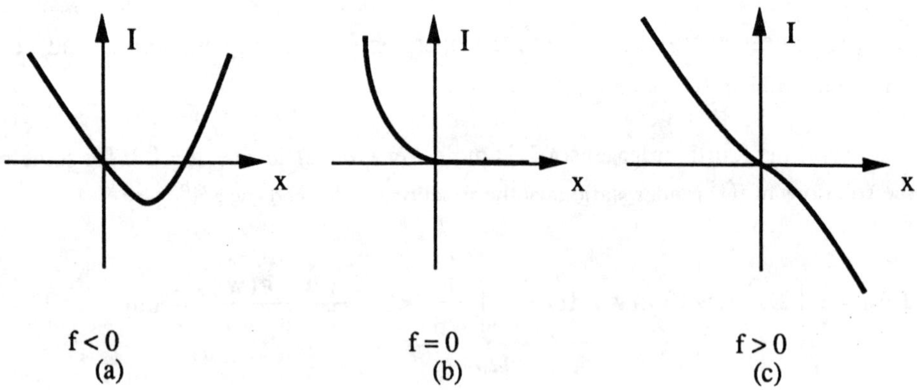


Figure 1 : Behaviour of function I for various values of f when $E = g^2$

Finally, if we consider the case when $E < g^2$, the picture is as follows :

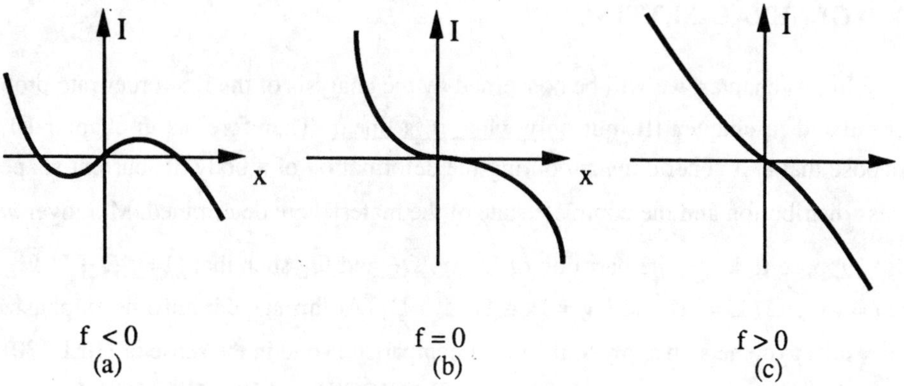


Figure 2 : Behaviour of function I for various values of f when $E < g^2$

Again figure 2 shows that uniqueness is lost in case (a) where $f < 0$, still prevails in case (b) with $f = 0$ and that existence is lost whenever $f > 0$.

This simple example demonstrates that at the failure of condition (III.2.11), not only uniqueness is lost, but there is the possibility that existence of solutions also fails. Let us point out that when $E = g^2$ , the existence depends on the sign of the data f. A similar situation has been observed (see [5]) for the existence of equilibrium configurations of a non-linear elastic body subjected to a system of applied forces and constrained to lie in a given region.

## III.3 HIGHER ORDER RATE PROBLEMS

So far, only the first order rate problem has been considered. It is quite clear that the exclusion of nonuniqueness for this problem, by no means implies exclusion of bifurcation for the evolution problem ($\mathcal{P}$). Indeed, bifurcation can manifest itself at a higher order. The second order problem has been considered by PETRIK and THERMANN [52] who showed that the exclusion condition is also sufficient for the uniqueness of the "regular" acceleration problem. Generalization of this result to any order for generalized standard materials has been obtained by SON and TRIANTAFYLLIDIS [64]. We send the interested reader to these references and also to the contributions of these authors to these lecture notes.

# IV. THE LINEAR RATE BOUNDARY VALUE PROBLEM

## IV.1 GENERAL SETTING

In this chapter, we will be concerned by the analysis of the first order rate problem as formulated in chapter III, but only when it is linear. Therefore, as in chapter III, let us suppose that at a generic instant during the deformation of a body, its current shape $\Omega$,the stress distribution and the complete state of the material are determined. Moreover we shall also suppose to know the partition of $\bar{\Omega}$ into $\bar{\Omega}_H$ and $\bar{\Omega}_E$ such that $\bar{\Omega} = \bar{\Omega}_H \cup \bar{\Omega}_E$ where $\Omega_H = \{x \in \Omega / \mathbb{L} = \mathbb{H}\}$ and $\Omega_E = \{x \in \Omega / \mathbb{L} = \mathbb{E}\}$. At this stage it must be emphasized that the results to come also apply to the linear comparison solid in the sense of HILL [30] and as generalized to nonsymmetric problems by RANIECKI and BRUHNS [54]. We will come back to this last point later.

For completeness, in order to handle more general situations, the body will be considered possibly heteregeneous in the sense that it is constituted of *two disjoint parts made of different materials*, named hereafter $\Omega_1$ and $\Omega_2$. This can represent for instance an inclusion in a matrix or a fiber in a composite. Generalization of the results to components with more than two parts is straightforward. Attention is then focused on the response of the body to an imposed infinitesimal change of external loading. This is represented by nominal traction rates $\dot{\mathbf{F}}$ acting on part $\partial_F\Omega$ of the boundary $\partial\Omega$ of the body (whose unit outward normal is denoted by $\mathbf{n}$), nominal body forces $\dot{\mathbf{f}}$per unit volume acting in its interior. On the remainder part $\partial_u\Omega$ of the boundary, the velocity $\mathbf{v}$ is prescribed as $\mathbf{v}^d$. *The materials constituting the structure have linear rate constitutive equations* in the form (see chapter III):

$$\dot{\mathbb{S}} = \mathbb{C} : \nabla \mathbf{v} \tag{IV.1.1}$$

The linear first order rate problem is then:

($\mathcal{LRP}$; t) *Find* $\mathbf{v}(x)$ *defined for* $x \in \Omega$ *satisfying* :

(a) the balance equations for continued quasi-static motions inside the body :

$$\begin{cases} \operatorname{div} \dot{\mathbb{S}}^T + [\dot{\mathbf{f}}_0 + \mathbb{h} . \mathbf{v}] = 0 & \text{in } \Omega_1 \text{ and } \Omega_2 \\ \dot{\mathbb{S}} - \dot{\mathbb{S}}^T = \sigma . (\nabla \mathbf{v})^T - \nabla \mathbf{v} . \sigma & \text{in } \Omega \end{cases} \tag{IV.1.2}$$

(b) <u>the linear rate constitutive equations</u> (IV.1.1).

(c) <u>the boundary conditions</u> :

$$\begin{cases} \dot{\mathbb{S}}^T . \mathbf{n} = \dot{\mathbf{F}} = \dot{\mathbf{t}} + \mathbb{k} . \mathbf{v} + \mathbf{n} . \Theta : (\nabla \mathbf{v})^T & \text{on } \partial_F\Omega \\ \mathbf{v} = \mathbf{v}^d & \text{on } \partial_u\Omega \end{cases} \qquad \text{(IV.1.3)}$$

(d) <u>the interfacial continuity requirements on the interface</u> I <u>between</u> $\Omega_1$ <u>and</u> $\Omega_2$. If the two parts are assumed perfectly bonded, ν denotes the normal to I (directed from $\Omega_1$ to $\Omega_2$ ) and [x] the jump of x across I, these requirements are :

$$\begin{cases} [\mathbf{v}] = \mathbf{0} \\ [\dot{\mathbb{S}}^T . \nu] = \mathbf{0} \end{cases} \quad \text{across I.} \qquad \text{(IV.1.4)}$$

Using the constitutive equations (IV.1.1) in (IV.1.2 - 4) the velocity field **v** then satisfies

$$\begin{cases} \operatorname{div}[\mathbb{C}_1 : \nabla \mathbf{v}] + \mathbb{h}_1 . \mathbf{v} + \dot{\mathbf{f}}_0 = \mathbf{0} & \text{in } \Omega_1 \\ \operatorname{div}[\mathbb{C}_2 : \nabla \mathbf{v}] + \mathbb{h}_2 \mathbf{v} + \dot{\mathbf{f}}_0 = \mathbf{0} & \text{in } \Omega_2 \\ [\mathbb{C}_1 : \nabla \mathbf{v} - \mathbf{n} . \Theta_1 : \nabla^T] . \mathbf{n} - \mathbb{k}_1 . \mathbf{v} - \dot{\mathbf{t}} = \mathbf{0} \text{ on } \partial_F\Omega \\ \mathbf{v} = \mathbf{v}^d & \text{on } \partial_u\Omega \\ [\mathbf{v}] = \mathbf{0} & \text{on I} \\ [(\mathbb{C} : \nabla \mathbf{v})^T . \nu] = \mathbf{0} & \text{on I} \end{cases} \qquad \text{(IV.1.5)}$$

## IV.2 GENERAL RESULTS FOR THE LINEAR CASE

### IV.2.1 General properties of the rate problem

Applying general results of modern theory of linear elliptic boundary value problems, it can be shown that ($\mathcal{L}\mathcal{R}\mathcal{P}$; t) is well posed (see the definitions and results recalled in the Appendix) in the sense that :

i) *the problem has a finite number of linearly independent solutions,*

ii) *these solutions depend continuously on the data* ,

iii) *the data must satisfy at most a finite number of linearly independent compatibility conditions* ,

if and only if the three following conditions hold :

a) the constitutive operators $\mathbb{C}_1$ and $\mathbb{C}_2$ are *elliptic* in the closures of $\Omega_1$ and $\Omega_2$ respectively,

b) the boundary conditions on $\partial\Omega$ complement the constitutive operator $\mathbb{C}_1$ at every point $\mathbf{x}_0$ of the boundary $\partial\Omega$ (they are said to satisfy the *boundary complementing condition*. at $\mathbf{x}_0$ ),

c) the constitutive operators $\mathbb{C}_1$ and $\mathbb{C}_2$ satisfy the *interfacial complementing condition* with respect to the interfacial conditions at every point $\mathbf{x}_0$ of the interface I.

Applications of this theory of elliptic boundary value-problems to mechanical problems have been made by THOMPSON [68] for the traction problems in finite elasticity, by SIMPSON and SPECTOR [61] for non-linear elasticity and by BENALLAL, BILLARDON and GEYMONAT [6,8,9] in the small strain range and for uncoupled isothermal conditions.

As already mentioned above, these results can be applied to the linear comparison solid. For sake of simplicity, let us consider here a homogeneous body $\Omega$ (constituted of only one material) such that $\Omega = \Omega_p \cup \Omega_e$ with

$$\Omega_p = \{\mathbf{x} \in \Omega \,/\, f = 0\} \tag{IV.2.1}$$

$$\Omega_e = \{\mathbf{x} \in \Omega \,/\, f < 0\} \tag{IV.2.2}$$

where f denotes the yield function (see (I.1.10)). The linear comparison solid associated to the real solid at hand is defined (see HILL [30]) by the constitutive relations

$$\dot{\mathbb{S}} = \mathbb{C}_e : \nabla \mathbf{v} \quad \text{in } \Omega_e \tag{IV.2.3}$$

$$\dot{\mathbb{S}} = \mathbb{C}_p : \nabla \mathbf{v} \quad \text{in } \Omega_p \tag{IV.2.4}$$

where, with the notations of chapter III,

$$\mathbb{C}_e = \mathbb{E} + \mathfrak{T} \tag{IV.2.5}$$

$$\mathbb{C}_p = \mathbb{L} + \mathfrak{T} \tag{IV.2.6}$$

It is concluded from the above results that the first order rate problem corresponding to this linear comparison solid is well posed if and only if :

a) the constitutive operators $\mathbf{C}_e$ and $\mathbf{C}_p$ are *elliptic* in the closures of $\Omega_e$ and $\Omega_p$ respectively,

b) the boundary conditions on $\partial\Omega$ complement the constitutive operator $\mathbf{C}$ at every point $\mathbf{x}_0$ of the boundary $\partial\Omega$.,

c) the constitutive operators $\mathbf{C}_e$ and $\mathbf{C}_p$ satisfy the *interfacial complementing condition* at every point $\mathbf{x}_0$ of the elastic-plastic boundary which plays here the role of an interface. (It is implicitely assumed here that the elastic-plastic boundary is compactly contained in $\Omega$.)

**Remark** :

Without going to technical discussions (see for that AGMON, DOUGLIS and NIRENBERG [4], GEYMONAT [28], LIONS and MAGENES [42], MARSDEN and HUGHES [46] the results given above apply rigorously in adequate solutions spaces,when the boundary $\partial\Omega$ is smooth (this eliminates corners and so on) and when the closures of $\partial_F\Omega$ and $\partial_u\Omega$ do not intersect. Also, in case of heteregeneous media, the interface should not intersect with the boundary. The results apply for example to an annulus where $\partial_u\Omega$ is the outer boundary and $\partial_F\Omega$ the inner one. They are of course also valid when only nominal tractions rates (or only velocities) are applied on $\partial\Omega$. The difficulty lies with the regularity of solutions and probably, the results can be extended by modifying the solution spaces but such results do not exist by now. The situation calls for further research to seek if these difficulties are just technical or are of physical interest. Presumably, the modifications in the spaces needed correspond to known asymptotic solutions near corners, cracks , etc.

We have given the general conditions under which the linear rate-problem is well posed. When it is so, it has a finite number of linearly independent solutions which depend continuously on the data. Hence, they can be interpreted as diffuse modes of deformation. We have also mentioned that the results can be applied with some minor changes to the linear comparison solid corresponding to the real solid at hand. The results are then also useful for standard results in bifurcation theory obtained with the piecewise-linear constitutive equations described in chapter I (e.g. see HUTCHINSON [35], HUTCHINSON, MILES [36]).

### IV.2.2 - Statements of the conditions and mechanical interpretations

In the following, we describe in details the aforementioned conditions and try to give mechanical interpretations for them. Before doing so, let us notice that the three conditions are local : consequently, they may be, in principle, easily implemented in finite element calculations for instance. Moreover, the ellipticity and the interfacial complementing conditions are expressed for an infinite body which is homogeneous for the first condition and made of two different materials for the second one. Besides, the boundary complementing condition is written for a homogeneous half space.

#### IV.2.2.1 The ellipticity condition.

$\mathbb{C}_1$ and $\mathbb{C}_2$ are *elliptic* in the closures of $\Omega_1$ and $\Omega_2$ if and only if

$$\det(\mathbf{N} \,.\, \mathbb{C}_1 \,.\, \mathbf{N}) \neq 0 \text{ for every vector } \mathbf{N} \neq \mathbf{0} \text{ and every point } \mathbf{x} \text{ of } \bar{\Omega}_1 \tag{IV.2.7}$$

$$\det(\mathbf{N} \,.\, \mathbb{C}_2 \,.\, \mathbf{N}) \neq 0 \text{ for every vector } \mathbf{N} \neq \mathbf{0} \text{ and every point } \mathbf{x} \text{ of } \bar{\Omega}_2 \tag{IV.2.8}$$

They are said to be properly elliptic if moreover,the equations in the complex variable $\tau$

$$\det[(\mathbf{k} + \tau\,\mathbf{n}) \,.\, \mathbb{C}_1 \,.\, (\mathbf{k} + \tau\,\mathbf{n}))] = 0 \tag{IV.2.9}$$

$$\det[(\mathbf{k} + \tau\,\mathbf{n}) \,.\, \mathbb{C}_2 \,.\, (\mathbf{k} + \tau\,\mathbf{n}))] = 0 \tag{IV.2.10}$$

have for every linearly independent vectors **k** and **n**, exactly three roots $\tau_j$ with positive imaginary parts. It must be noticed here that all real elliptic operators are properly elliptic (see AGMON, DOUGLIS and NIRENBERG [4] and LIONS and MAGENES [42] ).

The ellipticity condition is very classical in many problems of continuum mechanics. Its failure under homogeneous conditions is related to the occurence of deformation modes involving jumps of the velocity gradient across singular surfaces (e.g. HADAMARD [28], HILL [32], MANDEL [45], RICE [56], RUDNICKI, RICE [57], RICE, RUDNICKI [58]) and has served for long time as a localization criterion. It has also been related, in the dynamical context, to *stationary acceleration waves* (e.g HILL [32], MANDEL [45], RICE [56]). It is a necessary and sufficient condition for these last modes and waves to occur under isothermal condition (see BORRE, MAIER [21]).

Notice finally that only the coefficients of higher order terms appear in the ellipticity conditions and in particular, $\mathbb{h}_1$ and $\mathbb{h}_2$ do not appear.

### IV.2.2.2 The boundary complementing condition

Let $\mathbf{x}_0$ be a point of the boundary where the unit outward normal is denoted by $\mathbf{n}$. The boundary complementing condition can be expressed as an algebraic relationship between the coefficients of the constitutive operator $\mathbb{C}_1$ and those of the boundary operator given by (IV.1.3) or (IV.1.4) at $\mathbf{x}_0$. Rather, we will give an alternative differential statement which in the same time allows its mechanical interpretation. The condition is phrased in terms of an associated problem in the half space $\mathcal{D} = \{\mathbf{x} / \mathbf{n}.(\mathbf{x}_0 - \mathbf{x}) < 0\}$ and requires that for every vector $\mathbf{k}$ tangent to the boundary at $\mathbf{x}_0$, the *only solution* to the linear boundary value-problem

$$\begin{cases} \mathrm{div}[\, \mathbb{C}_1(\mathbf{x}_0) : \nabla \mathbf{v}\,] = 0 & \text{in } \mathcal{D} \\ [\, \mathbb{C}_1(\mathbf{x}_0) : \nabla \mathbf{v} - \mathbf{n} \,.\, \mathbb{S}(\mathbf{x}_0) : (\nabla \mathbf{v})^{\mathrm{T}}\,] \,.\, \mathbf{n} = 0 & \\ \text{or} \quad \mathbf{v} = \mathbf{0} & \end{cases} \left.\vphantom{\begin{matrix} a \\ b \end{matrix}}\right\} \text{on } \partial\mathcal{D} \qquad \text{(IV.2.11)}$$

of the form $\mathbf{v}(\mathbf{x}) = \mathbf{w}\,(\mathbf{n} \,.\, (\mathbf{x}_0 - \mathbf{x}))\, \exp\{i\mathbf{k} \,.\, (\mathbf{x}_0 - \mathbf{x})\}$ with bounded $\mathbf{w}$ is the null solution $\mathbf{v} \equiv \mathbf{0}$.

One important remark to be made here is the fact that only the highest order terms of the differential operators (field and boundary operators) appear in the condition just as for the ellipticity condition ; hence, $\mathbb{k}$ and $\mathbb{h}$ play no role in this condition.

To see the mechanical implications of the complementing condition, let us assume that it has failed at a point of the boundary $\mathbf{x}_0$. Then, there exist a solution satisfying (IV.2.11), in the form

$$\mathbf{v}(\mathbf{x}) = \mathbf{w}(\mathbf{n} \,.\, (\mathbf{x}_0 - \mathbf{x}))\, \exp\{i\, \mathbf{k} \,.\, (\mathbf{x}_0 - \mathbf{x})\} \qquad \text{(IV.2.12)}$$

with bounded $\mathbf{w}$. Since this problem is linear, it is easy to check that $\mathbf{w}$ satisfies a linear system of ordinary differential equations of second order

$$\mathbb{A}_1.\mathbf{w}'' + \mathbb{B}_1.\mathbf{w}' + \mathbb{C}_1.\mathbf{w} = \mathbf{0} \qquad \text{(IV.2.13)}$$

where the components of the matrices $\mathbb{A}$, $\mathbb{B}$and $\mathbb{C}$ are given respectively by :

$$\mathbb{A}_j = \mathbf{n} \,.\, \mathbb{C}_j \,.\, \mathbf{n} \qquad \text{(IV.2.14)}$$

$$\mathbb{B}_j = i\,[\mathbf{n} \,.\, \mathbb{C}_j \,.\, \mathbf{k} + \mathbf{k} \,.\, \mathbb{C}_j \,.\, \mathbf{n}] \qquad \text{(IV.2.15)}$$

$$\mathbb{C}_j = -\,\mathbf{k} \,.\, \mathbb{C}_j \,.\, \mathbf{k} \qquad \text{(IV.2.16)}$$

The solutions of system (IV.2.13) depend on the roots of the characteristic equation of this system, namely:

$$\det\,[\tau^2\, \mathbb{A}_1 + \tau\ \mathbb{B}_1 + \mathbb{C}_1\,] = 0 \qquad \text{(IV.2.17)}$$

which, by simple inspection can be shown to be

$$\det\,[(\mathbf{k} - i\,\tau\,\mathbf{n}) \,.\, \mathbb{C}_1 \,.\, (\mathbf{k} - i\,\tau\,\mathbf{n})) = 0 \qquad \text{(IV.2.18)}$$

It follows then that in the elliptic regime (i.e. the properly elliptic regime according to the aforementioned remark), equation (IV.2.17) has three roots with negative real parts denoted hereafter by $\tau_j = -\,\alpha_j + i\,\beta_j$ ($j = 1, 2, 3$) with $\alpha_j < 0$. Then, in the case where $\tau_j$ are distinct (when this does not hold, classical arguments can be followed to construct the set of solutions), the general bounded solution of (IV.2.13) is

$$\mathbf{w}(\mathbf{n}.(\mathbf{x}_0 - \mathbf{x})) = \sum_{j=1}^{j=3} a_j\, \mathbf{w}_j\, e^{\{-\,\alpha_j\, \mathbf{n}.(\mathbf{x}_0 - \mathbf{x})\}}\ e^{\{i\,\beta_j\, \mathbf{n}.(\mathbf{x}_0 - \mathbf{x})\}} \qquad \text{(IV.2.19)}$$

where $a_j$ are arbitrary complex numbers and $\mathbf{w}_j$ the eigenvectors of the matrix $[\tau_j^2\ \mathbb{A}_1 + \tau_j\ \mathbb{B}_1 + \mathbb{C}_1]$, associated to the zero eigenvalue. It follows that the failure of the complementing condition leads to the occurence of deformation modes of the form

$$\mathbf{v}(\mathbf{x}) = \sum_{j=1}^{j=3} a_j\, \mathbf{w}_j\, e^{\{-\,\alpha_j\, \mathbf{n}.(\mathbf{x}_0 - \mathbf{x})\}}\ e^{\,i\,\{\beta_j\, \mathbf{n}.(\mathbf{x}_0 - \mathbf{x}) + \mathbf{k}.(\mathbf{x}_0 - \mathbf{x})\}} \qquad \text{(IV.2.20)}$$

These solutions (for a practical problem, one must take real parts of these solutions) represent deformation modes which are oscillations which decay exponentially inside the half space $\mathcal{D}$. It can be seen from (IV.2.20) that the roots $\tau_j$ are proportional to $(\mathbf{k}.\mathbf{k})$ so that the frequency of the oscillations is undetermined : it is arbitrarily small or large ; the shorter the wavelength of the mode, the smaller the zone beneath the surface where the mode is localized. Such *surface modes* have been studied in their own by SAWYERS [59] for nonlinear elastic materials and for general incompressible incrementally linear solids by HUTCHINSON, TVERGAARD [37]. Further, these modes have been exhibited explicitely for simple boundary value problems by HILL, HUTCHINSON [34], YOUNG [71] and NEEDLEMAN [49] for the plane strain and compression tests, by TRIANTAFYLLIDIS [69] for the bending test and by VARDOULAKIS, SULEM, GUENOT for the borehole problem [70].

The modes given by (IV.2.20) look like what RAYLEIGH has studied in elastodynamics and called surface waves (see LOVE [44] ) : in that sense, one may say that the failure of the complementing condition allows for the existence of *stationary surface waves*. This has already been pointed out by THOMPSON [68] for traction boundary conditions ; these waves are however also possible under some circumstances for boundary conditions of place.

The modes corresponding to the failure of the boundary complementing condition depend on the type of boundary conditions applied at $\mathbf{x}_0$. For boundary conditions of place at $\mathbf{x}_0$ ($\mathbf{x}_0 \in \partial_u\Omega$), the parameters $a_j$ must satisfy (see (IV.2.20)):

$$\sum_{j=1}^{j=3} a_j \, \mathbf{w}_j = \mathbf{0} \qquad \text{(IV.2.21)}$$

and consequently the boundary complementing condition fails in this particular case if and only if at least two of the $\mathbf{w}_j$ are linearly dependent.

Now, for nominal traction rate boundary conditions at $\mathbf{x}_0$ ($\mathbf{x}_0 \in \partial_F\Omega$), the parameters must satisfy:

$$[\, \mathbb{C}_1(\mathbf{x}_0) : \nabla\mathbf{v} - \mathbf{n} \,.\, \Theta(\mathbf{x}_0) : (\nabla\mathbf{v})^{T} ] \,.\, \mathbf{n} = 0 \qquad \text{on } \partial\mathcal{D} \qquad \text{(IV.2.22)}$$

Combining (IV.2.20) and (IV.2.22), one gets a system of three algebraic equations and the boundary complementing conditions fails when non trivial solutions are possible for this system, that is if and only if its determinant is zero.

### IV.2.2.3 The interfacial complementing condition

This condition is similar to an extent to the boundary complementing condition. Here again, we present it as a differential statement but now in terms of an associated problem in the whole space. Let $\mathbf{x}_0$ be a point of the interface I where the unit outward normal to I, directed from $\Omega_2$ to $\Omega_1$ is denoted by $\nu$. This condition requires that for every vector $\mathbf{k}$ tangent to the interface at $\mathbf{x}_0$, the only solution of the linear boundary value problem

$$\begin{cases} \operatorname{div}[\ \mathbb{C}_1(\mathbf{x}_0) : \nabla \mathbf{v}_1\ ] = \mathbf{0} & \text{everywhere in } \mathscr{D}_1 \\ \operatorname{div}[\ \mathbb{C}_2(\mathbf{x}_0) : \nabla \mathbf{v}_2\ ] = \mathbf{0} & \text{everywhere in } \mathscr{D}_2 \\ \mathbb{C}_1(\mathbf{x}_0) : \nabla \mathbf{v}_1 . \nu\ = \mathbb{C}_2(\mathbf{x}_0) : \nabla \mathbf{v}_2 . \nu & \text{in } \partial\mathscr{D}_2 \\ \mathbf{v}_1\ = \mathbf{v}_2 & \text{in } \partial\mathscr{D}_2 \end{cases} \tag{IV.2.23}$$

in the form

$$\mathbf{v}_1(\mathbf{x}) = \mathbf{w}_1(\nu.(\mathbf{x}_0 - \mathbf{x})) \exp\{i\ \mathbf{k}.(\mathbf{x}_0 - \mathbf{x})\} \text{ in } \mathscr{D}_1 = \{\mathbf{x}/\ \nu.(\mathbf{x}_0 - \mathbf{x}) > 0\} \tag{IV.2.24}$$

$$\mathbf{v}_2(\mathbf{x}) = \mathbf{w}_2(\nu.(\mathbf{x}_0 - \mathbf{x})) \exp\{i\ \mathbf{k}.(\mathbf{x}_0 - \mathbf{x})\} \text{ in } \mathscr{D}_2 = \{\mathbf{x}/\ \nu.(\mathbf{x}_0 - \mathbf{x}) < 0\} \tag{IV.2.25}$$

with bounded $\mathbf{w}_1$ and $\mathbf{w}_2$, is the null solution $\mathbf{v}_1 = \mathbf{v}_2 \equiv \mathbf{0}$. The mechanical interpretation of this condition can be drawn in the same lines as that for the complementing condition given above. Indeed, let $\mathbf{x}_0$ be a point where this condition has failed. It follows that system (IV.2.23) has solutions of the form (IV.2.24) and (IV.2.25). Combining (IV.2.25), (IV.2.24) and (IV.2.23), one obtains that the $\mathbf{w}_j$ (j = 1, 2) satisfy systems of the form (IV.2.13) where the involved matrices are still given by (IV.2.14), (IV.2.15) and (IV.2.16) (with index 2 replacing 1 when necessary). Defining by $\tau_m = -\alpha_m + i\ \beta_m$(m =1, 2, 3) and $\kappa_m = -\gamma_m + i\ \delta_m$(m =1, 2, 3) the three roots with negative real parts of the respective charasteristic equations, and using the same developments as for the boundary, it is found that the general solution of (IV.2.23) with bounded $\mathbf{w}_j$ is:

$$\mathbf{v}_1(\mathbf{x}) = \sum_{j=1}^{j=3} a_j\, \mathbf{u}_j\, e^{\{-\alpha_j \nu.(\mathbf{x}_0 - \mathbf{x})\}}\ e^{i\ \{\beta_j \nu.\ (\mathbf{x}_0 - \mathbf{x}) + \mathbf{k}.\ (\mathbf{x}_0 - \mathbf{x})\}} \tag{IV.2.26}$$

$$\mathbf{v}_2(\mathbf{x}) = \sum_{j=1}^{j=3} b_j\, \mathbf{w}_j\, e^{\{-\gamma_j \nu\ .\ (\mathbf{x}_0 - \mathbf{x})\}}\ e^{i\ \{\delta_j \nu.\ (\mathbf{x}_0 - \mathbf{x}) + \mathbf{k}.\ (\mathbf{x}_0 - \mathbf{x})\}} \tag{IV.2.27}$$

where $a_j$ and $b_j$ are arbitrary complex numbers, and $\mathbf{u}_j$ (resp. $\mathbf{w}_j$ ) the eigenvectors of the matrices $[\tau_j^2\ \mathbb{A}_1 + \tau_j\ \mathbb{B}_1 + \mathbb{C}_1]$ (resp $[\kappa_j^2\ \mathbb{A}_2 + \kappa_j\ \mathbb{B}_2 + \mathbb{C}_2]$ ) associated to the zero eigenvalue.Here again, these modes are oscillations which decay exponentially perpendicularly to the interface between $\mathcal{D}_1$ and $\mathcal{D}_2$. Their frequencies are also unspecified and are then arbitrarily small or large. For high frequencies, the deformation is here localized beneath the interface in $\mathcal{D}_1$and $\mathcal{D}_2$ and in that sense, these modes represent *interfacial modes of deformation*. Such interfacial modes have been studied in their own by BIOT [18, 19]. Further, these modes look like the waves that propagate along interfaces in elastic media. As for the boundary, we will say that the failure of the interfacial complementing condition allows the occurence of *stationary interfacial waves* such as *Stonely* or *Love waves* (see LOVE [44], ACHENBACH [1]). In order that these interfacial modes appear, the parameters must satisfy

$$\sum_{j=1}^{j=3} a_j \mathbf{u}_j = \sum_{j=1}^{j=3} b_j \mathbf{w}_j \qquad \text{(IV.2.28)}$$

$$\sum_{j=1}^{j=3} a_j \left[\tau_j \nu \,.\, \mathbb{C}_1 \,.\, \nu + i\, \nu \,.\, \mathbb{C}_1 \,.\, \mathbf{k}\right] . \mathbf{u}_j = \sum_{j=1}^{j=3} b_j \left[\kappa_j \nu \,.\, \mathbb{C}_2 \,.\, \nu + i\, \nu \,.\, \mathbb{C}_2 \,.\, \mathbf{k}\right] . \mathbf{w}_j \qquad \text{(IV.2.38)}$$

This is a system of six algebraic linear homogeneous equations which has non-trivial solutions if and only if its deteminant is zero. This determinantal equation expresses then the failure of the interfacial complementing condition when the characteristic equations (IV.2.13) have distinct roots. When the characteristic roots are multiple, classical arguments can be used to construct the general solution to (IV.2.23).

### IV.2.2.4 Remarks

The complementing conditions have a theoretical interest in the elliptic regime in that the boundary value problem is already ill-posed in the non-elliptic one. However, one may wonder if in this non-elliptic regime, surface modes or interfacial modes are available (i.e. modes which decay inside the body). The answer is dependent on the type of loss of ellipticity. Indeed, ellipticity can fail in many ways (for an illustration, see HILL, HUTCHINSON [34] for a case where it is possible to have a transition from elliptic to parabolic or hyperbolic regimes).

## CONCLUSIONS

Though not developed in the paper, we have mentioned that the results given in the last chapters can be used to describe bifurcation and localization phenomena. Indeed, the ellipticity condition can be used to describe localization in the bulk of the solid whereas the boundary complementing conditions can be used to describe localization at the boundary or interfaces of the solid.

1) The ill-posedness which has been shown to accompany these localization phenomena implies that no attempt to have a reliable numerical solution can be successful. The only way to eliminate the associated numerical difficulties is to regularize the problem at hand and try to transform it into a well-posed one.This is the subject of an extensive current research and the recent interest in generalized continuum theories has taken its source there. Also, the absence of characteristic or internal length in classical continuum theories leads to the fact that the size of the bulk localized zone, the wavelength and the depth of the surface or interface modes are left unspecified and arbitrary.

In all the regularization procedures proposed so far, an internal length is introduced in a way or another. In the micropolar continuum (MUHLAUS, VARDOULAKIS [47], DE BORST [22], the additional degree of freedom (mico-rotation) brings in an internal scale which prevents the strain rate to become discontinuous under certain loading types (STEINMANN, WILLAM [66]). The same feature is attached to higher gradient theories (TRIANTAFYLLIDIS, AIFANTIS [3]), to nonlocal theories written either in a gradient format (LASRY, BELYTSCHO [39], DE BORST, MULHAUS [23], MULHAUS, AIFANTIS [48]) or in an integral one (PIJAUDIER-CABOT, BAZANT [53]). Another possibility explored by NEEDLEMAN [51], LORET, PREVOST [43], SLUYS, DE BORST [63] and SIMO [62], is the incorporation of rate effects. Let us examine now briefly these regularization procedures.

a) Roughly speaking, for nonlocal theories of the integral type, the linear boundary value problem ($\mathcal{LRP}$; t) described in chapter IV (for partial differential operators) is transformed into a linear boundary value problem for pseudo-differential operators, the type of which is (we limit the example to small strains) :

$$(\mathcal{LRP};\ t;\ NL_\eta)\quad \begin{cases} -\ \mathrm{div}\ \{\varphi_\eta * \mathbb{L} : \varepsilon(\mathbf{v}_\eta)\} = \mathbf{f} \quad \text{in } \Omega \\ \text{Boundary conditions} \end{cases}$$

where $*$ denotes the convolution, $\varphi_\eta$ a smooth approximation of the Dirac distribution $\delta$ and $\eta$ the caracteristic length (BENALLAL, GEYMONAT [14]).
Under some assumptions ( $\mathbb{L}$ is elliptic, ...), one can prove that for $\eta \to 0$, the solution(s) $\mathbf{v}_\eta$ of ($\mathcal{LRP}$; t; $NL_\eta$) approach the solution(s) of

$$(\mathcal{LRP};\ t) \qquad \begin{cases} -\operatorname{div}\{\ \mathbb{L} : \varepsilon(\mathbf{v})\} = \mathbf{f} & \text{in } \Omega \\ \text{Boundary conditions} \end{cases}$$

i.e. the solution(s) corresponding to the classical continuum. However, when $\mathbb{L}$ is not elliptic, the problem ($\mathcal{LRP}$; t; $\eta$) can still be well posed.

b) Roughly speaking, for higher gradient theories, the linear boundary value problem ($\mathcal{LRP}$; t) described in chapter IV (for partial differential operators) is transformed into a boundary value problem for higher order partial differential operators, the type of which is (we limit the example to small strains)

$$(\mathcal{LRP};\ t;\ HG_\eta) \qquad \begin{cases} -\operatorname{div}\{\ \mathbb{L} : \varepsilon(\mathbf{v}_\eta)\} + \eta\ \Delta^2(\mathbf{v}_\eta) = \mathbf{f} & \text{in } \Omega \\ \text{Boundary conditions} \end{cases}$$

A simple singular perturbation analysis (e.g. LIONS [41]) shows that under some assumptions ( $\mathbb{L}$ is coercive, ...) and for $\eta \to 0+$, the solution $\mathbf{v}_\eta$ of ($\mathcal{LRP}$; t; $HG_\eta$) approaches the solution of ($\mathcal{LRP}$; t), i.e. the solution corresponding to the classical continuum. However, when $\mathbb{L}$ is not elliptic, the problem ($\mathcal{LRP}$; t; $HG_\eta$) can be still well posed here again. An interesting remark of TRIANTAFYLLIDIS, AIFANTIS [3] is that, in the case of nonlinear elasticity, the solution of ($\mathcal{LRP}$; t; $HG_\eta$) has a "smooth singularity" in the direction of localization.

c) In the case of rate-dependent regularization, it is shown that the rate problem (and also the higher order rate problems) has a unique solution (see BENALLAL [10]). As in the previous cases, one can prove that under some suitable assumptions, the rate-dependent solution tends to the rate-independent solution. Let us notice here that this procedure is one of the keys to the proof of the fundamental result of JOHNSON [38] recalled in chapter II.

2) In all the previous regularization procedures, an internal length is introduced in a way or another. A possible justification of the choice of that internal length can be obtained by a suitable homogenization procedure.(e.g. AIFANTIS [2], FRANCFORT, MÜLLER [26], ...). Let us explicitely remark that in the case of brittle partially damageable materials

FRANCFORT, MARIGO [25] exhibited a time evolving zone where the sound and the damaged material are finely mixed. This is obtained without any homogenization procedure but only looking for stable generalized solutions.

3) Thermomechanical couplings were omitted for simplicity though their inclusion presents no difficulty The roles of thermal effects and thermomechanical couplings in localization phenomena in rate-independent materials can be found in BENALLAL [11, 12].

4) In all the previous chapters, we have only considered quasi-static situations. Some similar considerations could have been developed for the dynamic case : for instance, to the loss of ellipticity corresponds the loss of hyperbolicity, .... For some results in this direction, see SCHAEFFER [60] , BENALLAL [13], ...

5) The ill-posed character of the boundary value problem described above makes it a very difficult task to identify the constitutive behaviour in the range of ill-posedness (e.g. the softening range for associated plasticity,...). Indeed in this range the volume element in the core of the specimen can no more be considered homogeneous but it behaves as a structure. However, the given explicit forms for general constitutive properties and boundary conditions in the theoretical analysis so far developped can be used precisely in this range to take into account the experimental informations on the localization of the deformation. For some results in this direction see BILLARDON, DOGHRI [17], BERTHAUD, BILLARDON, COMI [15] ...

## REFERENCES


1. ACHENBACH, J. D. : Wave propagation in elastic media, North-Holland, Amsterdam, New York-Oxford, 1984.
2. AIFANTIS, E. C. : On the Microstructural Origin of Certain Inelastic Models, Trans ASME, J. Eng. Mat. Tech., **106**, pp. 326-320, 1984.
3. TRIANTAFYLLIDIS, N., AIFANTIS, E. C. : A gradient approach to localization of deformation - I. Hyperelastic materials, J. Elasticity, **16**, pp. 225-237, 1986.
4. AGMON, S., DOUGLIS, N., NIRENBERG, L. : Estimates near the boundary for solutions of partial differential equations II, Comm. pure and Appl. Math., **12**, pp. 623-727, 1964.

5. BAIOCCHI, G., BUTTAZZO, G., GASTALDI,F., TOMARELLI, F.: General existence theorems for unilateral problems in continuum mechanics, Arch. Rat. Mech. Anal., **100**, pp.149-189, 1988.

6. BENALLAL, A., BILLARDON, R. and GEYMONAT, G.: Some mathematical aspects of the damage softening rate problem, in : Strain localization and size effect, Ed. J. MAZARS & Z.P. BAZANT, Elsevier Applied Science, pp.247-258, 1989.

7. BENALLAL, A., BILLARDON, R. and GEYMONAT, G.: Conditions de bifurcation à l'intérieur et aux frontières pour une classe de matériaux non standards, C. R. Acad. Sci., Paris, t.**308**, série II, pp. 893-898, 1989.

8. BENALLAL, A., BILLARDON, R., GEYMONAT, G.: Phénomènes de localization à la frontière d'un solide, C. R. Acad. Sci. Paris, t. **310**, série II, pp. 679-684, 1990.

9. BENALLAL, A., BILLARDON, R., GEYMONAT, G.: Localization phenomena at the boundaries and interfaces of solids, in Constitutive Laws for Engineering Materials, Ed Desai et al, ASME Press, pp.387-390, 1991.

10. BENALLAL, A. : Thermoviscoplasticité et endommagement des structures, Thèse de doctorat d'état, Université Paris 6, 1989.

11. BENALLAL, A. : Quelques remarques sur le rôle des couplages thermomécaniques dans les phénomènes de localisation, C. R. Acad. Sci., Paris, t.**312**, Série II, pp. 117-122, 1991.

12. BENALLAL, A. : On localization phenomena in thermoelastoplasticity, Arch. Mech., **44**, pp. 15-29, 1992.

13. BENALLAL, A. : Ill-posedness and localization in solids structures, in Computational Plasticity, Ed. D.R.J. OWEN et al., Pineridge Press, pp. 581-592, 1992.

14. BENALLAL, A., GEYMONAT, G. : in preparation.

15. BERTHAUD,Y., BILLARDON, R., COMI C. : Identification of damageable elastic-plastic materials from localization measurements, in preparation.

16. BILLARDON, R. : Etude de la rupture par la mécanique de l' endommagement, Thèse de doctorat d'état, Université Paris 6, 1989.

17. BILLARDON, R., DOGHRI, I. : Prévision de l'amorçage d'une macro-fissure par localization de l'endommagement, C. R. Acad. Sci., Paris, t.**308**, série II, pp.347-352,1989.

18. BIOT, M. A. : Surface instability in finite anisotropic elasticity under initial stress, Proc. R. Soc. London, A **273**, pp.329-339, 1963.

19. BIOT, M. A. : Interface instability in finite elasticity under initial stress, Proc. R. Soc. London, A **273**, pp.340-344, 1963.

20. BLANCHARD D., LE TALLEC P.: Numerical analysis of the equations of small strains quasistatic elastoviscoplasticity, Numer. Math., **50**, pp. 147-169, 1986.

21. BORRE, G. and MAIER, G. : On linear versus non-linear flow rules in strain localization analysis, Meccanica, **24** , pp. 36-41, 1989.

22. DE BORST, R. : Simulation of strain localization : a reappraisal of the Cosserat continuum, Eng. Computations, **8**, pp. 317-332, 1991.

23. DE BORST, R., MULHAUS, H. B. : Gradient dependent plasticity : formulation and algorithmic aspects, Int. J. Num. Meth. Eng., **35,** 1992 (in press).

24. ERICKSEN, J.L. : Equilibrium of bars, Journal of Elasticity, **5**, pp. 131-201, 1975.

25. FRANCFORT, G., MARIGO, J. J. : Mathematical Analysis of damage evolution in a brittle damaging continuous medium, in Mécanique, Modélisation Numérique et Dynamique des matériaux, Publications du LMA n° 124, CNRS, Marseille, pp. 245-276, 1991.

26. FRANCFORT, G., MÜLLER, S. : Combined Effects of Homogenization and Singular perturbations in Elasticity, Preprint n° 201, Institut für Angewandte Mathematik, Universität Bonn, 1992.

27. GASTALDI, F., TOMARELLI, F. : Some Remarks on Nonlinear and Noncoercive Variational Inequalities, Bollettino U. M. I., (7) **1**-B, pp. 143-165, 1987.

28. GEYMONAT, G. : Sui problemi ai limiti lineari ellittici, Ann. Mat. Pura Appl., s. IV, **69**, pp. 207-284, 1965.

29. HADAMARD, J.: Leçons sur la propagation des ondes et les equations de l'hydrodynamique, Paris, 1903.

30. HILL, R.: A general theory of uniqueness and stability for elastic plastic solids, J. Mech. Phys.Solids, **6**, pp. 236-249, 1958.

31. HILL, R.: Acceleration waves in solids, J. Mech. Phys.Solids, **10**, pp.1-16, 1962.

32. HILL, R. : Uniqueness criteria and extremum principles in self-adjoint problems of continuum mechanics, J. Mech. Phys. Solids, **10**, pp. 185-194, 1962.

33. HILL, R. : Aspects of invariance in solids mechanics, Advances in Applied Mechanics, Vol.18, Academec Press, pp. 1-75, 1978.

34. HILL, R. , HUTCHINSON, J. W. : Bifurcation phenomena in the plane tension test, J. Mech. Phys. Solids, **23**, pp. 239-264, 1975.

35. HUTCHINSON, J. W. : Plastic buckling, Advances in applied mechanics, Vol. 14, Academic Press, pp.67-144, 1974.

36. HUTCHINSON, J. W., MILES, J. P. : Bifurcation analysis of the onset of necking in an elastic-plastic cylinder under uniaxial tension, J. Mech. Phys. Solids, **22**, pp.61-71, 1974.

37. HUTCHINSON, J. W. , TVERGAARD, V.: Surface instabilities on statically strained plastic solids, Int. J. Mech. Sci., **22**, pp.339-354, 1980.

38. JOHNSON, C. : On Plasticity with hardening, J. Math. Anal. Appl., **62**, pp. 325-336, 1978.

39 LASRY, D. , BELYTSCKO, T. : Localization limiters in transient problems, Int. J. Solids Struct., **24**, pp. 581-597, 1988.

40. LIONS, J.L.: Quelques méthodes de résolution de problèmes aux limites non linéaires, Dunod, Paris, 1969.

41. LIONS, J.L. : Perturbations singulières dans les problèmes aux limites et en contrôle optimal, L. N. in Mathematics, vol. 323, Springer-Verlag, Berlin, 1973.

42. LIONS, J.L. AND MAGENES, E. : Problèmes aux limites non homogènes, Tome 1, Dunod, Paris, 1964.

43. LORET, B., PREVOST, J.H. : Dynamic strain localization in elasto (visco) plastic solids, Part I -II, Comp. Meth. Appl. Mech. Engineering, **83**, pp. 247-294, 1990

44. LOVE, A. E. H. : A treatise on the mathematical theory of elasticity, New York-Dover Publications, 1944.

45. MANDEL, J.: Ondes plastiques dans un milieu indéfini à trois dimensions, J. Mécanique, **1**, pp. 3-30, 1962.

46. MARSDEN, J.E., HUGHES, T. J. R. : Mathematical foundations of elasticity, Prentice Hall, Englewoods Cliffs, New Jersey, 1983.

47. MULHAUS, H.B., VARDOULAKIS, I. : The thickness of shear bands in granular materials, Geotechnique, **37**, pp. 271-283, 1987.

48. MULHAUS, H. B., AIFANTIS, E. C. : A variational principle for gradient plasticity, Int. J. Solids Struct., 28, pp. 845-857, 1991.

49. NEEDLEMAN, A. : Non-normality and bifurcation in plane strain tension and compression, J. Mech. Phys. Solids, **27**, pp.231-254, 1979.

50. NEEDLEMAN, A., ORTIZ, M. : Effect of boundaries and interfaces on shear band localization, Int. J. Solids Structures, **28**, pp.859-877, 1991.

51. NEEDLEMAN, A. : Material rate dependence and mesh sensitivity in localization problems, Comp. Meth. Appl. Mech. Eng., **67**, pp. 64-85, 1988.

52. PETRIK, H., THERMANN, K.: Second order bifurcation in elastic-plastic solids,J. Mech.Phys. Solids, **33**, pp. 577-593, 1985.

53. PIJAUDIER-CABOT, G., BAZANT, Z. P. : Nonlocal damage theory, ASCE J. Eng. Mech., **112**, pp. 1512-1533, 1987.

54. RANIECKI, B. , BRUHNS, O. T. : Bounds to bifurcation stresses in solids with non-associated plastic flow law at finite strain, J. Mech. Phys. Solids, **29**, pp. 153-172, 1981.

55. RICE, J. R. : Plasticity and soil mechanics, Ed. A. C. PALMER, Department of Engineering, University of Cambridge, England, 1973.

56. RICE, J. R. : The localization of plastic deformation, in : Theoretical and Applied Mechanics , Ed. W. T. KOITER , North Holland, pp. 207-220, 1976.

57. RUDNICKI , J. W., RICE, J. R. : Conditions for the localization of deformation in pressure-sensitive dilatant materials, J. Mech. Phys. Solids, **23**, pp. 371-394, 1975.

58. RICE, J. R , RUDNICKI, J. W. : A note on some features of the theory of localization of deformation, Int. J. Solids Struct., **16**, pp. 597-605, 1980.

59. SAWYERS, K. N. : in : Finite Elasticity, AMD, 27, Ed. R. S. RIVLIN, A.S.M.E. New-York, 1977.

60. SCHAEFFER, O. G. : Instability and ill-posedness in the deformation of granular materials, Int. J. Num. Anal. Meth. Geomech., **14**, pp.253-278, 1990.

61. SIMPSON, SPECTOR : Some necessary conditions at an internal boundary for minimizers in finite elasticity, J. of Elasticity, **26**, pp. 203-, 1991.

62. SIMO, J. C. : Some aspects of continuum damage mechanics and strain softening, in Strain localization and size effect, Ed. J. MAZARS & Z. P. BAZANT, Elsevier Applied Science, pp. 247-258, 1989.

63. SLUYS, J. L., DE BORST, R. : Strain softening under dynamic loading conditions in : Computer-Aided Analysis and Design of Concrete Structures, Eds N. BICANIC & H. A. MANG, Pineridge Press, pp. 1091-1104, 1990.

64. SON Q. NGUYEN, TRIANTAFYLLIDIS, N. : Plastic bifurcation and post-bifurcation analysis for generalized standard continua, J. Mech. Phys. Solids, **37**, pp. 545-566, 1989

65. SUQUET, P. : Discontinuities and Plasticity, in Non-Smooth Mechanics and Applications, C.I.S.M. Lectures Notes, vol. 302, Ed. J. J.MOREAU and P. D. PANAGIOTOPOULOS, Springer-Verlag, 1988

66. STEINMANN, P. , WILLAM, K. : Localization in Micropolar elastoplasticity, in Constitutive Laws for Eng. Materials, Ed DESAI et al, ASME Press, pp. 387, 1991.

67. THOMAS, T. Y. : Plastic flow and fracture of solids, Academic Press, 1961.

68. THOMPSON, J. L. : Some existence theorems for the traction boundary value problem of linearized elastostatics, Arch. Rat. Mech. Anal., **32**, pp. 369-399, 1969.

69. TRIANTAFYLLIDIS, N. : Bifurcation phenomena in pure bending, J. Mech. Phys.Solids, **28**, pp. 221-245, 1980.

70. VARDOULAKIS, I., SULEM, J., GUENOT, A. : Borehole instabilities as bifurcation phenomena, Int. J. Rock. Mech. Min. Sci. & Geomech. Abstract, **25**, pp. 159-170, 1988.

71. YOUNG, N. J. B. : Bifurcation phenomena in the plane compression test, J. Mech. Phys.Solids, **24**, pp.77-, 1976.

## APPENDIX : RIESZ-FREDHOLM THEORY FOR LINEAR OPERATORS

In this appendix, classical results on Fredholm theory are gathered. The starting point is the situation which prevails in finite dimensional vector spaces.

Let $\mathcal{E}$ and $\mathcal{F}$ be finite dimensional complex vector spaces and let T be a linear operator from $\mathcal{E}$ to $\mathcal{F}$. An elementary result of linear algebra asserts that

$$\dim \mathcal{E} = \text{rank}(T) + \dim \text{Ker}(T) \tag{A.1}$$

If one defines the quotient space $\text{Coker}(T) = \mathcal{F}/\text{Im}(T)$ , then $\dim \text{Coker}(T) = \text{codim Im}(T) =$ rank (T) and (A.1) means that

$$\dim \text{Ker}(T) - \text{codim Im}(T) = \dim \mathcal{E} - \dim \mathcal{F} \tag{A.2}$$

so that the left hand side is indeed independent of the linear operator T. This left hand side is called *the index of the operator* T and denoted by ind(T).

A very special case is when $\mathcal{E} = \mathcal{F}$ (or generally when $\dim \mathcal{E} = \dim \mathcal{F}$) and so ind(T) = 0 which means that T is one-to-one if and only if it is surjective (and thus bijective). In this case, it is well known that for the equation

$$\mathbf{T} \,.\, \mathbf{x} = \mathbf{f} \tag{A.3}$$

the following *alternative* holds :

*either* det T ≠ 0 and then for all **f** ∈ $\mathcal{E}$, the equation (A.3) has a unique solution **u** ∈ $\mathcal{E}$

*or* det T = 0 and so the corresponding homogeneous equation (**f** = 0) has a finite number (say n) of linearly independent solutions and in that case equation (A.3) is solvable if and only if **f** satisfies the same number of linearly independent compatibility conditions which are : **f** ∈ $(\ker T^*)^{\perp}$.

Finally, the above solutions depend continuously on the data **f** as the inverse of a linear operator is continuous in finite dimensional situations.

These results do not hold anymore in infinite dimensional space as a linear continuous opeartor can be one-to-one but not surjective and vice-versa. Besides the inverse of a linear operator is not necessarily continuous.

For these reasons, conditions for which the finite dimensional situation still prevails in the case of infinite dimension are fundamental. Such an extension is the so called Riesz-Fredholm theory that we are going to sketch in the following.

The Fredholm alternative is the first result in this direction.

Proposition 1. : Let K be a linear continuous and compact operator in the Banach space $\mathcal{E}$. Then :

(i) Ker(I–K) has finite dimension

(ii) Im(I–K) is closed and more precisely Im(I–K) = $(\text{Ker}(I-K^*))^{\perp}$

(iii) Ker(I–K) = {0} if and only if Im(I–K) = $\mathcal{E}$

(iv) dim Ker(I–K) = dim Ker (I–K*)

The classical Fredholm Alternative concerns the equation

$$\mathbf{u} = \mathrm{K}.\, \mathbf{u} + \mathbf{f} \tag{A.4}$$

and Proposition 1 means that the following *alternative* holds :

*either* for all **f** ∈ $\mathcal{E}$, the equation (A.4) has a unique solution **u** ∈ $\mathcal{E}$

*or* the corresponding homogeneous equation (**f** = 0) has a finite number (say n) of linearly independent solutions and in that case equation (A.4) is solvable if and only if **f** satisfies the same number of linearly independent compatibility conditions which are **f** ∈ $(\text{Ker}(I-K^*))^{\perp}$.

Notice the similarity with the finite dimensional case. Indeed the finite dimensional situation holds for a larger class of linear operators called Fredholm operators.

Let $\mathcal{E}$ and $\mathcal{F}$ be Banach spaces and denote by $\mathcal{E}'$ and $\mathcal{F}'$ their (strong) duals and by $\mathcal{L}(\mathcal{E},\mathcal{F})$ the vector space of linear continuous operators from $\mathcal{E}$ to $\mathcal{F}$. Let $T \in \mathcal{L}(\mathcal{E},\mathcal{F})$. Then the kernel

$$\text{Ker } T = \{\mathbf{x} \in \mathcal{E}\ ; T.\mathbf{x} = 0\}$$

is a closed subspace of $\mathcal{E}$ but the range

$$\text{Im } T = \{\mathbf{y} \in \mathcal{F};\ T.\mathbf{x} = \mathbf{y} \text{ for some } \mathbf{x} \in \mathcal{F}\}$$

is generally not a closed subspace of $\mathcal{F}$. However, as a consequence of Banach's theorem one can prove that if the quotient space $\mathcal{F}/\text{Im } T$ has finite dimension, i.e. Im T has finite codimension in $\mathcal{F}$, then Im T is a closed subspace of $\mathcal{F}$.
We can now state the fundamental definition of Fredholm operator

Definition : The operator $T \in \mathcal{L}(\mathcal{E},\mathcal{F})$ is called a Fredholm operator when :

(i) dim Ker T is finite ,

(ii) Im(T) is closed and has finite codimension , i.e. $\mathcal{F}/\text{Im}(T)$ has finite dimension.

The index of a Fredholm operator is defined by

$$\text{ind}(T) = \dim \text{Ker}(T) - \dim (\mathcal{F}/\text{Im}(T))$$

The set of all Fredholm operators from $\mathcal{E}$ to $\mathcal{F}$ is a subspace of $\mathcal{L}(\mathcal{E},\mathcal{F})$ denoted $\Phi(\mathcal{E},\mathcal{F})$ and $\Phi(\mathcal{E})$ when $\mathcal{E} = \mathcal{F}$.

When $T \in \Phi(\mathcal{E},\mathcal{F})$ we say that the problem :

($\mathcal{P}$) Given $\mathbf{f} \in \mathcal{F}$ find $\mathbf{u} \in \mathcal{E}$ such that $T\mathbf{u} = \mathbf{f}$ .

is *well posed* in the sense that:

i) *the problem has a finite number of linearly independent solutions* (dim Ker T is finite ),

ii) *these solutions depend continuously on the data* (Im(T) is closed ),

iii) *the data must satisfy at most a finite number of linearly independent compatibility conditions* ( $\mathcal{F}/\text{Im}(T)$ has finite dimension.).

The most useful properties of Fredholm operators are summarized in the following theorem.

<u>Theorem.</u> (i) <u>If</u> $T \in \Phi(\mathcal{E},\mathcal{F})$ <u>and</u> K <u>is compact from</u> $\mathcal{E}$ <u>to</u> $\mathcal{F}$ <u>then</u> $(T+K) \in \Phi(\mathcal{E},\mathcal{F})$ <u>and</u> $\text{ind}(T+K) = \text{ind}(T)$

(ii) <u>If</u> $T \in \Phi(\mathcal{E},\mathcal{F})$ <u>then there exists</u> $\eta > 0$ <u>such that when</u> $S \in \mathcal{L}(\mathcal{E},\mathcal{F})$ <u>and</u> $\|T - S\| < \eta$ <u>then</u> $S \in \Phi(\mathcal{E},\mathcal{F})$ <u>and</u> $\text{ind}(T) = \text{ind}(S)$.

(iii) <u>If</u> $T \in \Phi(\mathcal{E},\mathcal{F})$ <u>and</u> $S \in \Phi(\mathcal{F},\mathcal{G})$ <u>then</u> $S \circ T \in \Phi(\mathcal{E},\mathcal{G})$ <u>and</u>

$$\text{ind}(S \circ T) = \text{ind}(S) + \text{ind}(T)$$

(iv) <u>If</u> $S \in \Phi(\mathcal{E},\mathcal{F})$ <u>then</u> $S^T \in \Phi(\mathcal{F}',\mathcal{E}')$ <u>and</u> $\text{ind}(S^T) = -\text{ind}(S)$.

Property (i) means that for the well-posedness of $(\mathcal{LRP}; t)$ one can neglect all the compact perturbations (such as, in general, lower order terms in partial differential equations). The interest of higher order gradient theories lies mainly on this property.
Property (ii) means that if $(\mathcal{LRP}; t_0)$ is well-posed and if the coefficients depends smoothly from t, then $(\mathcal{LRP}; t)$ is also well-posed for $|t - t_0| < \delta$ for some $\delta$ small enough.
Property (ii) means moreover that if $(\mathcal{LRP}; t)$ is well-posed then a good numerical approximation is also well-posed with the same index.

# BIFURCATION AND STABILITY OF TIME-INDEPENDENT STANDARD DISSIPATIVE SYSTEMS

**Q. S. Nguyen**
**Ecole Polytechnique, Palaiseau, France**

## ABSTRACT

This paper presents a general course on stability and bifurcation of dissipative systems based upon energetic considerations. For time-independent systems, it provides an unified framework for the study of quasi-static evolutions and of stability or bifurcation problems in a variety of interesting applications. The theoretical presentation is complemented by a number of simple analytical examples.

The first main issue to be addressed pertains to the buckling of elastic-plastic structures. Stability and bifurcation criteria are discussed for generalized standard models of plasticity in the light of Hill's results.

The second main issue to be addressed pertains to the stability and bifurcation of systems with internal damage or cracks. In brittle fracture or brittle damage, the evolution law of crack lengths or of damage parameters is time-independent as in plasticity and leads to a similar mathematical description of the quasi-static evolution of these systems. Stability and bifurcation analysis can be performed in the same spirit.

# 1.- INTRODUCTION

## 1.1.- Introduction

Our objective is to present here a **general course on stability and bifurcation of dissipative systems** based upon **energetic considerations**. This course provides an unified framework for the study of stability and bifurcation issues in a variety of interesting applications such as **plastic buckling, fracture and damage.**

The theoretical presentation will be given in a self-consistent manner and illustrated by a certain number of simple analytical examples.

This presentation consists of five chapters :

Chapter 1 is an introduction to stability and bifurcation problems in solids.
The objective of the course is first presented. A simple example of plastic buckling is considered in order to underline some principal features of plastic buckling compared to elastic buckling.

Chapter 2 is devoted to the constitutive equations of standard plasticity . Standard models of plasticity are first recalled. The incremental equations of the elastic plastic behaviour are discussed. The quasi-static evolution of elastic plastic structures submitted to a general loading path is considered.
A straightforward generalization of this discussion to the description of the quasi-static response of general standard dissipative systems is then given.

Chapter 3 is devoted to the study of the rate problem for a general time-independent dissipative system undergoing quasi-static evolution. Bifurcation analysis is then performed and the problem of dynamic stability of an equilibrium state is considered. General criteria of stability and non-bifurcation are formulated.

Chapter 4 is an application of Chapter 3 in the context of fracture mechanics, to the problem of multiple fracture. It illustrates the general theory in the case of **discrete variables.**

Chapter 5 is an application of Chapter 3 in the context of damage mechanics. Different models of total or partial brittle damage are considered. As an illustration of the case of **continuous variables**, special attention will be paid to the description of the evolution of damage zones.

## 1.2.- A simple example of plastic buckling

Mathematical modelling of plastic buckling is a difficult problem due to the complexity of the plastic constitutive equations. Historically, this problem has been studied since the early works of Considère (1891) and Von Karmann (1910). A complete comprehension of the problem was obtained only after Shanley's discussion (1947). A general formulation of stability and non-bifurcation criteria has been given by Hill (1958). Post-buckling analysis has been performed by Hutchinson (1973) for common structures like beams, plates and shells in compressive loadings.

### 1.2.1.- Shanley's column

Let us recall first Shanley's example of plastic buckling of a column :

A rigid $\perp$ frame ABCD is loaded by a vertical downward force $\lambda$ on the extremity D. Two elasto-plastic springs AE, BF are attached to points A and B in order to maintain the system in its vertical position while point C is constrained to move without friction on the Oy axis (Fig.1 (1)).

Our objective is to study all possible equilibrium branches when the load parameter $\lambda$ increases from 0. It is already clear that the branch $\theta = 0$ is a trivial branch.

This is an elastic-plastic system of two degrees of freedom : vertical displacement v of the point C and and rotation $\theta$ of the bar CD.

The elongations $e_i$ , i =1, 2 of the two springs can be obtained as a function of these parameters by the kinematic relations :

$$e_1 = v + \ell\theta \ , \quad e_2 = v - \ell\theta$$

for small rotations $\theta$.

It is assumed that linear kinematic hardening model is satisfied by the springs, i.e. the following classical equations hold for each spring :

$$\sigma = E( e - e^p )$$
$$f = ( \sigma - h\, e^p )^2 - k^2 \ \leq \ 0$$
$$\dot{e}^p \ = \ \mu \frac{\partial f}{\partial \sigma} \ , \ \mu \geq 0 \ , \ \mu\, f = 0$$

Stress and strain rates are related by :

$$\dot{\sigma}_i \ = \ E_i\, \dot{e}_i$$

with $E_i = E$ in the elastic regime
$E_i = E_T$ in the plastic regime

where $E_T = \frac{h}{E+h} E$ denotes the tangent modulus.

Static equilibrium of the system gives for small $\theta$ :

$$-\lambda = \sigma_1 + \sigma_2 \quad \text{( force equilibrium )}$$
$$-\lambda L\theta = \ell\,\sigma_2 - \ell\,\sigma_1 \quad \text{( moment equilibrium ).}$$

One obtains then :

$$-\dot{\lambda} = E_1\,\dot{e}_1 + E_2\,\dot{e}_2$$
$$-L(\dot{\lambda\theta}) = \ell\,E_2\,\dot{e}_2 - \ell\,E_1\,\dot{e}_1$$

substituting the kinematic relation for $e_i$ :

$$\dot{v} = -\dot{\lambda}\frac{1}{E_1 + E_2} + \frac{E_2 - E_1}{E_2 + E_1}\ell\dot{\theta}$$

$$(\dot{\lambda\theta}) = \frac{E_2 - E_1}{E_2 + E_1}\frac{\ell}{L}\dot{\lambda} + 4\frac{\ell^2}{L}\frac{E_1\,E_2}{E_1 + E_2}\dot{\theta}$$

If each spring has the same regime (elastic or plastic) during the loading process (assumption to be verified), the moduli $E_i$ are constants. The integration of the previous equations is then straightforward and gives a family of curves:

$$\lambda = \frac{\Delta\,\lambda_0 - \lambda_1\,\theta}{\Delta - \theta}$$

where $\lambda_0$ denotes a constant (initial load) and $\lambda_1$ and $\Delta$ are given by :

$$\Delta = \frac{\ell}{L}\frac{E_2 - E_1}{E_2 + E_1}\,, \quad \lambda_1 = 4\frac{\ell^2}{L}\frac{E_1\,E_2}{E_2 + E_1}$$

If $\Delta \neq 0$, the obtained curves $\lambda(\theta)$ represent a family of hyperbolae with horizontal and vertical asymptotes $\lambda = \lambda_1$ and $\theta = \Delta$ respectively.

Three situations are possible, according to the behaviour of the springs :

- If $E_1 = E_2 = E$, both springs are in the elastic regime. The obtained family of curves reduces to a horizontal line :

$$\lambda = \lambda_E = E\,\frac{2\ell^2}{L}$$

**$\lambda_E$ is the elastic buckling load** of the frame. This possibility occurs when $\lambda_E$ is low enough such that the yield condition is not yet attained in the springs (slender frame).
The frame buckles elastically if $\lambda_E < 2k$.

- If $E_1 = E_2 = E_T$, both springs are in the plastic regime. Since $\Delta = 0$, the horizontal line $\lambda$

$= \lambda_T = E_T \frac{2\ell^2}{L}$ is obtained. However, the assumption of plastic regime on both springs implies that $\dot{e}_i \leq 0$ , $i = 1,2$ $\forall\ \theta$ which is not possible unless $\theta = 0$.

- If $E_2 = E_T$ and $E_1 = E$, one spring is elastic and the other is plastic.

This situation leads to a family of hyperbolae of horizontal asymptote :

$$\lambda = \lambda_R = E_R \frac{2\ell^2}{L} \quad \text{with} \quad E_R = 2 \frac{E\, E_T}{E + E_T}$$

**$\lambda_R$ is by definition the critical load of reduced modulus.**

Here too, it is necessary to check the validity of the previous assumption which implies $\dot{e}_1 \geq 0$ and $\dot{e}_2 \leq 0$. At $\theta = 0$, this condition leads to :

$$\lambda_0 - \lambda_R - 2\, E_T\, \ell\Delta \geq 0\ , \quad \lambda_0 - \lambda_R + 2\, E\, \ell\Delta \leq 0$$

and thus :

$$\lambda_E \geq \lambda_0 \geq \lambda_T = E_T \frac{2\ell^2}{L}$$

**$\lambda_T$ is by definition the critical load of tangent modulus.**

Finally, in addition to the trivial equilibrium branches $\theta = 0$, one obtains non trivial branches which are the portions of hyperbolae situated in the region $\lambda_E \geq \lambda \geq \lambda_T$ (Fig.2 (1)).

### 1.2.2.- Remarks

- This discussion gives all possible branches of static solutions when the load parameter $\lambda$ increases from zero. In particular, it shows the set of **critical points of bifurcation** of the trivial branch where there is deviation from this branch to a non trivial branch.

The critical value $\lambda_T$ is the lowest critical load of bifurcation of the trivial branch. All points of this branch situated in the portion $\lambda_T \leq \lambda \leq \lambda_E$ are also critical bifurcation points. This situation is entirely new in respect to the classical theory of elastic buckling.

- It does not provide any information on the nature **(stable or instable)** of the different equilibria. For this, further study will be necessary.

- In this discussion, our attention was first focussed on the mechanical response of the system, in particular on displacements v, $\theta$ and on force-displacement curves $\lambda(\theta)$.

- Energetical aspects of the problem can also be studied easily.

In this example, the external load $\lambda$ derives from a potential $E_\lambda$ :

$$E_\lambda = \lambda( v + L \cos\theta ) \simeq \lambda\left( v + L\left( 1 - \frac{\theta^2}{2} \right)\right)$$

and the reversible energy stored in the system is the elastic energy of different springs :

$$E_i = \frac{1}{2} E (e_i - e_i^p)^2 + \frac{1}{2} h ( e_i^p )^2$$

The system is irreversible since plastic strains $e_i^p$ induce dissipation of energy in each spring. The associated dissipation power is $D_i = k \mid \dot{e}_i^p \mid$.

- Let us turn now our attention on the evolution of the plastic strains $e_i^p$ , $i = 1,2$.

The evolution of the plastic strains $e^p$ is shown in Fig.3 (1). In the same spirit as in Fig.2 (1), there is **symetry breaking** and **multiple bifurcation** in a continuous portion of the trivial branch $e_1^p = e_2^p$ which corresponds to the load interval $\lambda_E \geq \lambda \geq \lambda_T$.

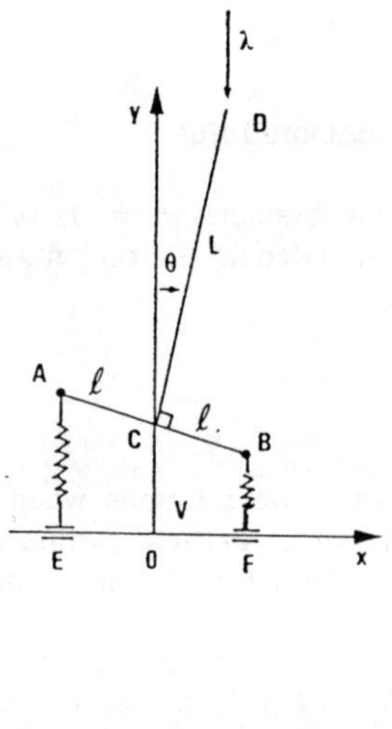


Fig. 1 (1) , Shanley's column

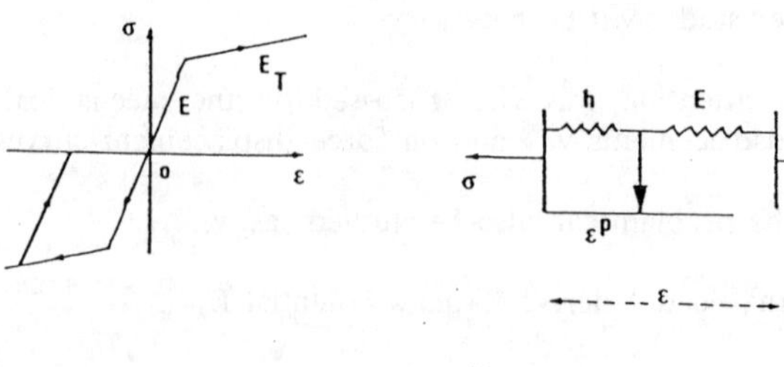

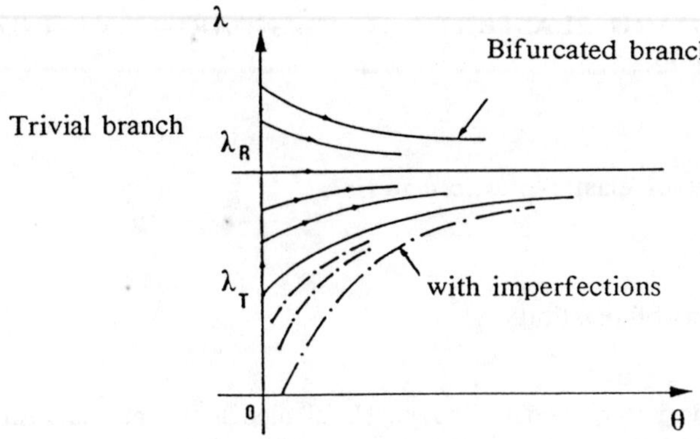


Fig. 2 (1) , Bifurcation diagram of Shanley's column. Trivial and bifurcated branches are presented in $\lambda$-$\theta$ plane. Responses in the presence of imperfections are also given.

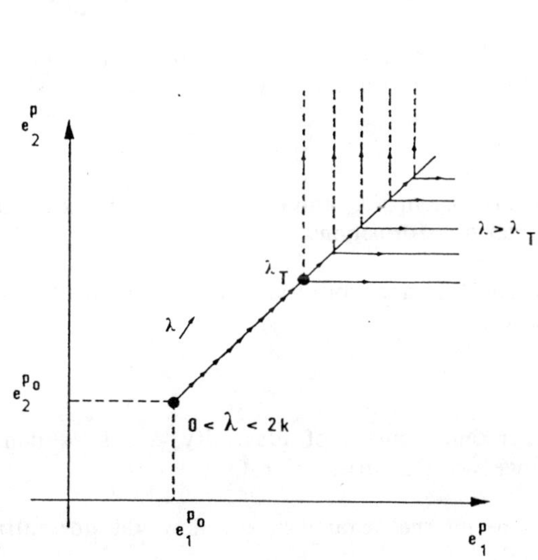


Fig. 3 (1) , Bifurcation diagram of the plastic strains

## 2.- FROM STANDARD PLASTICITY TO STANDARD TIME-INDEPENDENT DISSIPATIVE SYSTEMS

### 2.1.- Standard models of elastic-plastic materials

#### 2.1.1.- Basic equations of plasticity

Elastic plastic behaviour is now widely known. Usual models of **incremental plasticity** at small or finite strain can be introduced in several manners. Because of its large domain of validity, the following presentation will be adopted although it is not the simplest one for practical applications.

In a phenomenological approach, elastic plastic materials can be characterized by a set of **state variables** $(\epsilon, \alpha, T)$ where $\epsilon$ denotes the strain, T the temperature and $\alpha$ a set of internal parameters describing the irreversible behaviour of the material.

Let $w(\epsilon, \alpha, T)$ be the free energy per unit volume, the following **state equations** hold :

$$\sigma = \frac{\partial w}{\partial \epsilon} , \quad s = - \frac{\partial w}{\partial T} , \quad A = - \frac{\partial w}{\partial \alpha} \tag{2.1}$$

where $\sigma$ is the dual stress of the strain measure $\epsilon$, s the entropy per unit volume and A the force associated to the rate $\dot{\alpha}$ in the expression of the intrinsic dissipation :

$$d_i = \sigma \dot{\epsilon} - (\dot{w} + T\dot{s}) = A \dot{\alpha} \tag{2.2}$$

In the following discussion, the temperature T will be omitted since we are interested primarily in isothermal transformations.

The associated force A is not arbitrary. Physically admissible forces A must satisfy the yield condition of plasticity :

$$A \in C \tag{2.3}$$

where C is the present **domain of elasticity**, a convex domain in the force space. This domain may change when $\alpha$ varies , $C = C(\alpha)$.

The evolution of the internal parameters $\alpha$ follows the **normality law** :

$$\dot{\alpha} = N_C (A) \tag{2.4}$$

where $N_C (A)$ denotes an external normal to the convex C at the point A.

If A and $\dot{\alpha}$ are associated following the normality law (2.4), it follows that :

$$A\,\dot{\alpha} = \text{Max}\ A^*.\dot{\alpha} \quad \forall\ A^* \in C(\alpha) \tag{2.5}$$

This inequality is well known as Hill's maximum dissipation principle.

From (2.4) or (2.5), the dissipation $d_i$ can finally be expressed as a function $d(\alpha, \dot{\alpha})$, positively homogeneous of degree 1 and convex with respect to $\dot{\alpha}$ :

$$d(\alpha, m\,\dot{\alpha}) = m\,d(\alpha, \dot{\alpha}) \ , \ \forall \ m \geq 0. \tag{2.6}$$

such that :

$$A = \frac{\partial d}{\partial \dot{\alpha}} \tag{2.7}$$

When $\dot{\alpha} = 0$, the second member of (2.7) is still available if $\frac{\partial d}{\partial \dot{\alpha}}$ is understood as a sub-gradient, cf. Moreau [23].

Note that finally, the mechanical behaviour of an elastic plastic material is described by the expressions of two potentials, the thermodynamic potential $w(\epsilon, \alpha)$ , and the dissipative potential $d(\alpha, \dot{\alpha})$, cf. Germain [15].

For example, the usual model of perfect plasticity with Mises criterion is defined by $\alpha = \epsilon^p$ , $w = \frac{1}{2}(\epsilon-\alpha)\,E\,(\epsilon-\alpha)$ and $d = k\,|\dot{\alpha}|$. It follows from this expression of d that C is defined by the criterion $f = |A| - k \leq 0$ , C is independent of $\alpha$ if k is constant. The normality law (4) can also be written in a classical way as :

$$\dot{\alpha} = \mu\,\frac{\partial f}{\partial A} \ , \ f \leq 0 \ , \ \mu \geq 0, \ \ \mu\,f = 0. \tag{2.8}$$

In the same spirit, the usual Ziegler-Prager model of kinematic hardening is defined by the same function d and a different energy $w = \frac{1}{2}(\epsilon-\alpha).E.(\epsilon-\alpha) + \frac{1}{2}\,\alpha.h.\alpha$.

Note that if $\alpha = (\epsilon^p , \beta)$ and $w = w_1(\epsilon-\epsilon^p) + w_2(\beta)$ , the associated force A is $A = (\sigma, B)$ where $B = -\frac{\partial w_2}{\partial \beta}$ . If the elastic domain C can be expressed as an inequality $f(\sigma,B,\alpha) \leq 0$, the normality law (2.4) implies that

$$\dot{\epsilon}^p = \mu\,\frac{\partial f}{\partial \sigma} \ , \ \dot{\beta} = \mu\,\frac{\partial f}{\partial B} \tag{2.9}$$

and thus relations (2.4) requires a complete normality which is more restrictive than the so called associated models of plasticity in the literature. Normality is here fully assumed, for both plastic strain and internal parameters.

For example, the model of isotropic hardening is defined by the criterion
$f = |\sigma| + B - k \leq 0$ . From (2.9) , it follows that

$$\dot{\epsilon}^p = \mu \frac{\sigma}{|\sigma|} \ , \ \dot{\beta} = \mu = |\dot{\epsilon}^p|$$

$\beta$ is the equivalent plastic strain.

Such models will be denoted as **standard models** (in the sense of generalized standard models cf. Halphen & Nguyen [16]). This class of models is also compatible with general descriptions of metal plasticity, cf. Rice [31].

### 2.1.2.- Incremental relations

The incremental nature of the elastic plastic behaviour follows naturally from (2.4) and from the differentiation of state equations (2.1) with respect to a time-like parameter t.

Let us begin with the simplest case of **simple plastic potential,** in which the elastic domain C can be expressed by an inequality f $(\alpha, A) \leq 0$, f denotes a differentiable function. Our goal is to show that for a given state $(\epsilon,\alpha)$ , the stress rate $\dot{\sigma}$ and internal parameter rate $\dot{\alpha}$ (which are r.h.s. derivatives) can be expressed in terms of the strain rate $\dot{\epsilon}$ under the following **regularity assumption** of all state parameters :

The state $(\epsilon(t), \alpha(t))$ varies continuously such that their rate $(\dot{\epsilon}(t), \dot{\alpha}(t))$ are piecewise continuous. (2.10)

It follows from (2.1) that $\sigma$ and A must have the same regularity, i. e. $\dot{\sigma}$ and $\dot{A}$ are piecewise continuous. As a consequence of this regularity, the following equation holds for almost all t

$$(\mu \dot{f}) = \dot{\mu} f + \mu \dot{f} = 0 \tag{2.11}$$

$$\text{if } \mu > 0 \text{ then } \dot{f} = 0 \tag{2.12}$$

From (2.1), (2.8) and (2.12), one obtains then :

$$\mu = \frac{< f_{,A} \cdot w_{,\alpha\epsilon} \cdot \dot{\epsilon}>}{f_{,A} \cdot w_{,\alpha\alpha} \cdot f_{,A} - f_{,A} f_{,\alpha}} \tag{2.13}$$

where <.> denotes the positive part of a number. From (2.13), the expression of $\dot{\sigma}$ and $\dot{\alpha}$ can be derived explicitly. In particular, it is well known that a rate potential $U(\dot{\epsilon})$ exists such that

$$\dot{\sigma} = \frac{\partial U}{\partial \dot{\epsilon}} \tag{2.14}$$

with :

$$U = \frac{1}{2} \dot{\epsilon} . w_{,\epsilon\epsilon} . \dot{\epsilon} - \frac{1}{2} \frac{< f_{,A} \cdot w_{,\alpha\epsilon} , \dot{\epsilon} >^2}{f_{,A} \cdot w_{,\alpha\alpha} \cdot f_{,A} - f_{,A} f_{,\alpha}} \tag{2.15}$$

In the general case, the incremental behaviour cannot be written so easily. However, a general result can be given in the following way :

In the same spirit as (2.8) and (2.11), a general consequence of the normality law (2.4) and of the required regularity (10) is first derived :

**Proposition 1 (2)**

For any t , one obtains :

$$\dot{A} \cdot \dot{\alpha} = \dot{\alpha} \cdot d_{,\alpha} (\alpha, \dot{\alpha}) \tag{2.16}$$

$$\dot{A} \cdot \delta\alpha \leq \dot{\alpha} \cdot d_{,\alpha} (\alpha, \delta\alpha) \quad \forall \ \delta\alpha \in N_C (A) \tag{2.17}$$

Indeed, let A(t) be the force associated with $\alpha(t)$ by the normality law (2.4) with the required regularity. To simplify, the notations $\tau = t' - t$ , $A_\tau = A(t')$, $A = A(0)$ is introduced. From normality and convexity assumptions, inequality (2.5) is always satisfied and gives for any given $\delta\alpha$ :

$$g_\tau = d(\alpha_\tau , \delta\alpha) - A_\tau \cdot \delta\alpha \geq 0 \quad \forall \ \delta\alpha$$

On the other hand, $g_0 = 0$ if $\delta\alpha \in N_C (A)$. The r.h.s. derivative $\dot{g}$ is then necessarily non negative :

$$\dot{g} = \dot{\alpha} \, d_{,\alpha} (\alpha, \delta\alpha) - \dot{A} \cdot \delta\alpha \geq 0 \quad \forall \ \delta\alpha \in N_C (A).$$

while the l.h.s. derivative $\dot{g}-$ is non positive $\dot{g}- \leq 0$.

Thus inequality (2.17) is ensured.

To obtain (2.16), note that if A(t) and $\alpha(t)$ are derivable at time t ( i.e. if r.h.s. and l.h.s. derivatives are identical), then $\dot{g} = \dot{g}- = 0$ and thus (2.16) is true with $\delta\alpha = \dot{\alpha}$ .

If A(t) or $\alpha(t)$ is not derivable at time t, they are necessarily derivable at time t'> t near t because of the piecewise continuity assumption. Equality (2.16) is then obtained at time t' and also at time t when $t' \Rightarrow t$ by continuity.

If C does not depend on $\alpha$, equality (2.16) means that the rates $\dot{A}$ and $\dot{\alpha}$ are orthogonal. This result is well known in perfect plasticity, $\dot{\sigma} \, \dot{\epsilon}^p = \mu \dot{f} = 0$.

Note that, in the particular case of simple plastic potential i.e. when C can be expressed by a differentiable criterion $f(A,\alpha) \leq 0$ , equality (2.16) is equivalent to equality $\dot{f} = 0$ when $\dot{\alpha} \neq 0$ and inequality (2.17) is equivalent to inequality $\dot{f} \leq 0$ when $f = 0$.

From (2.1), (2.16) and (2.17), it follows that the rate $\dot{\alpha}$ can be expressed in terms of $\dot{\epsilon}$ as a solution of a variational inequality :

**Proposition 2 (2)**

The rate $\dot{\alpha}$ associated to the rate $\dot{\epsilon}$ is a solution of the following variational inequality :

$$\dot{\alpha} \in N_C(A) \text{ and satisfies } \forall \ \delta\alpha \in N_C(A) :$$

$$(\delta\alpha - \dot{\alpha}).( w_{,\alpha\alpha} . \dot{\alpha} + w_{,\alpha\epsilon} . \dot{\epsilon} ) + \dot{\alpha} \ (d_{,\alpha} (\alpha, \delta\alpha) - d_{,\alpha} (\alpha, \dot{\alpha} ) ) \geq 0 \qquad (2.18)$$

The reader may refer to Duvaut & Lions [10] for a general study on variational inequalities in physics and in mechanics.

Inequality (2.18) shows a priori that if the dissipative potential d depends on the present state, the term $\dot{\alpha} . d_{,\alpha} ( \alpha, \delta\alpha)$ is not symmetric with respect to the couple $\dot{\alpha}$ and $\delta\alpha$. This fact may induce potential difficulty. This variational inequality will be discussed later, in Chap. 3.

The incremental behaviour of an elastic plastic material is characterized by equations (2.18). For a given strain path $\epsilon(t)$ , this variational inequality allows to define an associated response $\alpha(t)$ from an initial given value $\alpha_0$ and to obtain the associated stress response $\sigma(t)$.

In the general case, it is not excluded that this behaviour of the material may present some singular features such as loss of uniqueness or unstable response.

## 2.2.- Quasi-static response of elastic-plastic structures

### 2.2.1.- Quasi-static evolution of an elastic-plastic structure

Let us consider now an elastic-plastic structure of volume $\Omega$ in the reference configuration (lagrangian description if finite displacement), submitted to a quasi-static loading controlled by a finite numbers of control parameters $\lambda_i$ , i = 1,n. This loading consists of applied volume force $F(\lambda)$ and surface force $T_d(\lambda)$ on a portion $S_T$ of the boundary $\partial\Omega$ and prescribed displacement $u_d(\lambda)$ on the complementary part $S_u$ . The function $\lambda(t)$ is given for $t \in [0,T]$. We are interested in the quasi-static response of this structure associated with the given loading.

The basic equations describing the quasi-static evolution of this structure are :

- The static conditions which can be expressed as virtual work equation :

$$\int_\Omega \sigma . \delta\epsilon \ d\Omega - \int_\Omega F \ \delta u \ d\Omega - \int_{S_T} T_d \ \delta u \ dS = 0$$

$$\forall \ \delta u = 0 \text{ on } S_u \qquad (2.19)$$

- The kinematic conditions $\mathbf{u} = \mathbf{u_d}$ on $S_u$

- The constitutive equations (1), (3), (4) at each point of $\Omega$.

- The initial conditions $\mathbf{u}(0) = \mathbf{u}_0$ , $\alpha(0) = \alpha_0$ .

The displacement field **u(t)** and the internal parameter field $\alpha(t)$ are the two principal unknowns of the evolution problem.

### 2.2.2.- Incremental response

The incremental nature of the constitutive equations leads to the formulation of the **rate problem** for the structure.

The present state i.e. the fields **u(t)**, $\alpha(t)$ and consequently also $\epsilon(t)$ **and** $\sigma(t)$ is assumed to be known in $\Omega$. With control parameters varying at rate $\dot{\lambda}$, the question is to determine the associated response $\dot{u}$ and $\dot{\alpha}$ .

This associated rate response can be obtained by the resolution of the rate problem which stems simply from the incremental form of the basic equations. These incremental equations are :

- The static rate equations :

$$\int_\Omega ( \dot{\sigma}\, \delta\epsilon + \sigma : \nabla\delta u .\, \nabla\dot{u} )\, d\Omega - \int_\Omega \dot{F}\, \delta u\, d\Omega - \int_{S_T} \dot{T}_d \,.\, \delta u\, dS = 0 \tag{2.20}$$

- The kinematic rate equations :

$$\dot{u} = \dot{u}_d \quad \text{on } S_u \tag{2.21}$$

- The rate constitutive equations : variational equation (2.18) and

$$\dot{\sigma} = w_{,\epsilon\epsilon} \,.\, \dot{\epsilon} + w_{,\epsilon\alpha} \,.\, \dot{\alpha} \quad \text{in } \Omega. \tag{2.22}$$

or , in the case of simple plastic potential, equations (2.8), (2.13) and (2.14), (2.15).

The rate problem is particularly simple in the case of simple plastic potential with the principal unknown $\dot{\mathbf{u}}$. Indeed, equations (2.20), (2.14) and (2.15) give :

$$\int_\Omega \left( \frac{\partial U}{\partial(\dot{\epsilon})}\, \delta\epsilon + \sigma : \nabla\delta u \,.\, \nabla\dot{u} \right) d\Omega - \int_\Omega \dot{F}\, \delta u\, d\Omega$$

$$- \int_{S_T} \dot{T}_d \,.\, \delta u\, dS = 0 \tag{2.23}$$

$\forall\ \delta u = 0$ on $S_u$ .

In the general case, it may be interesting to present the obtained equations in a more compact form in order to understand the mathematical nature of the rate problem. This is done in the following way (bold letters are related to fields with trivial significance) :

Let $\mathbf{E(u,\alpha,\lambda)}$ be the **total potential energy of the system** :

$$\mathbf{E(u,\alpha,\lambda)} = \int_\Omega w(\epsilon(u),\alpha)\, d\Omega - \int_\Omega F(\lambda)\, u\, d\Omega - \int_{S_T} T_d(\lambda)\, u\, dS \tag{2.24}$$

Equations (2.1), (2.19) lead to :

$$\boldsymbol{\delta u} \,.\, \mathbf{E}_{,\mathbf{u}} = \mathbf{0} \quad \forall\ \boldsymbol{\delta u} = \mathbf{0} \text{ on } S_u \tag{2.25}$$

and equations (2.20), (2.22) lead to :

$$\boldsymbol{\delta u} \,.\, (\, \mathbf{E}_{,\mathbf{uu}} \,.\, \dot{\mathbf{u}} + \mathbf{E}_{,\mathbf{u}\alpha} \,.\, \dot{\alpha} + \mathbf{E}_{,\mathbf{u}\lambda} \,.\, \dot{\lambda}\, ) = 0 \tag{2.26}$$

Let $\mathbf{D}(\alpha,\dot{\alpha})$ be the **dissipative potential of the system** :

$$\mathbf{D}(\alpha,\dot{\alpha}) = \int_\Omega d(\alpha,\dot{\alpha})\, d\Omega \tag{2.27}$$

Then equations (2.1), (2.18) lead to :

$$(\, \delta\alpha - \dot{\alpha}\, ) \,.\, (\, \mathbf{E}_{,\alpha\mathbf{u}} \,.\, \dot{\mathbf{u}} + \mathbf{E}_{,\alpha\alpha} \,.\, \dot{\alpha} + \mathbf{E}_{,\alpha\lambda} \,.\, \dot{\lambda}\, ) + \tag{2.28}$$
$$(\, \mathbf{D}_{,\alpha}(\alpha, \delta\alpha) - \mathbf{D}_{,\alpha}(\alpha, \dot{\alpha})\, ) \,.\, \dot{\alpha} \geq 0 \quad \forall\ \delta\alpha \in \mathbf{N_C\,(A)}\,.$$

Finally, the following proposition is obtained :

**Proposition 3 (2)**

The rate $(\dot{\mathbf{u}}\,,\, \dot{\alpha}\,)$ is a solution of the variational inequality :

$\dot{\mathbf{u}} \in \mathbf{V}$ and $\dot{\alpha} \in \mathbf{N_C\,(A)}$ and satisfy :

$$\boldsymbol{\delta u} \,.\, (\, \mathbf{E}_{,\mathbf{uu}} \,.\, \dot{\mathbf{u}} + \mathbf{E}_{,\mathbf{u}\alpha} \,.\, \dot{\alpha} + \mathbf{E}_{,\mathbf{u}\lambda} \,.\, \dot{\lambda}\, ) + \tag{2.29}$$
$$(\, \delta\alpha - \dot{\alpha}\, ) \,.\, (\, \mathbf{E}_{,\alpha\mathbf{u}} \,.\, \dot{\mathbf{u}} + \mathbf{E}_{,\alpha\alpha} \,.\, \dot{\alpha} + \mathbf{E}_{,\alpha\lambda} \,.\, \dot{\lambda}\, ) +$$
$$(\, \mathbf{D}_{,\alpha}(\alpha, \delta\alpha) - \mathbf{D}_{,\alpha}(\alpha,\dot{\alpha})\, ) \,.\, \dot{\alpha} \geq 0$$
$$\forall\ \boldsymbol{\delta u} = \mathbf{0} \text{ on } S_u \text{ and } \delta\alpha \in \mathbf{N_{C(A)}}$$

where $\mathbf{V}$ denotes the set of kinematically admissible rates of displacement.

The proof of this proposition is straightforward from (2.26) and (2.28).

### 2.2.3.- Remarks

Proposition 3 (2) presents a **mixed formulation** of the rate problem where the couple of rates ($\dot{\mathbf{u}}$ , $\dot{\boldsymbol{\alpha}}$) is the principal unknown.

In the particular case of simple plastic potential, it is noted from equation (2.23) that the internal parameter rate can be eliminated. Thus the rate problem can be formulated in terms of the principal unknown $\dot{\mathbf{u}}$ . Equation (2.23) corresponds to the **displacement rate formulation.**

Alternatively, it is possible to eliminate the displacement rate $\dot{\mathbf{u}}$ and retain as principal unknown the rate $\dot{\boldsymbol{\alpha}}$ . Such an idea is not new in Plasticity although it may appear to be curious and unusual (cf. Maier [21]).

To obtain the **internal parameter formulation** of the rate problem, it may be useful to note first that equation (2.25) gives an implicit representation of $\mathbf{u}$ in terms of $\boldsymbol{\alpha}$ and of $\boldsymbol{\lambda}$ :

$$\mathbf{u} = \mathbf{u}(\boldsymbol{\alpha},\boldsymbol{\lambda}) \tag{2.30}$$

when the quadratic form $E_{,uu}$ is positive definite :

$$\delta\mathbf{u} \,.\, \mathbf{E}_{,\mathbf{uu}} (\mathbf{u},\boldsymbol{\alpha},\boldsymbol{\lambda}) \,.\, \delta\mathbf{u} > 0 \quad \forall \;\; \delta\mathbf{u} \neq 0 \tag{2.31}$$

The physical significance of (2.31) is clear. This condition is related to the **elastic stability** of the present state of the system since all irreversible parameters are kept constant. In our discussion, condition (2.31) will be assumed to be always satisfied, at least in the vicinity of the present state.

The implicit representation (2.30) enables us to introduce the value of the potential energy of the system at equilibrium :

$$\mathbf{W}(\boldsymbol{\alpha},\boldsymbol{\lambda}) = \mathbf{E}(\mathbf{u}(\boldsymbol{\alpha},\boldsymbol{\lambda}),\boldsymbol{\alpha},\boldsymbol{\lambda}) \tag{2.32}$$

As a consequence of (2.25), it follows that :

$$\delta\alpha \,.\, \mathbf{W}_{,\boldsymbol{\alpha}} = \int_\Omega w_{,\alpha} (\epsilon,\alpha)\, \delta\alpha \;\; d\Omega \tag{2.33}$$

and thus the present force $\mathbf{A}$ is the general force associated with $\boldsymbol{\alpha}$ by the expression of the energy at equilibrium $\mathbf{W}$ :

$$\mathbf{A} = - \mathbf{W}_{,\boldsymbol{\alpha}} (\boldsymbol{\alpha},\boldsymbol{\lambda}) \tag{2.34}$$

It is thus shown that the unknown $\boldsymbol{\alpha}(\mathbf{t})$ satisfies formally an abstract differential equation with initial condition as follows :

**Proposition 4 (2)**

The unknown $\alpha$ is a solution of the following evolution equation :

Find $\alpha(t)$ such that :

$$\begin{aligned} A &= - W_{,\alpha}(\alpha,\lambda) \\ \dot{\alpha} &= N_C(A) \\ \alpha(0) &= \alpha_0 \end{aligned} \qquad (2.35)$$

where the expression of the energy at equilibrium **W** and the normal operator $N_C(A)$ to a variable convex $C(\alpha)$ play a fundamental role.

## 2.3.- Standard time-independent dissipative systems

### 2.3.1.- General framework

The evolution of an elastic plastic structure provides a good example of standard dissipative systems. Indeed, the previous equations enter into the framework of the quasi-static evolution of a large class of irreversible systems including not only incremental plasticity but also usual models of fracture, damage and friction.

The objective of this section is to give a **general presentation of standard time-independent dissipative systems** as a straightforward generalization of the equations of standard plasticity.

It is assumed that for such a system of material points, the set of variables $(u,\alpha,\lambda)$ represents state variables where **u** denotes the displacement of the system, $\alpha$ a set of local or global parameters and $\lambda$ are control variables of the external action (applied forces or displacements).

The evolution of parameters $\alpha$ characterizes all irreversible mechanisms of the considered system. In order to simplify the presentation, $\alpha$ are assumed to be elements of a vector space, $R_m$ in the discrete case or a functional space H in the continuous case, although this restriction is not essential as we will see in Chapter 5. If these parameters are kept constant, the evolution of the system will be purely elastic or reversible.

Displacement **u** must belong to a subset $U(\alpha,\lambda)$ of a vector space denoted as the set of kinematically admissible displacements. **U** depends on the present values of $\alpha,\lambda$.

If the system is not in static equilibrium, let $K_t$ be the present value of the kinetic energy of the system. The total reversible energy of the system is

$$\mathbf{K_t} + \mathbf{E}(\mathbf{u},\alpha,\lambda) \tag{2.36}$$

when $\mathbf{E}(\mathbf{u},\alpha,\lambda)$ represents as before the **thermodynamic potential** which is the sum of the free energy stored by all material elements and the potential energy of the external action associated with the present value of control $\lambda$.

The part of energy which is dissipated in the system can be defined from the **dissipative potential** of the system $\mathbf{D}(\alpha,\dot{\alpha})$. If $\mathbf{A}$ is the force parameter associated to the rate $\dot{\alpha}$ in the expression of the dissipation :

$$P_d = \mathbf{A} \cdot \dot{\alpha} \tag{2.37}$$

by definition, the following relation holds :

$$\mathbf{A} = \mathbf{D}_{,\dot{\alpha}} \tag{2.38}$$

In quasi-static evolution, the static equilibrium is always satisfied. In this case, it is assumed that the displacement solution $\mathbf{u}$ must realize a local minimum :

$$\mathbf{E}(\mathbf{u},\alpha,\lambda) = \underset{u^*}{\mathbf{Min}}\ \mathbf{E}(\mathbf{u}^*,\alpha,\lambda) = \mathbf{W}(\alpha,\lambda) \tag{2.39}$$

in order to avoid all risk of **elastic instability** of the present state. This assumption leads again to the implicit presentation $\mathbf{u} = \mathbf{u}(\alpha,\lambda)$ and to the introduction of the energy at equilibrium of the system $\mathbf{W}(\alpha,\lambda)$ with :

$$\mathbf{A} = -\mathbf{W}_{,\alpha} \tag{2.40}$$

For a given loading path controlled by the function $\lambda(t)$, the quasi-static evolution of the system is described by the following abstract differential equation :

$$\mathbf{W}_{,\alpha} + \mathbf{D}_{,\dot{\alpha}} = 0 \tag{2.41}$$

This abstract equation comprises the same number of equations and unknowns. For example if $\alpha$ is a vector of $R_m$ , a system of m differential equations of order 1 is obtained.

The general validity of equation (2.41) in Mechanics can be considered as classical, cf. the work of Biot [5] in visco-elasticity.

The behaviour is time-independent if the dissipative potential is positively homogeneous of degree 1 :

$$\mathbf{D}(\alpha, \mu\, \dot{\alpha}) = \mu\, \mathbf{D}(\alpha, \dot{\alpha}), \quad \forall\ \mu \geq 0 \tag{2.42}$$

If the disspative potential is also assumed to be **convex** with respect to $\dot{\alpha}$ , equation (2.38) or (2.41) is still valid for time-independent systems when $\mathbf{D}_{,\dot{\alpha}}$ is understood in the sense of sub-gradients (Moreau [23], Germain [15]). In this case, to the dissipative potential $\mathbf{D}(\alpha,\dot{\alpha})$ can be associated a domain of admissible forces $\mathbf{C}(\alpha)$ which is the set of sub-gradients of D at the point $\dot{\alpha} = \mathbf{0}$ such that the normality law :

$$\dot{\alpha} = N_C ( A ) \tag{2.43}$$

is satisfied for associated force and rate.

Finally for time-independent standard systems, evolution equation (2.41) with a prescribed initial condition $\alpha(0) = \alpha_0$ and convex dissipative potential can also be written under the equivalent form (2.35).

**Equations (2.35) represent the basic equations of evolution of time-independent standard systems.**

### 2.3.2.- Rate problem

Let us assume now that the present state is known. The rate problem consists in searching for the associated response $\dot{\alpha}$ as a function of $\dot{\lambda}$ .

Under the usual regularity assumption concerning $\alpha(t)$ given by (2.10), Proposition 1 (2) can be again applied to obtain the consequence of normality law (2.43) under the previous form (2.16) and (2.17) :

$$\dot{A} \cdot \dot{\alpha} = \dot{\alpha} \cdot D_{,\alpha} (\alpha, \dot{\alpha} ) \tag{2.44}$$

$$\dot{A} \cdot \delta\alpha \leq \dot{\alpha} \cdot D_{,\alpha} (\alpha,\delta\alpha) \quad \forall \ \delta\alpha \in N_C (A) \tag{2.45}$$

The following proposition is then obtained :

**Proposition 5 (2)**

The rate $\dot{\alpha}$ is a solution of the variational inequality :

$\dot{\alpha} \in N_C (A)$ and satisfies $\forall \ \delta\alpha \in N_C (A)$ :

$$( \delta\alpha - \dot{\alpha} ) \cdot ( W_{,\alpha\alpha} \cdot \dot{\alpha} + W_{,\alpha\lambda} \cdot \dot{\lambda} ) + \dot{\alpha} \cdot (D_{,\alpha} (\alpha, \delta\alpha ) - D_{,\alpha} (\alpha,\dot{\alpha} )) \geq 0 \tag{2.46}$$

Note that (2.46) is identical to (2.18). In (2.18), the strain history $\epsilon(t)$ plays the role of the control parameters.

### 2.3.3.- Illustration

Let us consider again the example of Shanley's column. In this elastic-plastic system, the displacement is $\mathbf{u} = (v,\theta)$ , the irreversible variables $\boldsymbol{\alpha} = (\epsilon_1^p , \epsilon_2^p )$ and the thermodynamic potential of the system is :

$$\mathbf{E} = \frac{1}{2} E ( v - \ell\theta - \epsilon_1^p )^2 + \frac{1}{2} h ( \epsilon_1^p )^2 +$$

$$\frac{1}{2} E ( v + \ell\theta - \epsilon_2^p )^2 + \frac{1}{2} h ( \epsilon_2^p )^2 + \lambda \left\{ v + L\left(1 - \frac{\theta^2}{2}\right)\right\}$$

Dissipative potential is :

$$\mathbf{D} = k \mid \dot\epsilon_1^p \mid + k \mid \dot\epsilon_2^p \mid$$

This function does not depend on the present value of $\alpha$. The associated domain of admissible forces C is :

$$C = \{ \mathbf{A} = (A_1 , A_2 ) \mid \; |A_i | \leq k , i = 1,2 \}$$

The static equations are :

$$\frac{\partial}{\partial v} \mathbf{E} = 0 \quad \text{( force equilibrium )}$$

$$\frac{\partial}{\partial \theta} \mathbf{E} = 0 \quad \text{( moment equilibrium )}$$

This two equations give v and $\theta$ in terms of $\epsilon_1^p$ and $\epsilon_2^p$ , $v = v(\epsilon_1^p , \epsilon_2^p , \lambda)$, $\theta = \theta(\epsilon_1^p ,\epsilon_2^p , \lambda)$ when $\lambda < \lambda_E$ and lead to the expression of energy at equilibrium $\mathbf{W}(\epsilon_1^p , \epsilon_2^p , \lambda)$.

## 3.- BIFURCATION AND STABILITY ANALYSES

In this chapter, bifurcation and stability analyses are performed in the general framework of time-independent standard dissipative systems.

The study of the rate problem provides the starting point for this discussion. The notion of critical points of bifurcation of a quasi-static branch is introduced. Hill's method leads to a criterion of non-bifurcation. The dynamic stability, i. e. stability in the sense of Liapunov of an equilibrium, is discussed.

### 3.1.- Analyses of the rate problem

Let us consider in full generality the evolution problem of a time-independent standard dissipative system :

$$\begin{aligned} A &= - W_{,\alpha} (\alpha,\lambda) \\ \dot{\alpha} &= N_C (A) \\ \alpha(0) &= \alpha_0 \end{aligned} \tag{3.1}$$

If the present state $\alpha(t)$ is assumed to be known, it has been established that the rate $\dot{\alpha}$ associated to the rate $\dot{\lambda}$ is a solution of the variational inequality :

$\dot{\alpha} \in N_C (A)$ and satisfies $\forall\ \delta\alpha \in N_C (A)$ :

$$\begin{aligned} &(\delta\alpha - \dot{\alpha}) \,.\, (W_{,\alpha\alpha} \,.\, \dot{\alpha} + W_{,\alpha\lambda} \,.\, \dot{\lambda}) + \\ &\qquad \dot{\alpha} \,.(D_{,\alpha} (\alpha,\delta\alpha) - D_{,\alpha} (\alpha,\dot{\alpha})) \geq 0. \end{aligned} \tag{3.2}$$

Concerning this variational inequality, the following results are obtained :

**Symmetry**

This variational inequality is not necessarily symmetric due to the last terms. Its symmetry is ensured only if :

$$\begin{aligned} &\delta\beta \,.\, D_{,\alpha} (\alpha, \delta\alpha) = \delta\alpha \,.\, D_{,\alpha} (\alpha, \delta\beta) \\ &\forall\ \delta\alpha, \delta\beta \in N_C (A) \end{aligned} \tag{3.3}$$

**Existence**

For a given arbitrary rate $\dot{\lambda}$ , equation (3.2) admits always at least one solution if the following positivity is satisfied :

$$\begin{aligned} &\delta\alpha \,.\, W_{,\alpha\alpha} \,.\, \delta\alpha + \delta\alpha \,.\, D_{,\alpha} (\alpha, \delta\alpha) > 0 \\ &\forall\ \delta\alpha \neq 0 \text{ and } \in N_C (A) \end{aligned} \tag{3.4}$$

For the proof of this statement, the reader may refer to Duvaut & Lions [10].

**Uniqueness**

The rate $\dot{\alpha}$ is also unique if the following positivity is satisfied :

$$\begin{aligned} &(\delta\beta-\delta\alpha) \,.\, W_{,\alpha\alpha} \,.\, (\delta\beta-\delta\alpha) + \\ &(\delta\beta-\delta\alpha) \,.\, (D_{,\alpha} (\alpha,\delta\beta) - D_{,\alpha} (\alpha,\delta\alpha)) > 0 \end{aligned} \tag{3.5}$$

$$\forall\ \delta\beta \neq \delta\alpha \in N_C (A)$$

It may be useful to note that condition (3.5) is more restrictive than (3.4).

## 3.2.- Bifurcation analysis

The notion of **point of bifurcation** of a quasi-static branch of responses in the **($\lambda$-$\alpha$)** space is a natural one (Fig.1 (3)) :

An equilibrium **($\alpha$,$\lambda$)** is a critical point of bifurcation if it corresponds to the intersection of at least two branches.

At a bifurcation point, the intersection of these two branches may be **angular** or **tangent** following the directions of their respective tangents. By definition, the bifurcation is angular if the tangents are different at this point.

It follows from this definition that the rate ($\dot{\alpha}$,$\dot{\lambda}$) is not unique at an angular bifurcation point and thus the following proposition is trivial :

**Proposition 1 (3)**

**The positivity condition (3.5) is a sufficient condition of non-bifurcation of angular type.**

Historically, a criterion of non-bifurcation in this sense was first studied by Hill [17] in Plasticity.

The notion of tangent bifurcation of different order must be introduced in order to study tangent bifurcation. Note that simple examples of tangent bifurcation can be constructed in Elastic Buckling. In Plastic Buckling, such bifurcation has been discussed by Triantafyllidis [36].

## 3.3.- Stability analysis

### 3.3.1.- Dynamic Stability

We are principally interested in the **stability of an equilibrium state.**

An equilibrium is stable in the **dynamic** sense, or in other words stable in the sense of Liapunov, if small perturbations of the equilibrium will result only small perturbed motions near the equilibrium.

To illustrate the concept, let us imagine that the considered system is in an equilibrium state **($\mathbf{u}$,$\alpha$,$\lambda$)** at time $\tau$ = 0. Perturbations are then applied to the system while $\boldsymbol{\lambda}$ remains constant on the time interval [0,$\theta$]. At time $\tau$ = $\theta$, all perturbations are removed, the system will evolve with a dynamic perturbed motion at the same control level $\lambda$ . This perturbed motion is characterized by the fields $(\mathbf{u}(\tau), \alpha(\tau))$ , $\tau \geq \theta$.

The considered equilibrium is stable if the subsequent states $(\mathbf{u}(\tau),\alpha(\tau))$ are sufficiently close to the equilibrium $(\mathbf{u},\alpha)$ for all $\tau > \theta$ when the perturbations are small enough.

As in elastic stability analysis, we try to apply a classical argument based upon Liapunov's method (or Lejeune-Dirichlet's method) to construct a sufficient condition of stability.

The quantity of the injected energy in the system by perturbations on time interval $[0,\tau]$, $W_{per}(\tau)$ , is first considered and introduced as a convenient measure of the perturbation and as a Liapunov functional.

The energy balance at time $\tau$ for the whole system is :

$$W_{per}(\tau) = K_\tau + E(u(\tau), \alpha(\tau),\lambda) + \int_0^\tau D(\alpha(z), \dot{\alpha}(z))\, dz \tag{3.6}$$

in which $K_\tau \geq 0$ denotes the kinetic energy.

By virtue of (2.39), the following inequality holds :

$$W_{per}(\tau) \geq W(\alpha(\tau), \lambda) + \int_0^\tau D(\alpha,\dot{\alpha})\, dz \tag{3.7}$$

The last term depends on the trajectory of $\alpha(\tau)$ . It is also clear that the dependence of $\mathbf{D}$ on $\alpha$ may play an important role.

Our attention focusses now on the last term and the notation $\Delta\alpha = \alpha(\tau) - \alpha(0)$ is introduced.

The following optimization problem is considered in order to obtain a lower bound of the r.h.s. of (3.7) (cf. also Nguyen & Radenkovic [25]) :

Among all admissible paths $\alpha(z)$, $z \in [0,\tau]$ realizing the same jump $\Delta\alpha = \alpha(\tau) - \alpha(0)$, determine the optimal ones which minimize the dissipated energy :

$$\underset{\alpha(z)}{\mathrm{Min}} \int_0^\tau D(\alpha(z), \dot{\alpha}(z))\, dz \tag{3.8}$$

Note that in order to be admissible, an arbitrary path must also obey the normality law (3.1).

A partial answer to this problem is given by the following results :

**Radial path lemma**

The radial path $\alpha(z) = \alpha + \frac{z}{\tau} \Delta\alpha$ minimizes the dissipated energy (3.8) :

$$\int_0^\tau D(\alpha(z), \dot{\alpha}(z)\, dz \geq \int_0^\tau D\left(\alpha + \frac{z}{\tau}\Delta\alpha, \Delta\alpha\right) dz \tag{3.9}$$

in the following cases :

i).- when the dissipated energy is a state function i.e. when there exists a function $W^d(\alpha)$ such that :

$$W^d(\alpha + \Delta\alpha) - W^d(\alpha) = \int_0^\tau D(\alpha, \dot{\alpha})\, dz \tag{3.10}$$

for all admissible paths.

ii).- when **C** is independent of $\alpha$ .

Part i) is trivial because the result is then path-independent. Part ii) results simply from the normality law.

Although the validity of the radial path lemma seems to be much larger than cases i) and ii), we will restrict our present discussion to these two principal situations which correspond to most models of brittle damage and brittle fracture.

For all **systems satisfying the radial path lemma**, the following estimate is then obtained :

$$W_{per}(\tau) \geq \Phi_\alpha(\Delta\alpha) = W(\alpha + \Delta\alpha, \lambda) + W_{dm}(\alpha, \Delta\alpha) \tag{3.11}$$

with :

$$W_{dm}(\alpha, \Delta\alpha) = \int_0^\tau D\left[\alpha + \frac{z}{\tau}\Delta\alpha\, ,\, \Delta\alpha\right] dz \tag{3.12}$$

and leads to the theorem :

**Generalized Lejeune-Dirichlet theorem**

If the function $\Phi_\alpha(\Delta\alpha)$ admits a strict (local) minimum at the point $\Delta\alpha = 0$ , then the considered equilibrium is stable in the dynamic sense.

The proof of this theorem follows the classical approach in standard textbooks on stability. It consists in taking as Liapunov functional the quantity $W_{per}$ $(\tau)$ which remains constant after time $\tau = \theta$ . The function $\Phi_\alpha$ provides the energy barrier necessary to ensure always small distances between the perturbed motions and the equilibrium position when the initial injected energy is sufficiently small.

For discrete systems, the validity of this theorem is clear. For continuous systems, the mathematical difficulty arising in the choice of suitable norms remains the same as in Elastic Stability (cf. Koiter [19]).

In the spirit of Elasticity, a criterion of second variations of energy can be associated to this theorem :

**Stability criterion**

**The positivity condition (3.4) is a sufficient condition of stability**

Indeed, let us verify that $\Delta\alpha = 0$ realizes a local minimum of $\Phi_\alpha$ under the positivity condition (3.4).

For this, we can consider the Taylor development of $\Phi_\alpha$ at the point $\Delta\alpha = 0$.
The first order term is :

$$\mathbf{W}_{,\alpha} \cdot \Delta\alpha + \mathbf{D}(\alpha, \Delta\alpha)$$

If $\Delta\alpha \in \mathbf{N_C}(A)$ this term is zero.

If $\Delta\alpha$ does not belong to $\mathbf{N_C}(A)$ , this term is strictly positive.

The second order term is :

$$\frac{1}{2}\left( \Delta\alpha \cdot \mathbf{W}_{,\alpha\alpha} \cdot \Delta\alpha \;+\; \Delta\alpha \cdot \mathbf{D}_{,\alpha}(\alpha, \Delta\alpha) \right)$$

If condition (3.4) is satisfied, it is then clear that $\Phi_\alpha(\Delta\alpha)$ admits a local minimum at point $\Delta\alpha = 0$.

### 3.3.2.- Remarks

- If $\mathbf{N_C}(A)$ is reduced to the vector $\mathbf{0}$, $\dot{\alpha} = 0$ the response is elastic. The considered equilibrium is stable by the elastic stability assumption (2.39).

- If the dependence $\mathbf{U}(\alpha,\lambda)$ is such that the partial derivations $E_{,u\alpha}$ , $E_{,u\lambda}$ can be introduced as in Chap.2, it may be interesting to keep the couple $(\mathbf{u},\alpha)$ as principal unknowns.

The function $\mathbf{u}(\alpha,\lambda)$ , given implicitly by equation (2.25), satisfies :

$$\delta\alpha \,.\, ( E_{,u\alpha} + E_{,uu}\, u_{,\alpha} ) = 0$$
$$\delta\lambda \,.\, ( E_{,u\lambda} + E_{,uu}\, u_{,\lambda} ) = 0$$

The assumption of elastic stability ensures that the operator $E_{,uu}$ is invertible and leads to the following formal expressions :

$$W_{,\alpha\alpha} = E_{,\alpha\alpha} - E_{,\alpha u} \,.\, E^{-1}_{,uu} \,.\, E_{,u\alpha} \qquad (3.13)$$

and

$$W_{,\alpha\lambda} = E_{,\alpha\lambda} - E_{,\alpha u} \,.\, E^{-1}_{,uu} \,.\, E_{,u\lambda} \qquad (3.14)$$

- In the same condition, it is not difficult to check that positivity condition (3.4) is also equivalent to the following one :

$$\forall\ 0 \neq \delta u \in V_0 \text{ and } 0 \neq \delta\alpha \in N_C\,(A) : \qquad (3.15)$$

$$\delta u \,.\, E_{,uu} \,.\, \delta u + 2\,\delta u \,.\, E_{,u\alpha} \,.\, \delta\alpha + \delta\alpha \,.\, E_{,\alpha\alpha} \,.\, \delta\alpha +$$

$$\delta\alpha \,.\, D_{,\alpha}\,(\alpha, \delta\alpha) > 0$$

and positivity condition (3.5) is also equivalent to the following one :

$$\forall\ \delta u \neq \delta v \in V_0 \text{ and } \delta\alpha \neq \delta\beta \in N_C\,(A) : \qquad (3.16)$$

$$(\delta u-\delta v) \,.\, E_{,uu} \,.\, (\delta u-\delta v) + 2\,(\delta u-\delta v) \,.\, E_{,u\alpha} \,.\, (\delta\alpha-\delta\beta) +$$

$$(\delta\alpha-\delta\beta) \,.\, E_{,\alpha\alpha} \,.\, (\delta\alpha-\delta\beta) + (\delta\alpha-\delta\beta) \,.\, (\, D_{,\alpha}\,(\alpha,\delta\alpha) - D_{,\alpha}\,(\alpha,\delta\beta)\,) \geq 0.$$

- In Incremental Plasticity and in the particular case of simple plastic potential, the rate of displacement $\dot{u}$ can be chosen as principal unknown. It is not difficult to establish that conditions (3.4) or (3.15) are also equivalent to Hill's stability criterion :

$$\int_{\Omega} \left\{ U(\dot{\epsilon}\,(\delta u)) + \frac{1}{2}\ \sigma : \nabla\delta u.\nabla\delta u \right\} d\Omega > 0 \qquad (3.17)$$

$$\forall\ 0 \neq \delta u \in V_0$$

while (3.5) or (3.16) are also equivalent to Hill's non-bifurcation criterion :

$$\int_{\Omega} \{(\, U_{,\dot{\epsilon}}\,(\delta u) - U_{,\dot{\epsilon}}\,(\delta v)\,) \,.\, (\,\dot{\epsilon}\,(\delta u) - \dot{\epsilon}\,(\delta v)\,) \qquad (3.18)$$

$$+\ \sigma :\ \nabla(\delta u - \delta v).\nabla(\delta u-\delta v)\ \}\, d\Omega > 0$$

$\forall\ \delta u \neq \delta v \in V_0$ .

- Note that stability in the **dynamic** sense is only obtained when the radial path lemma is satisfied, i.e. within the framework of assumptions i) or ii). In these two cases, the rate problem is also **symmetric**.

It is thus not proved that (3.4), (3.15) or (3.17) is strictly speaking a criterion of **dynamic** stability in all situations.

### 3.3.3.- Illustrations

*Example 1*

Let us consider the equilibrium of a stone on a slope with dry Coulomb friction. The stone of weight mg can be described by its abscissa x. If z = z(x) is the equation of the slope, the reversible energy is W = mgz(x) . The normal reaction is mg $\frac{dx}{ds}$ , the yield limit is mg $\frac{dx}{ds}$ tg $\phi$ . The slipping rate is $\frac{ds}{dx}\dot{x}$ , the dissipation is D = mg tg$\phi$ $|\ \dot{x}\ |$ and does not depend on the present abscissa x. The convex of admissible forces is C = { A | |A| $\leq$ mg tg$\phi$ } .

The present force A = - mg z′ attains the yield surface when $|z'|$ = tg$\phi$ . This equilibrium is stable if condition (3.4) is ensured i.e. when z" > 0 !

*Example 2*

The example of Shanley's column in Chap. 1 is considered again. When the vertical load $\lambda$ increases from 0, the trivial response is given by the vertical positions :

$$\theta = 0,\ \ v = -\frac{\lambda}{2k} + \alpha_1(\lambda)\ \ ,\ \alpha_1(\lambda) = \alpha_2(\lambda) = -<\frac{\lambda}{2} - k>\frac{1}{h}$$

Let us make a stability and bifurcation analysis of these equilibria. Since :

$$[\ E_{,uu}\ ] = \begin{bmatrix} 2\ell^2 k - \lambda L & 0 \\ 0 & 2k \end{bmatrix}$$

the elastic stability condition (2.31) is ensured when $\lambda < \lambda_E = 2\ell^2 \frac{k}{L}$ . The matrix $W_{,\alpha\alpha}$ is given by (3.13) :

$$[\ W_{,\alpha\alpha}\ ] = \begin{bmatrix} P & Q \\ Q & P \end{bmatrix}$$

with $2P = 2h - \frac{\lambda}{\lambda_E - \lambda} k$ , $2Q = \frac{\lambda}{\lambda_E - \lambda} k$. Its two associated eigenvalues are :

$$\mu_1 = h - \frac{\lambda}{\lambda_E - \lambda} k \text{ and } \mu_2 = h > 0.$$

Since $A_1 = A_2 = -k$ in the vertical position, $\mathbf{N_C}$ **(A)** is given by

$$\mathbf{N_C}\ \mathbf{(A)} = \{ (x,y) \in R_2 \mid x \leq 0, y \leq 0 \}$$

and corresponds to the subset of vectors of $R_2$ with non-positive components.

The condition of positivity (3.4) :

$$\mu_2 (x+y)^2 + \mu_1 (x-y)^2 > 0 \ \forall \ x \leq 0 , y \leq 0 , (x,y) \neq (0,0).$$

is satisfied if $\mu_1 > - \mu_2$ . The equality $\mu_1 = - \mu_2$ defines a critical value $\lambda_1 = \frac{2h}{k+2h} \lambda_E = \lambda_R$ .

The condition of positivity (3.5) :

$$\mu_2 (x+y)^2 + \mu_1 (x-y)^2 > 0 \ \forall \ (x,y) \neq (0,0)$$

is satisfied if $\mu_1 > 0$. The equality $\mu_1 = 0$ defines a critical value $\lambda_2 = \frac{h}{k+h} \lambda_E = \lambda_T$ .

Thus the straight position is stable if $\lambda < \lambda_R$ and is not a bifurcation point if $\lambda < \lambda_T$ .

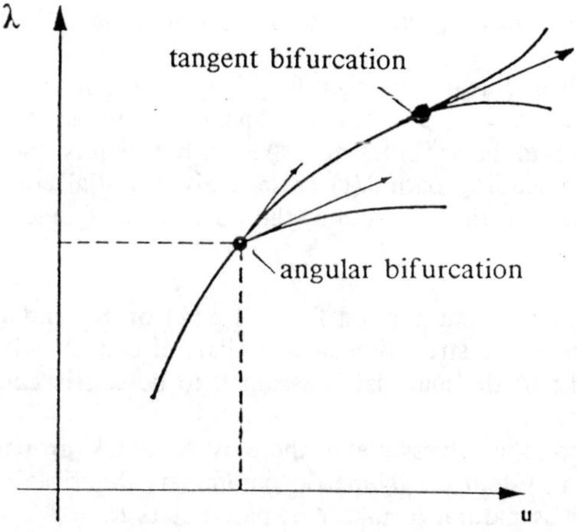


Fig. 1 (3) , Angular and tangent bifurcations

## 4. SOME APPLICATIONS IN BRITTLE FRACTURE

The study of crack propagation and stability is a classical problem in Fracture Mechanics. In brittle fracture, when Griffith's criterion of craçk propagation is assumed, a large body of research work has been dedicated to the bidimensional problem of one linear crack in plane strain, plane stress or antiplane shear (cf. Blum [6]). Its generalization to a system of interacting linear cracks has been discussed by Bazant & Ohtsubo [2], Nemat-Nasser, Sumi & Keer [24], Nguyen & Stolz [28]... The stability of a plane crack of arbitrary shape has been also discussed by numerous authors.

An elastic solid with Griffith cracks is a system undergoing irreversible transformation subsequent to crack propagation. Its quasi-static evolution enters completely in the framework of time-independent standard dissipative systems. For example, the quasi-static behaviour of a system of interacting linear cracks can be compared to the quasi-static behaviour of an elastic plastic structure with a finite number of d.o.f. : the equations of evolution of these systems are of the same mathematical nature, given formally by equations (2.35).

The objective of this chapter is to give some simple illustrations of the preceding results in the context of fracture.

### 4.1.- Quasi-static evolution of a system of interacting linear cracks

Consider (Fig.1 (4)) the following crack propagation problem in brittle fracture.

An elastic solid $\Omega$ with m linear cracks of length $\ell_i$ , i = 1,m is subjected to a loading path defined by load parameters $\lambda_\mu$ , $\mu$ = 1,n in plane strain, plane stress or antiplane shear. The deformation is assumed to be infinitesimal. We wish to study the quasi-static response $\boldsymbol{\ell}(t)$ associated with a given loading path $\boldsymbol{\lambda}(t)$ from a given initial configuration $\boldsymbol{\ell}_0$ . The cracks are assumed to propagate in their direction, the possibility of crack kinking is not considered here.

As usual, the load consists of an applied force $\mathbf{T_d}$ ($\lambda$) on $S_T$ and given displacement $\mathbf{u_d}$ ($\lambda$) on $S_u$ . Cracks surfaces $S_i$ are stress free with unilateral contact without friction. The volume force is here neglected and the material is assumed to be elastic and homogeneous.

The response of the solid is reversible if there is no crack propagation. The crack lengths $\boldsymbol{\ell}(t)$ represent clearly a system of geometric parameters describing the irreversible evolution of this system. Thus, it is natural to take $\boldsymbol{\ell}$ as parameters $\boldsymbol{\alpha}$.

The present volume of the cracked solid depends on the crack lengths $\Omega = \Omega_\ell$ .

The displacement field $\mathbf{u}$ must belong to the set of kinematically admissible displacements :

$$U(\ell,\lambda) = \{ u \mid u \text{ defined on } \Omega_\ell ,\ u = u_d \text{ on } S_u ,\ [u]_n \geq 0 \text{ on } S_i \} \tag{4.1}$$

If $w(\epsilon)$ denotes the elastic energy density per unit volume, the total energy of the system is :

$$E(u,\ell,\lambda) = \int_{\Omega_\ell} w(\epsilon(u))\, d\Omega - \int_{S_T} T_d\, u\, dS \tag{4.2}$$

At equilibrium, the displacement **u** must satisfy virtual work equation which can be written in the form of a variational inequality to take into account the unilateral contact without friction :

$$E_{,u}(u,\ell,\lambda) . \delta u \geq 0 \quad \forall\ \delta u = v - u ,\ v \in U(\ell,\lambda) \tag{4.3}$$

If energy density w is also strictly convex, it follows from (4.3) that the displacement at equilibrium $\mathbf{u} = \mathbf{u}(\ell,\lambda)$ brings the energy **E** to its minimum :

$$E(u(\ell,\lambda),\ell,\lambda) = \underset{v \in U(\ell,\lambda)}{\mathrm{Min}} E(v,\ell,\lambda) = W(\ell,\lambda) \tag{4.4}$$

It is important to note that, because of the strong dependence of U on $\ell$, the elimination of u is a necessity.

The generalized force $A_i$ :

$$A_i = - \frac{\partial W}{\partial \ell_i} \tag{4.5}$$

represents the **energy release rate** $G_i$ associated to the i-th crack length.

In brittle fracture, the Griffith criterion consists in adopting the following crack propagation law :

$$\begin{aligned} &\text{If } A_i < 2\gamma \quad \text{then } \dot{\ell}_i = 0 \quad \text{(no propagation)} \\ &\text{If } A_i = 2\gamma \quad \text{then } \dot{\ell}_i \geq 0 \quad \text{(possible propagation)} \end{aligned} \tag{4.6}$$

where $\gamma$ denotes a critical surface energy. Most often, $\gamma$ is a constant of the considered material. In some applications however, $\gamma$ may also be considered as a function of the effective propagated length $\Delta\ell_i = \ell_i - \ell_i^0$ in order to take into account the stabilizing effect due to the near-tip plastic strain.

If the convex domain of admissible forces **C** is introduced :

$$C = \{ A \mid A_i - 2\gamma\,(\Delta\ell_i) \leq 0 ,\ i=1,n \} \tag{4.7}$$

the normality law (2.43) :

$$\dot{\ell} = N_C(A) \tag{4.8}$$

is clearly satisfied. The associated dissipative potential is :

$$\mathbf{D}(\boldsymbol{\ell},\dot{\boldsymbol{\ell}}) = 2\gamma\,(\Delta\ell_i)\,\dot{\ell}_i \text{ if } \dot{\ell}_i \geq 0\,, \;= +\infty \text{ otherwise} \tag{4.9}$$

Note that the dissipated energy is here a state function with :

$$\mathbf{W}^d\,(\boldsymbol{\ell}) = \sum_i \int_0^{\ell_i} 2\gamma(\Delta\ell)\; d\ell$$

Thus, all the obtained results of Chapter 3 can be applied. More explicitly, let I be the set of indices of possible propagating cracks at time t i.e.:

$$A_i = 2\gamma\,(\Delta\ell_i) \quad \text{if } i \in I, \tag{4.10}$$

for these cracks, the Griffith criterion is attained. The set $\mathbf{N_C}$ **(A)** is simply :

$$\mathbf{N_C}\,(\mathbf{A}) = \{\, \delta\boldsymbol{\ell} \mid \delta\ell_i \geq 0 \text{ if } i\in \mathrm{I}\,, = 0 \text{ otherwise} \,\} \tag{4.11}$$

and leads to the following propositions :

**Proposition 1 (4)**

The rate $\dot{\boldsymbol{\ell}}$ is given by the following symmetric variational inequality :

$\dot{\ell}_i \geq 0$ and satisfy :

$$(\,\Phi_{,ij}\,.\,\dot{\ell}_j + W_{,i\mu}\,.\,\dot{\lambda}_\mu\,)(\,\delta\ell_i - \dot{\ell}_i\,) \geq 0 \tag{4.12}$$

**The total energy functional $\Phi$ = W + $W^d$ and the matrix $\Phi_{,ij}$ play a fundamental role.**

**Proposition 2 (4)**

The present state **(ℓ,λ)** is stable in the dynamic sense if the positivity condition (4.13) is satisfied :

$$\delta\ell_i\,.\,\Phi_{,ij}\,(\ell,\lambda)\,.\,\delta\ell_j > 0 \;\; \forall\; \delta\boldsymbol{\ell} \neq \mathbf{0}\,,\; \delta\ell_i \geq 0\;\;, i \in \mathrm{I} \tag{4.13}$$

**Proposition 3 (4)**

The present state **(ℓ,λ)** is not an angular bifurcation point if the matrix $\Phi_{,ij}\,(\ell,\lambda)$ , (i,j ∈ I) is positive definite.

Note that if $2\gamma$ is a constant, then the matrix [ $\Phi_{,ij}$ ] reduces to the matrix [ $W_{,ij}$ ]. In this case, the stability criterion for example can also be written as :

$$- \delta G \, . \, \delta \ell > 0 \tag{4.14}$$

for any $\delta\ell \neq 0$, $\delta\ell \geq 0$

where $\delta G = - W_{,\ell\ell} \, . \, \delta\ell$ .

## 4.2.- Illustrations

### 4.2.1.- Example 1

Let us consider the problem of propagation of a central crack in a thin plate under displacement control or force control of amplitude $\lambda$ as shown in Fig.2 (4). Under this symmetric loading and symmetric initial value $\ell_0$, the response can be symmetric with $\ell_1(\lambda) = \ell_2(\lambda)$ or bifurcates to an unsymmetric solution with $\ell_1(\lambda) \neq \ell_2(\lambda)$. Let us make a bifurcation and stability analysis of the trivial symmetric solution.

In order to obtain analytical results, a simple modelling of this structure as an assembly of two identical beams is first considered (cf. Pradeilles's thesis, to appear) as shown in Fig.3 (4).

- **Displacement control** :

$v(0) = \lambda$

If $0 < x < \ell_2$ : $v = \lambda + qx + rx^2 + sx^3$

$$q = 3\lambda \frac{\ell_2 - \ell_1}{2(\ell_2 \ell_1)} , \quad r = - \frac{3\lambda}{\ell_1 \ell_2} , \quad s = \lambda \frac{3\ell_2 + \ell_1}{2(\ell_1 \ell_2^3)}$$

If $-\ell_1 < x < 0$ : $v = \lambda + qx + rx^2 + s'x^3$ , $s' = -\lambda \dfrac{3\ell_1 + \ell_2}{2(\ell_1^3 \ell_2)}$

$$W(\ell_1 , \ell_2 , \lambda) = 3 \, EI \, \lambda^2 \, (\ell_1 + \ell_2)^3 \frac{1}{\ell_1^3 \ell_2^3}$$

$$[W_{,ij}] = 18 \, EI \, \lambda^2 \frac{\ell_1 + \ell_2}{\ell_1^5 \ell_2^5} \begin{bmatrix} Q & R \\ R & S \end{bmatrix}$$

with $Q = \ell_2^3 (\ell_1 + 2\ell_2)$, $R = \ell_1^2 \ell_2^2$ , $S = \ell_1^3 (2\ell_1 + \ell_2)$

When $2\gamma$ is assumed to be a constant, only this matrix contributes to bifurcation and stability analysis.

This matrix is always positive-definite, in particular when $\ell_1 = \ell_2 = \ell$. A trivial equilibrium is always stable and is not a bifurcation point.

- **Force control** :

$\lambda$ is now the applied force at x = 0

From the previous value of energy, it follows that :

$$W(\ell_1 , \ell_2 , \lambda) = - \frac{\lambda^2}{k} \quad \text{with } k = 3 \, EI \, (\ell_1 + \ell_2)^3 \frac{1}{(\ell_1 \ell_2)^3}$$

$$[W_{,ij}] = - \frac{2}{EI} \lambda^2 \frac{\ell_1 \ell_2}{(\ell_1 + \ell_2)^5} \begin{bmatrix} Q & R \\ R & S \end{bmatrix}$$

with $Q = \ell_2^3 (\ell_2 - \ell_1)$, $R = 2 \ell_1^2 \ell_2^2$ , $S = \ell_1^3 (\ell_1 - \ell_2)$

This matrix is always negative definite, in particular when $\ell_1 = \ell_2 = \ell$. A trivial equilibrium is always unstable with possible bifurcation .

- **mixed control** :

A spring of rigidity K is added at the central point to the structure and $\lambda$ is the total implied displacement :

$$W(\ell_1 ,\ell_2 , \lambda) = \frac{kK}{k+K} \lambda^2$$

$$[W_{,ij}] = 18 \, EI \, \lambda^2 \frac{K^2}{(k+K)^2} \frac{\ell_1 + \ell_2}{(\ell_1 \ell_2)^5} \begin{bmatrix} Q & R \\ R & S \end{bmatrix}$$

$$\text{with } Q = \left(2 + \frac{\ell_1}{\ell_2} - \frac{3k}{k+K}\right) \ell_2^4$$

$$R = \left(1 - \frac{3k}{k+K}\right) \ell_1^2 \ell_2^2$$

$$S = \left( 2 + \frac{\ell_2}{\ell_1} - \frac{3k}{k+K} \right) \ell_1{}^4$$

When $\ell_1 = \ell_2$ , the matrix $\begin{bmatrix} q & r \\ r & q \end{bmatrix}$ with q = 3K , r = K - 2k has to be considered.

Its eigenvalues are : $\mu_1 = 2(k+K) > 0$ , $\mu_2 = 2(2K - k)$ .

The present trivial equilibrium is not a bifurcation point if $\mu_2 > 0$ or if 2K > k i.e. if $\ell > \left( 12 \frac{EI}{K} \right)^{\frac{1}{3}}$ .

The present trivial equilibrium is stable if $\mu_1 + \mu_2 > 0$ . But this is always satisfied since $\mu_1 + \mu_2 = 6K > 0$.

These results show that symmetric equilibria are stable. However, for short cracks, bifurcation is possible to an unsymmetric response. There is no risk of bifurcation when the crack length $\ell$ is long enough.

Now, if the two dimensional problem of the propagation of a central crack in a plate is considered, it is not possible to obtain the explicit expression of the energy at equilibrium **W** . Numerical calculation of **u** and **W** , by the finite element method for example, will be necessary. Then, by finite difference method, $A_i$ can be approached as $-\frac{\Delta W}{\Delta \ell_i}$ and $W_{,ij}$ can be approched as $-\frac{\Delta A_i}{\Delta \ell_j}$ .

This elementary method can also be completed by a numerical computation of the analytical expression of the energy release rates $A_i$ in terms of J-integral as it is well known in fracture mechanics.

In the same spirit, the analytical expressions of $W_{,ij}$ in terms of new path independent J-type integrals have been also obtained (cf. Nguyen & Stolz [28]).

### 4.2.2.- Example 2

This example is a simple illustration of the **propagation of interface cracks** by delamination and its **interaction with possible local buckling** phenomena.

Let us consider the compression of a sheet with an interfacial crack as shown in Fig. 3 (4). In the spirit of the previous example, a simple modelling in the context of beam theory is here discussed.

The behaviour of this system of four beams can be modelled as follows :

A beam of small section S and length $\ell = \ell_1 + \ell_2$ is submitted to an axial force N. If $N \leq N_c = 4\pi^2 \frac{EI}{\ell^2}$, there is no buckling and its elastic energy, due to pure compression, is $\frac{1}{2} \frac{\ell}{ES} N^2$. If $N = N_c$, there is Euler buckling, an additional term $\Delta N_c$ due to the transverse displacement $v \neq 0$ exists, $\Delta$ denotes the length shortening of this beam. These results come directly from the expression of the elastic energy of a beam in non-linear theory :

$$\int_0^\ell \left\{ \frac{1}{2} ES \left[u' + \frac{1}{2} v'^2\right]^2 + \frac{1}{2} EI v''^2 \right\} dx$$

with $N = ES\left(u' + \frac{1}{2} v'^2\right) = N_c$ and $v = a\left(1 - \cos 2\pi \frac{x}{\ell}\right)$ in the buckled configuration.

A second beam of section A > S and length $\ell$ is in pure compression. Its elastic energy is $\frac{1}{2} EA \left(\frac{\Delta}{\ell}\right)^2 \ell$.

A third beam of section T = S + A and length $L-\ell$ is in pure compression, of energy $\frac{1}{2} ET \left(\frac{\delta}{L-\ell}\right)^2 (L-\ell)$.

The potential energy of this system under axial force $\lambda$ is :

$$E(\delta, \Delta, \ell_1, \ell_2, \lambda) = -\frac{1}{2} \frac{\ell}{ES} N_c^2 + \Delta N_c + \frac{1}{2} EA \left(\frac{\Delta}{\ell}\right)^2 \ell + \frac{1}{2} ET \left(\frac{\delta}{L-\ell}\right)^2 (L-\ell) - \lambda(\delta + \Delta)$$

The displacement parameters $\delta, \Delta$ can be eliminated by the equilibrium equations :

$$E_{,\Delta} = N_c + EA \frac{\Delta}{\ell} - \lambda = 0 \quad \text{and} \quad E_{,\delta} = E \frac{\delta}{L-\ell} - \lambda = 0$$

The potential energy at equilibrium is :

$$W(\ell_1, \ell_2, \lambda) = \frac{\lambda}{EA} 4\pi^2 EI \frac{1}{\ell} - \frac{1}{2E}\left(\frac{1}{S} + \frac{1}{A}\right) (4\pi^2 EI)^2 \frac{1}{\ell^3} - \frac{1}{2} \frac{\lambda^2}{EA} \ell - \frac{1}{2} \frac{\lambda^2}{ET} (L-\ell)$$

and leads to the following expression :

$$[W_{,ij}] = W_{,\ell\ell} \cdot \begin{bmatrix} 1 & 1 \\ 1 & 1 \end{bmatrix}$$

with $$W_{,\ell\ell} = 2\,N_c \frac{1}{EA\ell} \left\{ \lambda - N_c\, 3\left( 1 + \frac{A}{S} \right) \right\}$$

It is clear that the stability condition (13) is satisfied when $\lambda > N_c \ 3\left(1+\frac{A}{S}\right)$.

However, this matrix is not positive-definite and thus each equilibrium position may be a critical point of bifurcation. This is effectively true as one can see from the multiple possibility of rates $\dot{\ell}_1$ and $\dot{\ell}_2$ associated to a rate $\dot{\ell}$.

This results from the fact that energy **W** depends on $\ell_1$ and $\ell_2$ only by their sum $\ell$ in the proposed model.

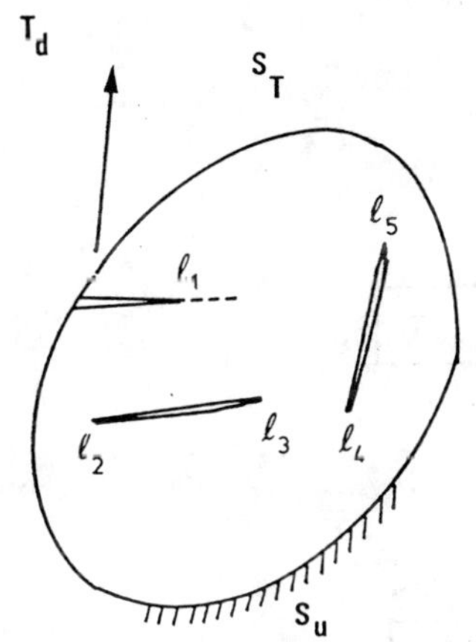


Fig. 1 (4) ,

A system of interacting linear cracks

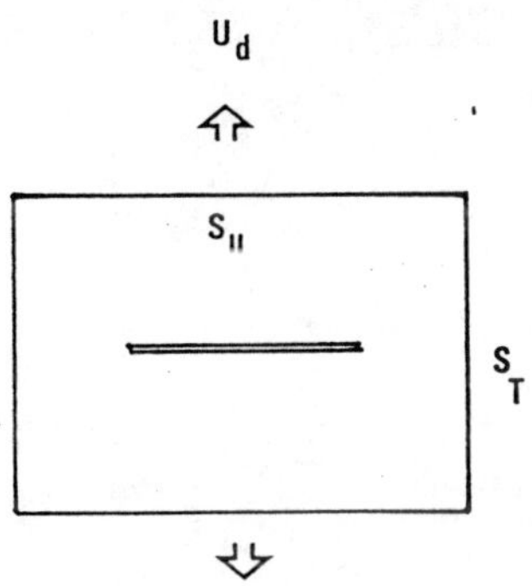


Fig. 2 (4) ,

Central crack in a thin plate

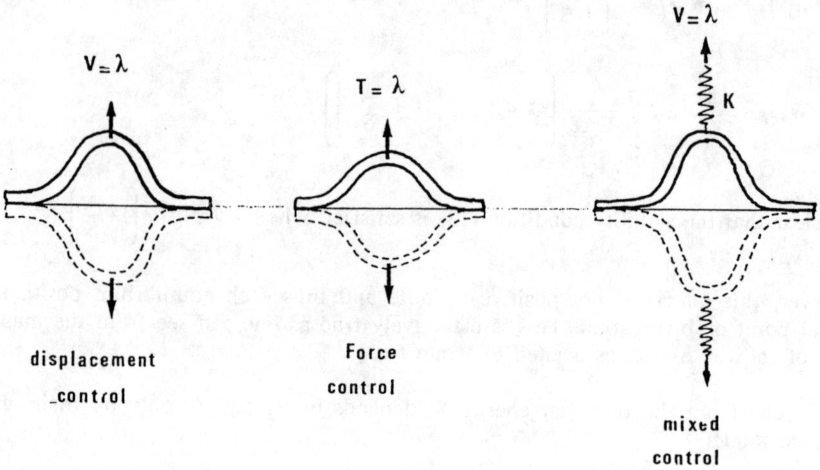


Fig. 3 (4) , Model of two beams

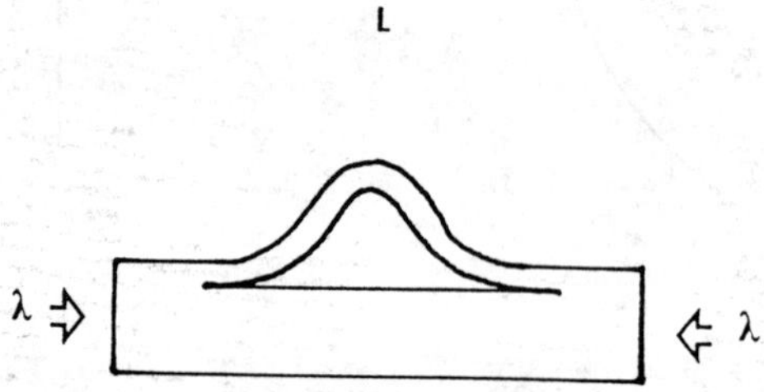


Fig. 4 (4) , Crack propagation by delamination and buckling

# 5.- SOME APPLICATIONS IN BRITTLE DAMAGE

This chapter introduces some illustrations of our general results in the framework of damage mechanics.

Models of brittle damage are first studied in order to illustrate their connections with incremental plasticity as it has been presented in Chapter 2.

Next, emphasis will be put on the problem of brittle and total damage (cf. Bui & Ehrlacher [8]), or brittle and partial damage (cf. Dems & Mroz [9], Marigo [22]) in order to study an interesting situation where the irreversible parameters are **geometric domains**. The derivation with respect to a domain is introduced. It is shown that the general theory can be then applied pratically without modification. Some examples of illustration (cf. Berest [4]) will be presented.

## 5.1.- Brittle Damage

### 5.1.1.- Models of brittle damage

Usual models of brittle damage enter strictly in the framework of the previous description (Chapter 2). The basic constitutive equations for an elastic material with brittle damage are thus given by the following equations :

$$\sigma = \frac{\partial w}{\partial \epsilon} \ , \ A = - \frac{\partial w}{\partial \alpha} \tag{5.1}$$

$$\dot{\alpha} = N_C (A) \tag{5.2}$$

Most often, $\alpha$ is a scalar representing the level of damage. As a consequence, the associated force is then a scalar.

For example, if the damage yield condition is :

$$A - A_c (\alpha) \leq 0 \ , \tag{5.3}$$

The normality law (5.2) is simply :

$$\dot{\alpha} \geq 0 \ \text{if} \ A = A_c \ , \ \dot{\alpha} = 0 \ \text{if} \ A < A_c \tag{5.4}$$

Note that a dissipative energy $w^d$ also exists :

$$w^d(\alpha) = \int_0^{\alpha} A_c(\beta)\, d\beta \tag{5.5}$$

All general results can be applied without modification. In particular, the positivity (3.4) is a sufficient condition for dynamic stability since the radial path lemma is automatically satisfied.

In particular, it is clear that :

An elastic material with brittle damage is **stable at the present state of stress** (stress control) if the matrix

$$[\,E''\,] = \begin{bmatrix} E_{,\epsilon\epsilon} & E_{,\alpha\epsilon} \\ E_{,\epsilon\alpha} & E_{,\alpha\alpha} \end{bmatrix} \tag{5.6}$$

defines a positive form on the set of elements $(\delta\epsilon,\delta\alpha) \neq (0,0)$ with non-negative component $\delta\alpha \geq 0$.

### 5.1.2.- Examples :

i).- if the special expression $w(\epsilon,\alpha) = \frac{1}{2}\, e^{-\alpha}\, \epsilon L \epsilon$ is assumed and $A_c$ is a constant, the stress-strain curve of the Fig.1a (5) is then obtained. This is a model of progressive damage, the material may be more or less damaged with a progressive evolution of the damage level.

The stability of this material is not granted since the matrix :

$$[\,E''\,] = \begin{bmatrix} L & -L\epsilon \\ -L\epsilon & \frac{1}{2}\,\epsilon L \epsilon \end{bmatrix}$$

does not define a positive form on the set $(\delta\epsilon, \delta\alpha) \neq (0,0)$ with non-negative component $\delta\alpha \geq 0$.

ii).- if the special expression $w(\epsilon,\alpha) = \frac{1}{2}\,(1-\alpha)\,\epsilon L \epsilon$ , with $0 \leq \alpha \leq 1$ , is assumed and $A_c$ is a constant $w_c$ , the stress-strain curve of the Fig.1b (5) is then obtained. This is the model of brittle and total damage in the sense of Bui [8].

iii).- if the special expression $w(\epsilon,\alpha) = (1-\alpha)\, U(\epsilon) + \alpha\, V(\epsilon)$ , $0 \leq \alpha \leq 1$ with $U = \frac{1}{2}\, \epsilon . L_1 . \epsilon$

and $V = \frac{1}{2}\,\epsilon.L_2.\epsilon$ , is assumed and $A_c$ is a constant, the stress-strain curve of the Fig.1c (5) is then obtained. This is the model of brittle and partial damage in the sense of Dems & Mroz [9], Marigo [22].

iv).- if the energy $w(\alpha,\epsilon)$ is obtained as the macro energy density associated with a bi-material model of energy $e(\alpha,\chi,\eta) = (1-\alpha)U(\chi) + \alpha V(\eta) + I(\alpha)$ at local strains $\chi$ and $\eta$ with local constraint $\epsilon = (1-\alpha)\chi + \alpha\,\eta$ , the stress-strain curve of the Fig.1d (5) may be obtained for a suitable choice of the dissipation. Here, U and V stand for the elastic energy of each phase and $I(\alpha)$ is the interaction energy between these phases.

Indeed, w is now given by the value of the lagrangian :

$$w = L(\alpha,\epsilon,\mu,\chi,\eta) = e(\alpha,\chi,\eta) - \mu[\,(1-\alpha)\chi + \alpha\eta - \epsilon\,] \tag{5.7}$$

with the associated state equations :

$$L_{,\mu} = 0\ ,\ L_{,\chi} = 0\ ,\ L_{,\eta} = 0 \text{ and } L_{,\epsilon} = \sigma \tag{5.8}$$

where $\mu$ denotes the lagrangian multipliar.

Irreversible force associated to $\alpha$ is now :

$$A = -\,L_{,\alpha} = U(\chi) - V(\eta) - \mu\,(\chi-\eta) \tag{5.9}$$

If the following expression of the dissipative potential is adopted :

$$\begin{aligned} d &= +\infty \quad \text{if} \quad \dot\alpha < 0 \\ d &= k\,\dot\alpha \quad \text{if} \quad \dot\alpha \geq 0 \end{aligned}$$

the Fig. 1d (5) is then obtained when I = 0.

*Exercice*

Discuss the stability of the previous material in a force control experiment.

## 5.2.- Solid with total or partial brittle damage

### 5.2.1.- Solid with total damage

As expected by the negative slope of the stress-strain curves, brittle models (i), (ii), (iii) are all highly unstable. More precisely, the stability analysis of the two models (ii) and (iii) can still be discussed **for a structure** in the framework of the present theory under some proper choices of the irreversible parameter $\boldsymbol{\alpha}$.

In this section, Bui's model of total damage (ii) is considered for a structure in order to illustrate the general theory when the irreversible parameter $\alpha$ is a part of the volume of the whole structure.

Let us consider an elastic solid of volume V in total and brittle damage. Since the material presents no rigidity beyond the yield limit in the damaged zone, the problem to be discussed is the evolution of an elastic structure with a variable domain of definition $V-\Omega$ which represents the undamaged part of the volume. **The damaged volume $\Omega$ represents the irreversible parameter $\alpha$ in this case.**

To simplify the presentation, it is assumed that the undamaged zone $V-\Omega$ is submitted to a force $T(\lambda)$ on a fixed part $S_T$ of the boundary $\partial\Omega$. The complementary part S is a weakened zone where the material may be damaged and represents the moving boundary between damaged and undamaged zones when damage propagates inside the structure (Fig.2 (5)).

The geometric domain $\Omega$ is a variable domain, its motion may be represented by its normal velocity denoted as $\dot{\Omega}$ (s) for $s \in S$ . This suggestive notation shows that if $\alpha = \Omega$ is a geometric domain, its velocity $\dot{\alpha}$ may be presented by a function $\dot{\Omega}(s) \geq 0 \ \forall \ s \in S$ , which is an element of a functional space $\mathbf{Z}(\alpha)$. Its associated force may be presented by an element of the dual space $\mathbf{Z}^*(\alpha)$.

The potential energy at equilibrium is :

$$\mathbf{W}(\Omega,\lambda) = \mathbf{E}(\mathbf{u}(\Omega,\lambda),\Omega,\lambda) = \underset{\mathbf{u}^* \in \mathbf{U}(\Omega,\lambda)}{\mathbf{Min}} \ \mathbf{E}(\mathbf{u}^*,\Omega,\lambda) \tag{5.10}$$

where $U(\Omega,\lambda)$ denotes the set of admissible displacements and :

$$\mathbf{E}(\mathbf{u},\Omega,\lambda) = \int_{V-\Omega} w(\epsilon(u)) \ dV - \int_{S_T} T(\lambda).u \ ds \tag{5.11}$$

## 2.2.- Derivation with respect to a domain

To follow the general description, it is necessary to make the derivation $\mathbf{W}_{,\Omega}$ and $\mathbf{W}_{,\Omega\Omega}$ by the techniques of derivation with respect to a domain.

For clarity, our discussion will be here limited to a bi-dimensional case, S is then a curve.

The dependence of $\mathbf{u}(\Omega,\lambda)$ on $\Omega$ can be understood in the following way :

Let **du** be the displacement increment associated to a variation of the domain $\mathbf{d\Omega}$ at each material point of $\Omega$, the normal rate $d\Omega(s)$ is oriented by the external normal n.

This field is given by the following proposition :

**Proposition 1 (5)**

The associated field **du(dΩ)** is a solution of the variational equation :

$\mathbf{du} \in \mathbf{dU}$ and satisfies $\forall\ \boldsymbol{\delta u} \in \mathbf{dU_0}$ :

$$\int_{V-\Omega} d\epsilon \,.\, \frac{\partial^2 w}{\partial\epsilon\partial\epsilon}\, \delta\epsilon \; dV + \int_S \sigma \,.\, \delta\epsilon \; d\Omega(s)\; ds = 0. \tag{5.12}$$

which is obtained simply by derivation with respect to Ω of the equilibrium equation.

In this proposition, dU denotes the set of admissible rates and $dU_0$ the set of virtual admissible rates.

From (5.11) and (5.10), it follows by derivation that :

$$\frac{\partial W}{\partial \Omega} \,.\, d\Omega = - \int_S w(\epsilon) \,.\, d\Omega \;\; ds \tag{5.13}$$

The second derivative of energy is obtained by derivation of (5.13) in a direction **dΩ′** :

$$d\Omega' \,.\, \frac{\partial^2 W}{\partial\Omega\partial\Omega'} \,.\, d\Omega = \int_S \sigma . d\epsilon' \; d\Omega \; ds + \int_S \left( w_{,n} + \frac{w}{R} \right) d\Omega \; d\Omega' \; ds \tag{5.14}$$

A more symmetric expression follows also from (5.12) and (5.14) :

$$d\Omega' \,.\, \frac{\partial^2 W}{\partial\Omega\partial\Omega'} \,.\, d\Omega = - \int_{V-\Omega} d\epsilon \,.\, \frac{\partial^2 w}{\partial\epsilon\partial\epsilon} \,.\, d\epsilon' \;\; dV \tag{5.15}$$

$$+ \int_S \left( w_{,n} + \frac{w}{R} \right) d\Omega \; d\Omega' \; ds$$

where **dϵ** and **dϵ′** are associated respectively to **dΩ** and **dΩ′** by (5.12), and R denotes the curvature of the curve S.

## 5.2.3.- Quasi-static evolution of damage

Relation (5.13) shows clearly that the force **A** associated with the rate $\dot{\Omega}$ is the function $G(s) = w(s),\ s \in S$ .

The natural propagation law :

$$\begin{array}{l} \text{If } w(s) < w_c \text{ then } \dot{\Omega}(s) = 0 \text{ (no propagation)} \\ \text{If } w(s) = w_c \text{ then } \dot{\Omega}(s) \geq 0 \text{ (possible propagation)} \end{array} \quad (5.16)$$

is associated with the dissipative potential :

$$\mathbf{D}(\Omega,\dot{\Omega}) = \int_S w_c\, \dot{\Omega}(s)\, ds \quad (5.17)$$

for $\dot{\Omega}(s) \geq 0 \ \forall \ s \in S$. The dissipated energy :

$$\mathbf{W^d}(\Omega) = w_c \,.\, \mathrm{Vol}(\Omega) \quad (5.18)$$

is proportional to the volume of the damaged zone.

The rate problem of propagation of the damaged zone $\Omega$ is given by variational inequality (2.2) which may also be written as :

$$\dot{\Omega}(s) \geq 0 \text{ on } S_c \text{ and satisfies } \forall\ \delta\Omega(s) \geq 0 \text{ on } S_c : \quad (5.19)$$

$$(\delta\Omega - \dot{\Omega}) \,.\, (\Phi,_{\Omega\Omega} \,.\, \dot{\Omega} + W,_{\Omega\lambda} \,.\, \dot{\lambda}) \geq 0$$

where $S_c$ denotes the portion of S on yield ( $w(s) = w_c$ ) and $\Phi = W + W^d$ :

$$d\Omega \,.\, \Phi,_{\Omega\Omega} \,.\, \delta\Omega = - \int_{V-\Omega} d\epsilon \,.\, w,_{\epsilon\epsilon} \,.\, \delta\epsilon \ dV + \int_S w,_n \ d\Omega\ \delta\Omega\ ds \quad (5.20)$$

and :

$$d\Omega \,.\, W,_{\Omega\lambda} = \int_S \sigma \,.\, \epsilon,_\lambda \ d\Omega\ ds \quad (5.21)$$

Stability and bifurcation again can be discussed as in previous chapters . For example, the following proposition is obtained :

**Proposition 2 (5)**

The present equilibrium is stable in the dynamic sense if the quadratic form $d\Omega \,.\, \Phi,_{\Omega\Omega} \,.\, d\Omega$ is positive definite on the set of admissible rates $\delta\Omega(s) \geq 0$ on $S_c$ .

### 5.2.4.- Remarks

- The set dU of admissible rates depends on $d\Omega$ if displacements $u_d$ are prescribed on a moving boundary S. Such a dependence will be illustrated later in an example.

- The case of solids with brittle and partial damage (iii) can be discussed in the same manner. This model leads also to the study of the motion of a surface of discontinuity S separating the undamaged zone V-$\Omega$ and the damage zone $\Omega$ (cf. Pradeilles & Stolz [30], Mroz & Dems [9]).

The expressions of the first and second derivatives of energy can be obtained in the same spirit.

- In the general case, it is established that :

The first derivative of energy W is :

$$\mathbf{W}_{,\Omega} \cdot \mathbf{d\Omega} = - \int_S [\, \mathbf{w - n \,.\, \sigma \,.\, u_{,n}} \,]\, \mathbf{d\Omega \;\; ds} = - \int_S \mathbf{G \; d\Omega \;\; ds} \tag{5.22}$$

which reduces to (5.13) in our particular case since $\sigma \,.\, n = 0$ on S.

In (5.22), the quantity $G = [\, w - n.\sigma.u_{,n} \,]$ is the energy release rate at a point of the moving surface S. It represents the local value of the force A associated with irreversible parameter $\alpha = \Omega$ .

The second derivative of energy $\Phi$ is :

$$\mathbf{d\Omega \,.\, \Phi_{,\Omega\Omega} \,.\, d\Omega} = \int_S \mathbf{- \, DG \,.\, d\Omega \;\; ds} \tag{5.23}$$

In this expression, DG denotes the variation of G following the motion $d\Omega$ of S.
This expression is clear from (5.14) for example.

In the spirit of (4.14), the stability criterion can also be written as :

$$\int_S \mathbf{-DG \,.\, d\Omega \;\; ds} > \mathbf{0} \tag{5.24}$$

for any $\mathbf{d\Omega} \neq \mathbf{O}$ such that $d\Omega(s) \geq 0$ on S .

- Simple examples illustrating the two models (ii) and (iii) have been given by Mroz & Dems [9], Berest & Fedelich [13], Marigo [22] (cylinders under pressure, flexion or torsion of beams..etc.) .

### 5.3.- Illustrative example : torsion of a brittle and elastic cylinder

This example is due to Berest and Fedelich [22].

Let us consider the torsion of a brittle elastic cylinder (model (iii) of total damage). If $\ell$ is the length , $\lambda\ell$ the torsion angle between the two extreme sections, we wish here to investigate the stability of the homogeneous equilibrium with the control parameter $\lambda$.

This solution is given by the displacement :

$$u_x = -\lambda\, yz \quad , \; u_y = \lambda\, xz \; , \; u_z = \lambda\, \eta(x,y)$$

where $\eta(x,y)$ denotes the warping function.
The associated potential energy is

$$\mathbf{E}(\boldsymbol{\eta},\boldsymbol{\Omega},\boldsymbol{\lambda}) = \lambda^2 \mu\ell \int_{V-\Omega} [(\eta,_x - y)^2 + (\eta,_y + x)^2\ ]\ dV$$

Equilibrium leads to :

$$\mathbf{E},_{\boldsymbol{\eta}} \cdot \delta\boldsymbol{\eta} = 2\lambda^2 \mu\ell \int_{V-\Omega} (\eta,_x - y)\ \delta\eta,_x + (\eta,_y + x)\ \delta\eta,_y = 0$$

and gives :

$$\Delta\eta = 0 \text{ in } V\text{-}\Omega \; , \; \frac{\partial\eta}{\partial n} = n_x\ y - n_y\ x \text{ on } S$$

For example, for a cylinder of circular section, these equations give the trivial equilibrium $\eta = 0$.

The dependence $d\eta(d\Omega)$ can be obtained directly from the derivation of previous equations with respect to $\Omega$ . For example, the variational form leads to :

$$\int_{V-\Omega} (\ d\eta,_x\ \delta\eta,_x + d\eta,_y\ \delta\eta,_y\ )\ dV +$$

$$\int_S [\ (\ \eta,_x - y)\ \delta\eta,_x + (\eta,_y + x)\ \delta\eta,_y\ ]\ d\Omega\ ds = 0$$

For a cylinder of circular section since $\eta = 0$ , this equation gives :

$$\int_{V-\Omega} ( d\eta_{,x}\ \delta\eta_{,x} + d\eta_{,y}\ \delta\eta_{,y} )\ dV + R \int_S d\Omega\ \delta\eta_{,s}\ ds = 0$$

and thus, after intergration by parts on the boundary of the last term :

$$\int_{V-\Omega} ( d\eta_{,x}\ \delta\eta_{,x} + d\eta_{,y}\ \delta\eta_{,y} )\ dV - R \int_S d\Omega_{,s}\ \delta\eta\ ds = 0$$

Finally, the following equations are obtained :

$$\Delta(d\eta) = 0 \text{ in } V\text{-}\Omega \quad , \quad \frac{\partial(d\eta)}{\partial n} = R\ d\Omega_{,s}$$

These equations can also be derived from a direct differentiation of the previous local equations with respect to $\Omega$.

The quadratic form $d\Omega \cdot \Phi_{,\Omega\Omega} \cdot d\Omega$ is :

$$2\ \lambda^2\ \mu\ell \left\{ - \int_{V-\Omega} [ (d\eta_{,x})^2 + (d\eta_{,y})^2 ] + \int_S R\ d\Omega^2\ ds \right\}$$

Let $\quad d\Omega(\theta) = da_0 + \sum_{j=1}^{\infty} ( da_j \cos j\theta + db_j \sin j\theta )$

be the Fourier expansion of $d\Omega$.

Then $d\eta(d\Omega)$ is a harmonic function in $\Omega$ with normal derivative on S ($r = R$) :

$$\frac{\partial(d\eta)}{\partial n} = \frac{1}{R} \sum_{j=1}^{\infty} j ( - da_j \sin j\theta + db_j \cos j\theta )$$

and thus :

$$d\eta(r,\theta) = \sum_{j=1}^{\infty} ( - da_j \sin j\theta + db_j \cos j\theta ) \left(\frac{r}{R}\right)^j$$

Finally, the following result is obtained :

$$d\Omega \,.\, \Phi,_{\Omega\Omega} \,.\, d\Omega = 2\,\lambda^2\,\mu\ell\,2\pi R^2 \left\{ da_0^{\,2} - \sum_{j=1}^{\infty} (j-1)\,[\,da_j^{\,2} + db_j^{\,2}\,] \right\}$$

this form is not positive definite on the set { $da_j$ , j = 0,1,.. } such that the rate $d\Omega(\theta)$ is non positive.

The criterion of stability is thus violated.

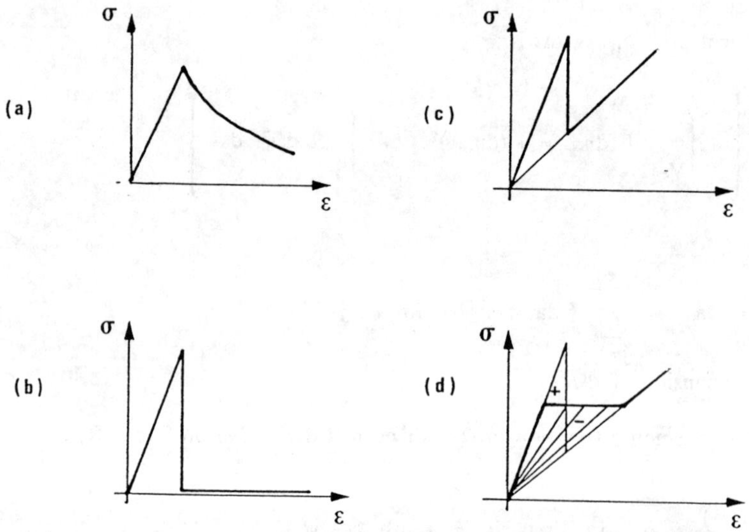


Fig. 1 (5) , Some models of brittle damage

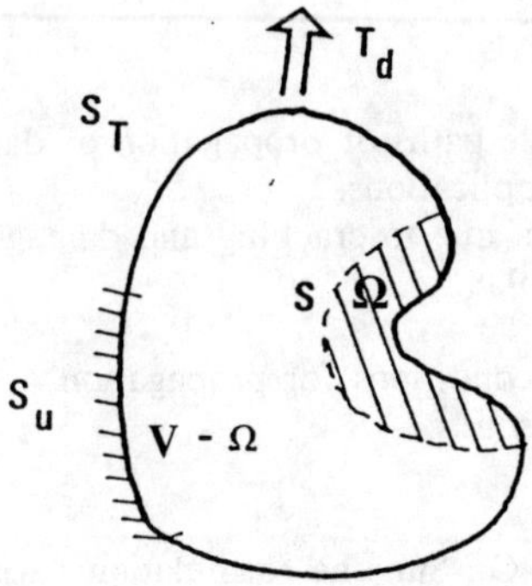


Fig. 2 (5) , Propagation of a damage zone

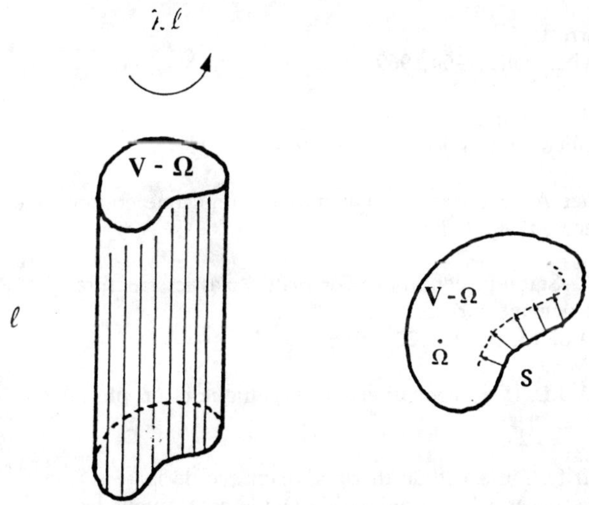


Fig. 3 (5) , Torsion of a brittle elastic cylinder

## REFERENCES

[1] Bazant Z.P. , Stable states and stable paths of propagation of damage zones : inelastic stability criteria with applications.
Strain localization and size effect due to cracking and damage.
France-US Workshop, Cachan 1988.

[2] Bazant Z.P., Ohtsubo H., Stability conditions for propagation of a system of cracks in brittle solids.
Adv. Civil Engng, 1977.

[3] Benallal A., Florez J. & Geymonat G., Sur une formulation variationelle à deux champs en élasticité couplée à l'endommagement.
C.R. Acad. Sciences, T305, 1987.

[4] Bérest P. , Problèmes de Mécanique associés au stockage souterrain.
Thèse de Doctorat, Paris, 1989.

[5] Biot M.A., Mechanics of incremental deformations
Wiley, 1965.

[6] Blum J., Fracture arrests.
Fracture, Liebowitz, vol 5,1-64,1969.

[7] Budiansky B., Elastic buckling.
Advances in Applied Mechanics, vol. 14, 1974.

[8] Bui H.D. & Ehrlacher A., Propagation dynamique d'une zone endommagée.
C.R. Acad. Sciences, t290, 1980.

[9] Dems K. & Mroz Z., Stability conditions for brittle plastic structures with propagating damage surfaces.
J. Struc. Mech., Vol 13, pp 95-122, 1985.

[10] Duvaut G. & Lions J.L., Les inéquations en mécanique et en physique.
Dunod, Paris, 1972.

[11] Ehrlacher A., Contribution à l'étude thermodynamique de la progression de fissure et à la mécanique de l'endommagement brutal.
Thèse, Paris, 1985.

[12] Fedelich B., Trajets d'équilibre des systèmes mécaniques dissipatifs à comportement indépendant du temps physique.
Thèse, Paris,1990.

[13] Fedelich B. & Bérest P., Torsion d'un cylindre élasto-fragile, Stabilité de l'équilibre.
Arch. Mech. Stos., Vol 40, 5-6, 1988.

[14] Florez J., Elasticité couplée à l'endommagement :
formulation, analyse théorique et approximation numérique.
These, Paris 1989.

[15] Germain P., Mécanique des Milieux Continus.
Masson, Paris, 1973.

[16] Halphen B. & Nguyen Q.S., Sur les matériaux standards généralisés.
J. de Mécanique, vol 1, 1975.

[17] Hill R., A general theory of uniqueness and stability in elastic-plastic solids.
J. Mech. Phys. Solids, 236-249, 1958.

[18] Hutchinson J., Plastic buckling.
Advances in Applied Mechanics. 1973.

[19] Koiter W.T., A basic open problem in the theory of Elastic Stability.
IUTAM Symposium "Applications of methods of functional analysis
to problems in mechanics". Marseille, 1975.

[20] Lemaitre J. & Chaboche J.L., Mécanique des Matériaux solides.
Dunod, Paris, 1985.

[21] Maier G. , A shake-down matrix theory allowing for workhardening
and second order geometric effect.
Symposium "Foundations of Plasticity", Varsovie, 1972.

[22] Marigo J.J., Modélisation et analyse qualitative de l'endommagement
des matériaux et des structures.
Cours "Problèmes non linéaires appliqués". EDF-INRIA-CEA.
Paris, 1990.

[23] Moreau J.J., On unilateral constraints, friction and plasticity.
Cours CIME, Bressanone 1973.
Edition Cremonese, Roma 1974.

[24] Nemat-Nasser, Sumi Y. & Keer L.M., Unstable growth of tensile cracks
in Brittle Solids.
Int. J. Solids & Structures, 1017-1035, 1980.

[25] Nguyen Q.S. & Radenkovic D., Stability of an equilibrium
in elastic plastic solids.
IUTAM Symposium "Application of methods of functional analysis to problems in Mechanics.
Marseille, 1975.

[26] Nguyen Q.S., Bifurcation et stabilité des systèmes irréversibles
obéissant au principe de dissipation maximale.
J. de Mécanique, Paris, 1984.

[27] Nguyen Q.S., Stabilité et bifurcation des systèmes dissipatifs standards

à comportement indépendant du temps physique.
C. R. Acad. Sciences, T310, 1990.

[28] Nguyen Q.S., Stolz C., Energy methods in fracture mechanics :
Stability, Bifurcation and Second variations.
IUTAM Symposium "Applications of multiple scaling in mechanics".
Masson, Paris 1987.

[29] Petryk H., On energy criteria of plastic stability.
Symposium Plastic instability. Considère memorial. ENPC, 1985.

[30] Pradeilles R. & Stolz C. , Sur le problème d'évolution des milieux à
endommagement partiel.
C.R. Acad. Sciences, 1991.

[31] Rice J.R., Inelastic constitutive relations for solids :
an internal variable theory and its application in metal plasticity.
J. Mech. Phys. Solids, vol 19, 1971.

[32] Sewell M.J. , A survey of plastic buckling.
Stability, Waterloo Press, 1972.

[33] Stolz C., Anélasticité et stabilité.
Thèse, Paris, 1987.

[34] Storakers B., Non linear aspects of delamination in structural members.
Theoretical and Applied Mechanics, ICTAM, Grenoble 1988.

[35] Suquet P., Plasticité et homogénéisation.
Thèse, Paris, 1982.

[36] Triantafyllidis N., On the bifurcation and post-bifurcation analysis
of elastic plastic solids under general pre-bifurcation conditions.
J. Mech. Phys. Solids, vol.31, 1983.

# THEORY OF BIFURCATION AND INSTABILITY IN TIME-INDEPENDENT PLASTICITY

**H. Petryk**
**Polish Academy of Sciences, Warsaw, Poland**

ABSTRACT

The theory is developed for constitutive rate equations which are arbitrarily nonlinear and can thus describe the important effect of formation of a yield-surface vertex. Basic elements of Hill's theory of bifurcation and stability in time-independent plastic solids are presented. In particular, the condition sufficient for uniqueness of a solution to the first-order rate boundary value problem, the stationary and minimum principles for velocities, the concept of a comparison solid and the primary bifurcation point are discussed. Distinction is emphasized between the conditions for uniqueness and for stability of equilibrium, and between the bifurcation point and an eigenstate. Several recent extensions of Hill's theory are next presented. It is shown how the property of the comparison solid required in the uniqueness criterion can be weakened and justified then on micromechanical grounds, without the need of complete specification of the macroscopic constitutive law. The question of non-uniqueness and instability in the post-critical range is examined, and respective theorems are formulated for discretized systems and for a certain class of continuous systems. The energy interpretation of the basic functionals and conditions in the bifurcation theory is given. Finally, a unified approach to various bifurcation and instability problems is presented which is based on the concept of instability of a deformation process and on the relevant energy criterion.

## INTRODUCTION

These lecture notes are aimed at presenting in a systematic way certain general results obtained in the area of uniqueness and stability, or bifurcation and instability, of quasi-static solutions for time-independent inelastic solids. The word "general" does not mean, of course, that the theory is universal and can be applied to any problem. For instance, the assumptions adopted exclude problems involving fracture or strong discontinuities in velocities, and most of the conclusions will be obtained for conservative loading and symmetric moduli only. The class of constitutive laws and problems under consideration is, however, still very broad, and the theory can be applied to a variety of phenomena as buckling, necking, snap-through, etc.

Constitutive framework is given in Chapter 1 and discussed further in Chapter 3, with the emphasis on the effect of changes in geometry and on incremental nonlinearity of the constitutive law. Fundamental results concerning uniqueness and bifurcation in velocities and stability of equilibrium, due to R.Hill, are presented in Chapter 2. Later chapters complement Hill's theory and are mainly based on the present author's own work during last several years. The classical assumption of a smooth yield surface is nowadays known to be an oversimplification in bifurcation problems, and for practical applications it is of interest whether the bifurcation criterion may be linearized also at a yield-surface vertex whose formation is predicted by micromechanical studies; this problem is discussed in Chapter 3. Stability of equilibrium and incremental uniqueness on post-critical paths where Hill's sufficiency conditions no longer apply are examined in Chapter 4. Chapter 5 shows that the basic conditions from the bifurcation theory can be given an energy interpretation by expressing them in terms of the second time derivative of the total energy functional. Finally, in Chapter 6 a unified approach to bifurcation and stability problems is presented which is based on the concept of instability of a quasi-static deformation process. The concept itself is discussed first, and then the energy criterion of instability of a process is introduced. It is demonstrated how various critical points connected with bifurcation, loss of stability of equilibrium or material instabilities are generated by the energy criterion as particular cases of the onset of instability of a deformation process.

No attempt is made here to review the related vast literature devoted to particular problems or specific material models, or to compare the theory with experimental or numerical results. References to the literature are intended primarily to indicate where theoretical results quoted below were obtained, and not how they relate to other investigations. The contents of the lectures decide that many important contributions but not directly to the general theory described below are not mentioned. Fortunately, the other Parts of this Volume are expected to fill this gap.

## 1. PRELIMINARIES

### 1.1 Notation. Deformation gradient and nominal stress

Lower case Latin subscripts varying from 1 to 3 denote vector or tensor components on a fixed orthogonal triad of unit vectors $\mathbf{a}^1$ , $\mathbf{a}^2$ ,$\mathbf{a}^3$ , and the usual summation convention will be adopted for repeated indices. Boldface letters will symbolize vectors or tensors themselves which often will be identified with a collection of their components, just to simplify the terminology. A dot over a symbol of a quantity denotes the material time derivative understood as a *forward* rate, i.e. the right-hand derivative with respect to a scalar time-like parameter $\theta$.

Two configurations of a continuous deformable material body are considered: the actual one is variable and is called *the current configuration* while the other is fixed and called *the reference configuration.* In many cases it is convenient to assume that these two configurations *momentarily* coincide. Denote by $\xi_i$ and $x_i$ the rectangular Cartesian coordinates of a material point in the reference and current configurations, respectively, on the common basis $\mathbf{a}^i$ . Infinitesimal elements of volume and of surface area in the *reference* configuration are denoted by $d\xi$ and $da$, respectively. Displacement and velocity components are $u_i = x_i - \xi_i$ and $v_i \equiv \dot{x}_i \equiv \partial x_i/\partial\theta^+$ , respectively. A spatial field will be distinguished from its value at some point by superimposing a tilde over the symbol.

The first weak variation (Gateaux differential) of a functional $J$ at its argument value $\tilde{\mathbf{v}}$ in the direction $\tilde{\mathbf{w}}$ is denoted by

$$\delta J(\tilde{\mathbf{v}}, \tilde{\mathbf{w}}) \equiv \frac{d}{d\gamma} J(\tilde{\mathbf{v}} + \gamma\tilde{\mathbf{w}})|_{\gamma=0} \,, \tag{1.1}$$

where $\tilde{\mathbf{w}} = \delta\tilde{\mathbf{v}}$ stands for an admissible variation of $\tilde{\mathbf{v}}$ and $\gamma$ is a real variable.

If the mapping $x_i(\xi_j)$ is differentiable then the formula $F_{ij} = \partial x_i/\partial\xi_j \equiv x_{i,j}$ defines *the deformation gradient*; henceforth the partial differentiation with respect to the rectangular Cartesian coordinates $\xi_i$ (but *not* $x_i$ ) is denoted by the corresponding index preceded by a comma. The stress measure which is work-conjugate to $F_{ij}$ is *the nominal stress* $N_{ij}$ , that is,

$$N_{ij}\, \dot{F}_{ji} \;=\; \text{the rate of deformation work per unit reference volume}\,. \tag{1.2}$$

This may be regarded as a definition of $N_{ij}$ ; equivalently, $N_{ij}$ are obtained by decomposing the load vector $dT_j$ (acting in the current configuration) on an infinitesimal surface element in the reference configuration, viz.

$$dT_j \;= n_i\, N_{ij}\, da, \tag{1.3}$$

where $n_i$ and $da$ denote the unit normal to the element and the element area, both taken in the reference configuration. Generally, the nominal stress is not symmetric. It is connected with the Cauchy stress $\boldsymbol{\sigma}$ and the Kirchhoff stress $\boldsymbol{\tau}$ by the formula

$$F_{ji}\, N_{ik} \;= \det(F_{pq})\, \sigma_{jk} \;= \tau_{jk}\,; \tag{1.4}$$

the transpose of $\mathbf{N}$ is also called the first Piola-Kirchhoff stress. Of course, if the reference and current configurations coincide then $F_{ij} = \delta_{ij}$ ($\delta_{ij}$ is the Kronecker symbol) and $N_{ij}$ reduces to $\sigma_{ij}$ .

The moment balance at the assumed absence of body moments requires the Cauchy stress to be symmetric, which is equivalent to

$$F_{ji}\, N_{ik} \;= F_{ki}\, N_{ij}\,. \tag{1.5}$$

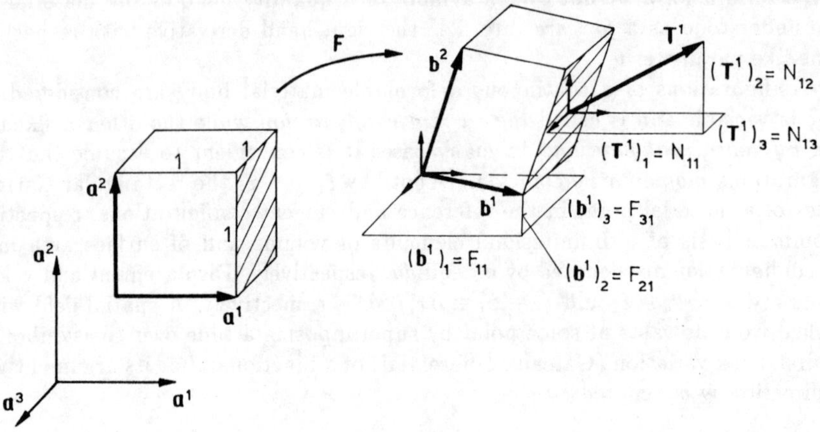


Fig. 1, Deformation gradient and nominal stress.

The deformation gradient and nominal stress as particular measures of deformation and stress are directly connected with typical experimental measurements. Consider in the reference configuration a unit cube of the edges defined by the triad of base vectors $\mathbf{a}^i$ (Fig. 1). The cube is thought as an element of a specimen in an initial state. After a (macroscopically) *uniform* deformation the cube becomes a parallelepiped with the edges $\mathbf{b}^i$ such that $(\mathbf{b}^i)_j = F_{jk}(\mathbf{a}^i)_k = F_{ji}$; the former equality follows from the definition of $F_{ij}$ and the latter from $(\mathbf{a}^i)_k = \delta_{ik}$ .It becomes evident that measurements of the components of $\mathbf{b}^i$ provide a direct way to determine the deformation gradient in an experiment.

The surface tractions are assumed to be (macroscopically) uniformly distributed on each of the parallelepiped faces. The resultant load vector on the parallelepiped face in the current configuration which in the *reference* configuration had the outward normal $\mathbf{a}^i$ is denoted by $\mathbf{T}^i$ . In agreement with (1.3), the nominal stress $N_{ij}$ can be directly defined as the j-th component of the load vector $\mathbf{T}^i$ (Fig. 1); the components can be measured experimentally. Thus

$$F_{ji} = (\mathbf{b}^i)_j \quad N_{ij} = (\mathbf{T}^i)_j \tag{1.6}$$

where $\mathbf{b}^i$ and $\mathbf{T}^i$ have a clear mechanical interpretation.

**Remark**

Since the basis $\mathbf{a}^i$ has been assumed fixed and common for all configurations, it is not necessary here to distinguish between components connected with the reference or current configurations. However, it should be remembered that both the deformation gradient and nominal stress are in fact based partly on each of the two configurations (they are so-called two-point tensors). This becomes essential when the question of objectivity (or frame-indifference) is examined which, however, is not discussed here.

**1.2. Incremental effect of changes in geometry**

The theory where changes in geometry of a deformed material body are disregarded is customarily called the small strain theory. Unfortunately, the terminology may suggest that the converse implication is valid, that is, that all changes in geometry may be disregarded provided only that the strains are known to be small. That supposition is evidently incorrect in a study of buckling of slender structures but might seem to be justified for solid bodies where the rotations are of order of the strains. However, even then the supposition is generally incorrect if the bifurcation and stability phenomena are examined, as R.Hill clearly demonstrated in the cycle of papers in 1957-1959 [1÷5]. In the analysis of such phenomena the *incremental* effect of changes in geometry has to be taken into account. This can be done by using various variables but Hill has shown that the resulting formulae have particularly concise form when expressed in terms of the rates of the deformation gradient and nominal stress, $\dot{F}_{ij}$ and $\dot{N}_{ij}$ . The rate $\dot{F}_{ij}$ of the deformation gradient is, of course, nothing else than the velocity gradient $v_{i,j}$ evaluated in the *reference* configuration. The Eulerian strain rate $\dot{\varepsilon}_{ij} = \frac{1}{2}(\partial v_i/\partial x_j + \partial v_j/\partial x_i)$ and material spin $\omega_{ij} = \frac{1}{2}(\partial v_i/\partial x_j - \partial v_j/\partial x_i)$ can be expressed in terms of $\dot{F}_{ij}$ on substituting $\partial v_i/\partial x_j = \dot{F}_{ik}F^{-1}_{kj}$ , where $\mathbf{F}^{-1}$ denotes the inverse of $\mathbf{F}$.

To illustrate the incremental effect of change in geometry, choose the reference configuration to coincide momentarily with the current configuration so that $x_i(\boldsymbol{\xi}) = \xi_i$ , $F_{ij} = \delta_{ij}$ and the current strains and rotations are exactly zero. In that special case $\dot{F}_{ij}$ coincides with the spatial velocity gradient $\partial v_i /\partial x_j$ . Consider the cube discussed above under a current stress $N_{ij} = \sigma_{ij}$ and suppose that $v_{i,j}$ and constitutively related $\dot{\sigma}_{ij}$ are known. From (1.4) it follows that in the reference configuration we have

$$\dot{N}_{ij} = \dot{\sigma}_{ij} + v_{k,k}\,\sigma_{ij} - v_{i,k}\,\sigma_{kj} \tag{1.7}$$

From (1.6) we obtain that the components $(\dot{\mathbf{T}}^i)_j$ of the rate of the load vector $\mathbf{T}^i$ on the cube face are given by the right-hand expression in (1.7); in the small strain theory the last two terms would be absent. This does not necessarily mean that it is these two terms which represent precisely the effect of changes in geometry. For, $\dot{\sigma}_{ij}$ is influenced by the material spin and can thus hardly be regarded as a proper generalization of the stress-rate from the small strain theory which is related constitutively to $\dot{\varepsilon}_{ij}$ and not to $\omega_{ij}$ . Rather, an objective stress flux should be regarded as such a generalization but then a problem arises which (if any) of the many proposals known from the literature is the most appropriate . Leaving this problem aside, consider one of the most popular, the Zaremba–Jaumann flux of Kirchhoff stress, defined by

$$\overset{\circ}{\tau}_{ij} = \dot{\tau}_{ij} + \tau_{ik}\,\omega_{kj} - \omega_{ik}\,\tau_{kj}\,. \tag{1.8}$$

Substitution of (1.8) into (1.7) at $F_{ij} = \delta_{ij}$ yields

$$\dot{N}_{ij} = \overset{\circ}{\tau}_{ij} - \dot{\varepsilon}_{ik}\,\sigma_{kj} - \sigma_{ik}\,\omega_{kj}\,. \tag{1.9}$$

If the relationship $\overset{\circ}{\tau}_{ij}\ (\dot{\boldsymbol{\varepsilon}})$ is taken of the same form as the incremental constitutive relationship from the small strain theory then the last two terms in (1.9) (multiplied by

$d\theta$) represent the effect of changes in geometry on the increment $d\mathbf{T}^i$ of the load vector associated with a given deformation increment. It is clear that if stress-rates are of order of strain-rate × stress, that is, if the instantaneous stiffness moduli are of order of the stresses, then that incremental geometric effect is *in general* not negligible. It is recalled that one arrives at that conclusion despite the fact that instantaneous strains and rotations are exactly zero.

**Exercise**

Examine the effect of changes in geometry on the load increment in pure tension and simple shear.

### 1.3 Constitutive framework for time-independent inelastic solids

A general assumption is made that mechanical properties of the material in isothermal deformation do not depend in any way on a natural time. Under restriction to piecewise smooth deformation paths, it is also assumed that the stresses vary in time in a continuous and piecewise smooth manner. It must be noted, however, that that natural assumption excludes rigid-plastic solids. The constitutive relationship between the (right-hand) rates of stress and strain, no matter in which variables it is originally formulated to ensure objectivity of the material description, can be expressed as

$$\dot{N}_{ij} = \dot{N}_{ij}(\dot{\mathbf{F}},\mathbf{H}), \tag{1.10}$$

where the parameter $\mathbf{H}$ symbolizes the influence of the deformation history prior to the current instant, including the current stress and deformation. For instance, $\mathbf{H}$ may be taken as a set $\{F_{ij}\ , \sigma_{ij}\ , \alpha_K\}$ , where $\alpha_K$ are internal state variables; arbitrary "hardening" or "softening" characteristics are allowed. If the current state of the material is regarded as known then the symbolic parameter $\mathbf{H}$ will be omitted for simplicity. At a given state, the function $\dot{N}_{ij}(\dot{\mathbf{F}})$ is assumed to be single-valued, continuous and piecewise smooth (with a possible singularity at $\dot{F}_{ij} = 0$). The following function

$$U(\dot{\mathbf{F}},\mathbf{H}) = \frac{1}{2}\dot{N}_{ij}(\dot{\mathbf{F}},\mathbf{H})\,\dot{F}_{ji} \tag{1.11}$$

will play a fundamental role in the theory of bifurcation and instability.

For the material to be time-independent, the dependence of $\dot{N}_{ij}$ on $\dot{F}_{kl}$ must be positively homogeneous of degree one, i.e.

$$\dot{N}_{ij}(\gamma\dot{\mathbf{F}},\mathbf{H}) = \gamma\dot{N}_{ij}(\dot{\mathbf{F}},\mathbf{H}) \ \text{ for every } \gamma > 0 \tag{1.12}$$

but otherwise may be arbitrarily nonlinear (piecewise linear in particular cases). Consequently, $U(\dot{\mathbf{F}})$ is homogeneous of degree two. By the Euler theorem on homogeneous functions, at every point of differentiability of $\dot{\mathbf{N}}(\dot{\mathbf{F}})$ the equations (1.10) can be written down in the form

$$\dot{N}_{ij} = C_{ijkl}(\dot{\mathbf{F}},\mathbf{H})\,\dot{F}_{lk}\ ,\ C_{ijkl} \equiv \partial\dot{N}_{ij}/\partial\dot{F}_{lk} \tag{1.13}$$

where, at a given state, $C_{ijkl}$ depend only on the direction of strain rate; the dependence of $C_{ijkl}$ on $\dot{\mathbf{F}}$ may be discontinuous and nonlinear. Later, of primary importance will be *the tangent moduli* $C_{ijkl}(\dot{\mathbf{F}}^0,\mathbf{H})$ evaluated at the actual velocity gradient $\dot{\mathbf{F}}^0$ directed tangentially to a given deformation path.

Suppose that at a certain given state the incremental constitutive relationship is specified in the form

$$\overset{\circ}{\tau}_{ij} = L_{ijkl}(\dot{\varepsilon})\,\dot{\varepsilon}_{kl} \tag{1.14}$$

e.g. as an adaptation of a constitutive relationship from the small strain theory. We assume that $L_{ijkl} = \partial\overset{\circ}{\tau}_{ij}/\partial\dot{\varepsilon}_{kl}$ and $L_{ijkl} = L_{ijlk}$. Then the moduli $C_{ijkl}(\dot{\mathbf{F}})$ can be found from the following relationship

$$\det(F^{-1}_{rs})\,F_{ip}\,F_{kq}\,C_{pjql}(\dot{\mathbf{F}}) = L_{ijkl}(\dot{\varepsilon}) - \frac{1}{2}(\sigma_{jk}\,\delta_{il} - \sigma_{ik}\,\delta_{jl} + \sigma_{il}\,\delta_{jk} + \sigma_{jl}\,\delta_{ik}), \tag{1.15}$$

where $\boldsymbol{\sigma}$ and $\mathbf{F}$ are the current Cauchy stress and deformation gradient, and $\dot{\varepsilon}_{ij} = \frac{1}{2}(\dot{F}_{ik}F^{-1}_{kj} + \dot{F}_{jk}F^{-1}_{ki})$. The derivation of (1.15) is omitted here; all needed transformations are given in [6]. The stress-dependent right-hand terms in (1.15) are associated with the *incremental* effect of changes in geometry discussed in the preceding subsection, while the left-hand multiplicative term dependent on the deformation gradient represents the influence of a *finite* deformation between the reference and current configurations.

Following Hill [5] [6], we shall later introduce an additional assumption that the constitutive rate equations (1.10) admit a potential, viz.

$$\dot{N}_{ij} = \frac{\partial U}{\partial \dot{F}_{ji}}, \qquad C_{ijkl} = \frac{\partial^2 U}{\partial \dot{F}_{ji}\partial \dot{F}_{lk}}, \tag{1.16}$$

where $U$ necessarily satisfies (1.11) on account of the homogeneity of the constitutive equations and is continuously differentiable by the assumed continuity of $\dot{N}_{ij}(\dot{\mathbf{F}})$. Of course, (1.16) implies that

$$C_{ijkl} = C_{klij} \tag{1.17}$$

whenever $C_{ijkl}$ are well defined, and conversely, differentiation of (1.11) by $\dot{F}_{ji}$ shows that (1.17) implies (1.16) . Existence of the potential and diagonal symmetry of the moduli are preserved when (2.1) is expressed in terms of rates of any *work-conjugate* pair of stress and strain measures [6]. In particular, we have (1.17) if and only if the moduli in (1.14) satisfy $L_{ijkl} = L_{klij}$ . For elastic materials, this symmetry property is a consequence of existence of a strain energy potential.

**Example: Classical elastic-plastic solids**

For the classical models of incremental elastic-plastic response, the moduli in the constitutive rate equations (1.14) have the following form (cf. [7])

$$L_{ijkl} = \begin{cases} L^p_{ijkl} & \text{if} \quad \lambda_{ij}\dot{\varepsilon}_{ij} > 0 \quad \text{(loading)} \\ L^e_{ijkl} & \text{if} \quad \lambda_{ij}\dot{\varepsilon}_{ij} < 0 \quad \text{(unloading)} \end{cases} \tag{1.18}$$

with

$$L^p_{ijkl} = L^e_{ijkl} - g^{-1}\lambda^p_{ij}\lambda_{kl}\,. \tag{1.19}$$

$\lambda_{ij}$ is the (outward) normal to the current smooth yield surface in the *strain* space, $g$ is a positive scalar parameter, and $\lambda^p_{ij}$ defines the direction of the so-called plastic parts $\dot{\varepsilon}^p_{ij}$ and $\overset{\circ}{\tau}{}^p_{ij}$ of the strain-rate and stress-rate when $\lambda_{ij}\,\dot{\varepsilon}_{ij} > 0$, viz.

$$\overset{\circ}{\tau}{}^p_{ij} = \overset{\circ}{\tau}_{ij} - L^e_{ijkl}\dot{\varepsilon}_{kl} = (-g^{-1}\lambda_{kl}\dot{\varepsilon}_{kl})\lambda^p_{ij}\,, \tag{1.20}$$

$$\dot{\varepsilon}^p_{ij} = \dot{\varepsilon}_{ij} - M^e_{ijkl}\overset{\circ}{\tau}_{kl} = (g^{-1}\,\lambda_{pq}\dot{\varepsilon}_{pq})M^e_{ijkl}\lambda^p_{kl} \tag{1.21}$$

and $L^e_{ijkl}$ and $M^e_{ijkl}$ are elastic moduli and compliancies, respectively, with the assumed symmetry $L^e_{ijkl} = L^e_{klij}$ . It can be seen that $L^p_{ijkl} = L^p_{klij}$ if and only if $\lambda^p_{ij}$ has the direction of $\lambda_{ij}$ , that is, if the normality flow rule holds. Consequently, the normality flow rule is here necessary and sufficient for existence of the constitutive potential (1.16). Note that no distinction has been made between "hardening" and "softening" materials which themselves are not measure-invariant concepts [7].

**Example: Elastic-plastic materials with a discrete set of internal plastic deformation mechanisms**

Consider constitutive rate equations for crystals deformed plastically by multislip, or more generally, for solids with a finite number $N$ of internal mechanisms of rate-independent inelastic deformation. The constitutive framework for such materials at finite strain under the assumption of the "normality structure" was given by Hill and Rice [8] and Sewell [9], generalizing earlier theories of Koiter [10], Mandel [11] and Hill [12]. It is assumed that

$$\dot{\gamma}_K \geq 0\,, \quad f_K \leq 0\,, \quad f_K\,\dot{\gamma}_K = 0 \quad \text{(no sum)}\,, \tag{1.22}$$

$$\lambda_{Kij} = \partial f_K / \partial \varepsilon_{ij}\,, \tag{1.23}$$

$$\overset{\circ}{\tau}_{ij} = L^e_{ijkl}\,\dot{\varepsilon}_{kl} - \lambda_{Kkl}\,\dot{\gamma}_K\,, \tag{1.24}$$

$$\dot{f}_K = \lambda_{Kkl}\,\dot{\varepsilon}_{kl} - g_{KL}\,\dot{\gamma}_L \tag{1.25}$$

Upper case lower indices varying from 1 to $N$ refer to quantities related to a specific plastic deformation mechanism; the summation convention is adopted here for those indices except when an indication "no sum" appears. $\dot{\gamma}_K$ denotes a scalar measure of the rate of activity of the $K$-th mechanism and $f_K(\varepsilon_{ij})$ is the respective smooth yield function in strain space, functionally dependent on the deformation history, where $\varepsilon_{ij}$ is a strain measure based on the current configuration as reference. For instance, in crystals $\dot{\gamma}_k$ may stand for the rate of shearing on $K$-th slip system, with $f_K = \tau_K - \tau^c_K$ , where $\tau_K$ is the *generalized* resolved shear stress on that system and $\tau^c_K$ is its critical value. If $f_K = 0$ then (1.22) implies $\dot{f}_K \leq 0$ and $\dot{f}_K\,\dot{\gamma}_K = 0$ (no sum) while $\lambda_{Kij}$ defined by (1.23) represents a normal to the $K$-th yield surface in strain space, directed outward from the elastic domain. To ensure that $\tau_{ij}(\dot{\varepsilon}_{kl})$ is a single-valued function, one assumes that [8] [9]

$$(g_{KL}) \quad \text{is positive definite}\,, \tag{1.26}$$

where $(g_{KL})$ is the matrix for *potentially active* mechanisms. Then one easily shows that the constitutive rate equations can be written down in the potential form (1.16) provided that

$$L^e_{ijkl} = L^e_{klij}\,, \qquad g_{KL} = g_{LK}\,. \tag{1.27}$$

In fact, this follows from the explicit expression for the moduli

$$L_{ijkl}(\dot{\varepsilon}) = L^{e}_{ijkl} - \bar{g}^{-1}_{KL}(\dot{\varepsilon})\,\lambda_{Kij}\,\lambda_{Lkl} \tag{1.28}$$

where the submatrix $\bar{g}_{KL}(\dot{\varepsilon},\mathbf{H})$ of $g_{KL}(\mathbf{H})$ corresponds only to the mechanisms which are active for that strain-rate $\dot{\varepsilon}$, and $\bar{g}^{-1}_{KL}$ is its inversion.

For N = 1 we recover the constitutive rate equations for the classical elastic-plastic solids obeying the normality flow rule.

**Remark**

The relationship between the first-order rates of stress and strain is nonlinear but homogeneous. This is not so for higher-order rates. Let $\mathbf{H}$ be identified with $\{F_{ij}, \sigma_{ij}, \alpha_K\}$ and suppose that the constitutive function $\dot{N}_{ij} = \dot{N}_{ij}(\dot{F}_{kl}, F_{kl}, \sigma_{kl}, \alpha_K)$ is "sufficiently smooth" in its arguments. Then straightforward time differentiation of $\dot{N}_{ij}$ followed by substitution of (1.13) yields

$$\ddot{N}_{ij} = C_{ijkl}(\dot{\mathbf{F}},\mathbf{H})\ddot{F}_{lk} + \frac{\partial \dot{N}_{ij}}{\partial F_{kl}}(\dot{\mathbf{F}},\mathbf{H})\dot{F}_{kl} + \frac{\partial \dot{N}_{ij}}{\partial \sigma_{kl}}(\dot{\mathbf{F}},\mathbf{H})\dot{\sigma}_{kl} + \frac{\partial \dot{N}_{ij}}{\partial \alpha_K}(\dot{\mathbf{F}},\mathbf{H})\dot{\alpha}_K\,. \tag{1.29}$$

In a time-independent material, $\dot{\alpha}_K$ is a homogeneous function of degree one of the strain-rate but we may assume that $\dot{\alpha}_K$ is independent of higher-order rates of the deformation gradient. It follows that all right-hand terms in (1.29), except the leading one, are independent of $\ddot{F}_{lk}$. Now, let a difference between the quantities corresponding to two different continuations of the deformation process be denoted by a prefix $\Delta$. Then from (1.29) we obtain

$$\Delta\ddot{N}_{ij} = C_{ijkl}(\dot{\mathbf{F}},\mathbf{H})\,\Delta\ddot{F}_{lk} \quad \text{if} \quad \Delta\dot{F}_{ij} = 0\,. \tag{1.30}$$

That procedure could formally be repeated to draw a conclusion that the constitutive relationship between the *differences* of the stress- and deformation-rates of any order is linear with the same coefficients defined as the actual moduli (1.13) provided that there is no difference in all lower-order rates. This observation [13] is essential for studying ***regular*** higher-order bifurcations. However, the underlying regularity assumptions should always be carefully verified for the problem at hand since they need not always be satisfied. For instance, (1.30) is invalid in the case of so-called "neutral" loading in the classical elastic-plastic solids [14].

### 1.4 **Incremental boundary value problem**

We are concerned with isothermal deformations of a time-independent solid body subject to a quasi-static loading program. The body may be inhomogeneous but any material property is always assumed to vary at least piecewise smoothly with place. Suppose that the body occupied in a given reference configuration a spatial domain $V$ bounded by a piecewise smooth surface $S$. Let $\lambda = \lambda(\theta)$ denote a scalar loading parameter which varies continuously in time (infinitely slowly with respect to a natural time); it will be convenient to keep the parameters $\theta$ and $\lambda$ as being distinct. An initial equilibrium state of the body at a certain value of $\lambda$ is regarded as known. We shall assume that on a nonzero part $S_u$ of $S$ the displacements $u_j = \bar{u}_j(\boldsymbol{\xi},\lambda)$ are controlled so that rigid-body motions at fixed $\lambda$ are kinematically excluded. The part $S_u$ might be taken different for different components in order to include the mixed boundary conditions but is regarded as known and independent of $\lambda$.

The standard boundary value problem is posed by assuming that on a part $S_T$ of $S$, complementary to $S_u$, the nominal surface tractions $T_j = \overline{T}_j(\boldsymbol{\xi}, \lambda)$ (per unit reference area) are controlled, while in V the nominal body forces $b_j = \bar{b}_j(\boldsymbol{\xi}, \lambda)$ (per unit reference volume) are controlled; note that $\lambda$ need not be a load multiplier. The loading functions $\bar{u}_j, \overline{T}_j, \bar{b}_j$ are assumed to be at least piecewise smooth with respect to their arguments and to exhibit no strong discontinuities other than across material surfaces or lines. In a more general case, $T_j$ and $b_j$ need not be controlled directly but may depend on the actual deformation process.

Suppose that a theoretical quasi-static response of the body to the loading program is known, at least in principle, up to a current equilibrium state of the body at a certain value of $\lambda$. A convenient starting point to formulate the incremental boundary value problem is the principle of virtual power, which on account of (1.2) and of the *nominal* character of $T_j$ and $b_j$ can be written down in the form

$$\int_V N_{ij} w_{j,i}\, d\xi = \int_V b_j w_j\, d\xi + \int_{S_T} T_j w_j\, da \quad \text{for every } \tilde{\mathbf{w}} \in \mathcal{W}. \tag{1.31}$$

$\mathcal{W}$ denotes the linear space of admissible variations $\delta\tilde{\mathbf{v}} \equiv \tilde{\mathbf{w}}$ of velocities at a fixed value of the loading parameter $\lambda$; the fields from $\mathcal{W}$ are defined in the reference configuration and vanish on the part $S_u$ of the body surface $S$ where displacements and velocities have prescribed values. Different regularity restrictions could be imposed here; if not stated otherwise, we shall assume that $\mathcal{W}$ consists of all continuous and piecewise continuously twice differentiable vector fields on $\overline{V} = V \cup S$ which vanish on $S_u$ . Since $S_u$ has been assumed to be independent of $\lambda$, $\mathcal{W}$ is also independent of $\lambda$. It is well known that fulfillment of the principle of virtual power in the form (1.31) is equivalent to fulfillment of the standard conditions of equilibrium in the current configuration.

Consider now the first-order rate boundary value problem, formulated at the current equilibrium state. Admissible velocity fields $\tilde{\mathbf{v}}$ must take on $S_u$ the prescribed values $\bar{v}_j = (\partial\bar{u}_j/\partial\lambda)\,\dot{\lambda}$; moreover, $\tilde{\mathbf{v}}$ are assumed to be continuous and piecewise continuously twice differentiable. Let $\tilde{\mathbf{v}}^k$ be such a fixed field; the class $\mathcal{V}$ of kinematically admissible velocity fields is defined by

$$\mathcal{V} = \{\tilde{\mathbf{v}} : \tilde{\mathbf{v}} = \tilde{\mathbf{v}}^k + \tilde{\mathbf{w}}\,,\ \mathbf{v}^k = \bar{\mathbf{v}} \text{ on } S_u,\ \tilde{\mathbf{w}} \in \mathcal{W}\}. \tag{1.32}$$

From a mathematical point of view it would be natural to take $\tilde{v}_j$ from a wider space, known as the Sobolev space $H^1(V)$, but we prefer to deal with solutions in the classical sense. Note that any functional defined on $\mathcal{V}$ can equivalently be regarded as a certain other functional defined on the *linear* space $\mathcal{W}$.

Suppose that there are no moving strong discontinuities in **N**, **b** and **T**. Since the integration domains $V$ and $S_T$ are fixed, differentiation of (1.31) with respect to a time-like parameter $\theta$ at *fixed* $w_j$ yields

$$\int_V \dot{N}_{ij} w_{j,i}\, d\xi = \int_V \dot{b}_j w_j\, d\xi + \int_{S_T} \dot{T}_j w_j\, da \quad \text{for every } \tilde{\mathbf{w}} \in \mathcal{W}. \tag{1.33}$$

Assuming that inertia forces are neglected, a solution to the first-order rate boundary value problem, or simpler: *a solution in velocities*, or *a first-order solution*, is defined as a velocity field $\tilde{\mathbf{v}} \in \mathcal{V}$ such that the stress-rate field related constitutively to $\tilde{\mathbf{v}}$ through (1.10) satisfies the variational equality (1.33). By the standard transformation with the help of Green's theorem, (1.33) is written equivalently as

$$\int_V (\dot{N}_{ij,i} + \dot{b}_j) w_j \, d\xi + \int_{S_D} [\dot{N}_{ij}] n_i w_j \, da + \int_{S_T} (\dot{T}_j - \dot{N}_{ij}\nu_i) w_j \, da = 0 \quad \text{for every } \tilde{\mathbf{w}} \in \mathcal{W}, \tag{1.34}$$

where $\nu_i$ denotes the unit outward normal to the body surface, $S_D$ stands for the internal surface(s) of discontinuity in the stress-rate, $n_i$ is the unit normal to $S_D$ , and [ ] denotes a jump across $S_D$ with the former term of the difference taken from the side of $S_D$ into which $n_i$ is pointed. Hence, under the assumed regularity restrictions, fulfillment of (1.33) is equivalent to fulfillment of the set of conditions of continuing equilibrium in the form

$$\dot{N}_{ij,i} + \dot{b}_j = 0 \quad \text{in } V \setminus S_D , \qquad [\dot{N}_{ij}] n_i = 0 \quad \text{on } S_D , \qquad \dot{N}_{ij}\nu_i = \dot{T}_j \quad \text{on } S_T . \tag{1.35}$$

Higher-order rate problems can in principle be formulated in an analogous way by successive differentiation of (1.33) with respect to time provided the problem is assumed *a priori* to be "sufficiently regular". Also "regular" moving surfaces of discontinuity in stress-rates can be allowed for by using the differentiation rules for integrals defined on varying domains (e.g. [15]) and the respective compatibility conditions on the interfaces (e.g. [16]). We shall not trouble to write down explicitly the resulting formulae. However, it must be remarked that such a formal analysis is much less satisfactory than in the case of the first-order problem since higher-order rate fields may be discontinuous and have other singularities absent in velocity fields.

## 2. ELEMENTS OF HILL'S THEORY OF BIFURCATION AND STABILITY

### 2.1 Condition for uniqueness of first-order solution

The question of uniqueness of a response of the material body to the quasi-static loading program is at the first instance reduced to the problem of uniqueness of a solution in velocities. As a starting point we can adopt the rate form (1.33) of the principle of virtual power. On substituting (1.32) we can rewrite (1.33) equivalently as

$$\int_V \dot{N}_{ij} (\Delta v_j)_{,i} \, d\xi = \int_V \dot{b}_j \Delta v_j \, d\xi + \int_{S_T} \dot{T}_j \Delta v_j \, da \quad \text{for every } \tilde{v}_j^* \in \mathcal{V},\ \Delta v_j = v_j^* - v_j ; \tag{2.1}$$

If $\tilde{v}_j^*$ is another first-order solution then the equality in (1.33) must hold also when the fields $\tilde{v}_j$ and $\tilde{v}_j^*$ are interchanged, that is, if $\dot{N}_{ij}$ , $\dot{b}_j$ and $\dot{T}_j$ are replaced by $\dot{N}_{ij}^*$ , $\dot{b}_j^*$ and $\dot{T}_j^*$ which correspond to the velocity field $\tilde{v}_j^*$ . By subtracting these two equalities, we obtain the identity

$$\int_V \Delta\dot{N}_{ij} (\Delta v_j)_{,i} \, d\xi = \int_V \Delta\dot{b}_j \Delta v_j \, d\xi + \int_{S_T} \Delta\dot{T}_j \Delta v_j \, da \tag{2.2}$$

valid for a pair of first-order *solutions*, where the differences of corresponding quantities are denoted by the prefix $\Delta$.

Hence [5] [17], *uniqueness of a first-order solution is ensured* (i.e. bifurcation in velocities is excluded) *when*

$$\int_V \Delta\dot{N}_{ij}(\Delta v_j)_{,i}\, d\xi > \int_V \Delta\dot{b}_j \Delta v_j\, d\xi + \int_{S_T} \Delta\dot{T}_j \Delta v_j\, da \tag{2.3}$$

*for every pair of kinematically admissible and distinct velocity fields from* $\mathcal{V}$. This is the most general form of the uniqueness criterion for velocities; in particular cases it can further be specified. It may be remarked that (2.3) ensures uniqueness also in a wider class of generalized solutions to the first-order rate boundary value problem (in the sense of (1.33)) not necessarily belonging to $\mathcal{V}$. An alternative general formulation of the uniqueness criterion for solids obeying the maximum dissipation principle has been given by Nguyen [18] but it is not discussed here.

Further discussion will for simplicity be limited to the standard boundary value problem, i.e. to controlled nominal loads; an extension to configuration-dependent *conservative* loading will be indicated later. In the standard case, $\dot{b}_j = \dot{\bar{b}}_j = (\partial\bar{b}_j/\partial\lambda)\dot{\lambda}$ in $V$ and $\dot{T}_j = \dot{\overline{T}}_j = (\partial\overline{T}_j/\partial\lambda)\dot{\lambda}$ on $S_T$ are prescribed independently of the actual velocity field, so that $\Delta\dot{b}_j$ and $\Delta\dot{T}_j$ vanish and the uniqueness criterion (2.3) reduces to [5]

$$\int_V \Delta\dot{N}_{ij}\Delta v_{j,i}\, d\xi > 0 \tag{2.4}$$

for every pair of distinct velocity fields from $\mathcal{V}$.

It is worth pointing out that the criterion (2.4) is applicable also to a certain wider class of surface loading than the standard one. Namely, suppose that $\dot{b}_j$ are prescribed while $\dot{T}_j$ depend on the actual velocity field in such a manner that

$$\int_{S_T} \Delta\dot{T}_j \Delta v_j\, da \leq 0 \tag{2.5}$$

for all pairs of admissible velocity fields from $\mathcal{V}$. It is clear that (2.4) contradicts (2.2) also for that kind of loading. For instance, (2.5) holds for the unilateral frictionless contact with a rigid tool if the actual contact pressure happens to vanish over the contact surface (e.g. [6]).

Let the current configuration be taken as reference; then from (1.9) we obtain

$$\Delta\dot{N}_{ij}\Delta\dot{F}_{ji} = \Delta\overset{\circ}{\tau}_{ij}\Delta\dot{\varepsilon}_{ij} - \sigma_{kj}\left(\Delta\dot{\varepsilon}_{ji}\,\Delta\dot{\varepsilon}_{ik} + 2\Delta\omega_{ji}\Delta\dot{\varepsilon}_{ik} + \Delta\omega_{ji}\Delta\omega_{ik}\right); \tag{2.6}$$

On substituting (2.6), the condition (2.4) takes the form

$$\int_V \Delta\overset{\circ}{\tau}_{ij}\Delta\dot{\varepsilon}_{ij}\, d\xi - \int_V \sigma_{kj}(\Delta\dot{\varepsilon}_{ji}\Delta\dot{\varepsilon}_{ik} + 2\Delta\omega_{ji}\Delta\dot{\varepsilon}_{ik} + \Delta\omega_{ji}\,\Delta\omega_{ik})\, d\xi > 0 \tag{2.7}$$

If the constitutive relationship $\overset{\circ}{\tau}_{ij}(\dot{\varepsilon})$ is defined as a generalization of an incremental constitutive law from the small strain theory then the second integral in (2.7) which is linearly dependent on the current Cauchy stresses represents the effect of changes in geometry on the uniqueness criterion. It is evident that in general the effect is not negligible.

**Example: A pointwise condition sufficient for uniqueness in the case of predominantly tensile stresses** [19]

Suppose that uniqueness is examined at the state of predominantly tensile stress, that is, when at each point of the body the principal Cauchy stresses $\sigma_i$ satisfy the inequalities

$$\sigma_1 + \sigma_2 > 0, \qquad \sigma_2 + \sigma_3 > 0, \qquad \sigma_3 + \sigma_1 > 0. \tag{2.8}$$

Then, for given $\Delta\dot{\varepsilon}_{ij}$, the right-hand integrand in (2.7) attains a maximum value when

$$\Delta\omega_{pr} = \Delta\dot{\varepsilon}_{pr}\frac{\sigma_p - \sigma_r}{\sigma_p + \sigma_r}, \qquad \text{(no sum)} \tag{2.9}$$

where the components are taken on the principal axes of $\boldsymbol{\sigma}$. Fulfillment of the inequality (2.7) is thus ensured if

$$\Delta\overset{\circ}{\tau}_{ij}\Delta\dot{\varepsilon}_{ij} > \sum_{r=1,2,3}\sigma_r(\Delta\dot{\varepsilon}_{rr})^2 + \sum_{\substack{r,p=1,2,3\\ r\neq p}}\frac{\sigma_p^2+\sigma_r^2}{\sigma_p+\sigma_r}(\Delta\dot{\varepsilon}_{pr})^2 \tag{2.10}$$

holds inside the body. It follows that fulfillment of the pointwise condition (2.10) in the body is sufficient for uniqueness in velocities.

Similarly, if at uniaxial tension we have

$$\Delta\overset{\circ}{\tau}_{ij}\Delta\dot{\varepsilon}_{ij} > \sigma_1((\Delta\dot{\varepsilon}_{11})^2 + 2(\Delta\dot{\varepsilon}_{12})^2 + 2(\Delta\dot{\varepsilon}_{13})^2) \tag{2.11}$$

then uniqueness of the solution in velocities is ensured with accuracy to rigid body rotations about the tensile axis.

## 2.2 Stationarity and minimum principles for velocities

Consider the standard boundary value problem where $\dot{b}_j$ in $V$ and $\dot{T}_j$ on $S_T$ are prescribed, and suppose that the constitutive rate equations (1.10) admit a velocity-gradient potential (1.16). Then [5] *a velocity field* $\tilde{\mathbf{v}} \in \mathcal{V}$ *is a first-order solution if and only if it assigns to the functional*

$$J(\tilde{\mathbf{v}}) = \int_V (U(\nabla\mathbf{v}) - \dot{\bar{b}}_j v_j)\, d\xi - \int_{S_T} \dot{\bar{T}}_j v_j\, da\,, \qquad \tilde{\mathbf{v}} \in \mathcal{V}\,, \tag{2.12}$$

*a stationary value*, in the sense that its first weak (Gateaux) variation vanishes in any direction, viz. (cf. (1.1))

$$\delta J(\tilde{\mathbf{v}}, \tilde{\mathbf{w}}) = 0 \quad \text{for every } \tilde{\mathbf{w}} \in \mathcal{W}\,, \tag{2.13}$$

where $\tilde{\mathbf{w}} = \delta\tilde{\mathbf{v}}$ stands for an admissible variation of $\tilde{\mathbf{v}}$. Equivalence between (2.13) and (1.33) is immediate since from (1.16) we have $\delta U(\nabla\mathbf{v}, \nabla\mathbf{w}) = \dot{N}_{ij}(\nabla\mathbf{v})w_{j,i}$.

Suppose now that the uniqueness condition (2.4) holds. Observe that (2.4) can be rewritten as

$$\delta J(\tilde{\mathbf{v}}^*, \tilde{\mathbf{v}}^* - \tilde{\mathbf{v}}) - \delta J(\tilde{\mathbf{v}}, \tilde{\mathbf{v}}^* - \tilde{\mathbf{v}}) > 0 \quad \text{for every } \tilde{\mathbf{v}}^*, \tilde{\mathbf{v}} \in \mathcal{V}, \quad \tilde{\mathbf{v}}^* \neq \tilde{\mathbf{v}}, \tag{2.14}$$

which is nothing else than the condition of strict convexity of the functional (2.12) in $\mathcal{V}$. It follows [5] [17] [6] that *in the range where* (2.4) *holds, the unique solution* $\tilde{\mathbf{v}}^0$ *assigns to the functional* $J(\tilde{\mathbf{v}})$ *a strict and absolute minimum value*, viz.

$$J(\tilde{\mathbf{v}}) > J(\tilde{\mathbf{v}}^0) \quad \text{for every } \tilde{\mathbf{v}} \in \mathcal{V}, \quad \tilde{\mathbf{v}} \neq \tilde{\mathbf{v}}^0 . \tag{2.15}$$

### 2.3 Incrementally linear comparison solid

The uniqueness criterion (2.4) may be difficult to be applied directly to practical problems. Hill [4] [5] has proposed the following "linearizing device" to make it more easy to handle, although in general at the cost of overestimating the possibility of bifurcations. Consider a hypothetical material body of the same configuration and Cauchy stress distribution under the current loading as those for the actual inelastic body, but of a different incremental constitutive law. Namely, at each material point the hypothetical solid is assumed to be incrementally *linear*, with the constitutive relationship

$$\dot{N}^L_{ij} = C^L_{ijkl}(\mathbf{H})\dot{F}_{lk} , \tag{2.16}$$

The moduli $C^L_{ijkl}$ are dependent on place in general; the moduli field $\tilde{C}^L_{ijkl}$ may be discontinuous, for instance, across the elastic-plastic interface.

Suppose that $C^L_{ijkl}$ are defined in such a way that

$$\Delta\dot{N}_{ij}\Delta\dot{F}_{ji} \geq C^L_{ijkl}\Delta\dot{F}_{ji}\Delta\dot{F}_{lk} \tag{2.17}$$

for every pair of velocity gradients whose difference $\Delta\dot{\mathbf{F}}$ generates the difference $\Delta\dot{\mathbf{N}}$ of the stress-rates which are first found from (1.10) for each velocity gradient separately and then subtracted. The incrementally linear solid of the instantaneous moduli $C^L_{ijkl}$ satisfying (2.17) is called the *comparison solid.* Since diagonally antisymmetric part of $C^L_{ijkl}$ gives no contribution to the right-hand side of (2.17), henceforth we assume that $C^L_{ijkl} = C^L_{klij}$ . If the potential (1.16) exists then (2.17) means that the *difference* $U - U^L$ of the two potentials is a convex function of $\dot{\mathbf{F}}$, where $U^L(\dot{\mathbf{F}}) \equiv \frac{1}{2}C^L_{ijkl}\dot{F}_{ji}\dot{F}_{lk}$ is the quadratic potential for the comparison solid. (2.17) is then called the *relative convexity property.*

Define the quadratic functional

$$I^L(\tilde{\mathbf{w}}) = \int_V C^L_{ijkl}w_{j,i}w_{l,k}\, d\xi , \quad \tilde{\mathbf{w}} \in \mathcal{W} . \tag{2.18}$$

If (2.17) holds then the integrand in (2.4) is not smaller in value that the integrand in (2.18) with $w_{j,i} = \Delta v_{j,i}$ . It follows [5] that *fulfillment of the condition*

$$I^L(\tilde{\mathbf{w}}) > 0 \quad \text{for every } \tilde{\mathbf{w}} \in \mathcal{W}, \quad \tilde{\mathbf{w}} \neq \tilde{\mathbf{0}}, \tag{2.19}$$

*excludes bifurcation in velocities in the actual incrementally nonlinear solid.* [1)]

The problem lies in optimal choice of the comparison solid, in the sense of making the difference between the integrands in (2.4) and (2.18) as small as possible. In certain cases (see the examples below) this can be decided by fulfillment of (2.17) for $C^L_{ijkl}$ equal to the moduli from a specific constitutive cone for the actual nonlinear law, where the "stiffness" of the material is the lowest. This means that we have equality in (2.17) whenever both velocity gradients lie within that cone. If (1.16) holds then this is equivalent to saying that the convex function $U - U^L$ vanishes in that cone. Of course, in an elastic zone in $V$ where $U$ is quadratic, $U^L = U$ trivially.

From (1.9) or (2.6) it can be seen that (2.17) is equivalent to

$$\Delta\overset{\circ}{\tau}_{ij}\Delta\dot{\varepsilon}_{ij} \geq L^L_{ijkl}\Delta\dot{\varepsilon}_{ij}\Delta\dot{\varepsilon}_{kl} \tag{2.20}$$

where $L^L_{ijkl}$ are related to $C^L_{ijkl}$ by the formula analogous to (1.15). This observation can be extended to an incremental constitutive relationship expressed in terms of the rates of any pair of work-conjugate measures of strain and stress [6].

**Examples of the comparison solid for elastoplastic models**

As shown by Hill [4], for the classical elastic-plastic solids obeying the normality flow rule (i.e. $\lambda^p_{ij} = \lambda_{ij}$ in (1.19)), the relative convexity property (2.20) is satisfied if we define $L^L_{ijkl}$ as

$$L^L_{ijkl} = L^p_{ijkl} \tag{2.21}$$

at any material element stressed to the current yield point; of course, $L^L_{ijkl}$ in the elastic zone can be taken as the actual elastic moduli. This means that in the plastic zone the comparison solid is defined by the moduli from the loading branch.

For the classical elastic-plastic solids but without normality, Raniecki [20,21] has shown that (2.20) is satisfied when in the plastic zone we define

$$L^L_{ijkl} = L^e_{ijkl} - \frac{1}{4gr}(\lambda^p_{ij} + r\lambda_{ij})(\lambda^p_{kl} + r\lambda_{kl}), \qquad r > 0. \tag{2.22}$$

Contrary to the above case with normality, this comparison material does not follow from linearization of the constitutive relationship about a principal (loading) solution.

For the material model defined by (1.22) - (1.27) with the moduli (1.28), Sewell [9] has proved that (2.20) is ensured if the comparison moduli are taken from the so-called total loading branch, that is, if

$$L^L_{ijkl} = L^e_{ijkl} - g^{-1}_{KL}\lambda_{Kij}\lambda_{Lkl}, \tag{2.23}$$

where the subscripts $K, L$ run over all potentially active mechanisms of plastic deformation.

---

[1)] Fulfillment of (2.19) ensures also uniqueness of a solution in quasi-static accelerations to a *regular second-order* rate boundary value problem [14].

**Remark**

There is an open problem in which circumstances (2.19) excludes also higher-order bifurcations. This delicate question has been discussed when (1.30) holds [13][22] or need not hold [14] but always under certain simplifying assumptions (sometimes introduced tacitly) concerning regularity of the bifurcating branch. Since the latter is unknown, no definite conclusion of analogous generality as in the first-order case is available at present, to the writer's knowledge.

### 2.4 Primary bifurcation point

The discussion above was concerned with conditions *sufficient for uniqueness* of a solution in velocities. Now, we will discuss a condition sufficient for bifurcation. Consider the critical instant beyond which the condition (2.19) fails for the first time along a given deformation path. Generally, that critical instant determines *a lower bound* to the first value of $\theta$ at which bifurcation in velocities is possible. Below we shall specify the circumstances in which the critical instant coincides with the primary bifurcation point.

It is natural to assume [2)] that at the critical instant the functional (2.18) is just positive semi-definite on $\mathcal{W}$, in the sense that

$$\begin{cases} I^L(\tilde{\mathbf{w}}) \geq 0 & \text{for every} \quad \tilde{\mathbf{w}} \in \mathcal{W}, \\ I^L(\tilde{\mathbf{w}}^*) = 0 & \text{for some} \quad \tilde{\mathbf{w}}^* \in \mathcal{W}, \ \tilde{\mathbf{w}}^* \neq \tilde{0}. \end{cases} \tag{2.24}$$

$I^L$ must be stationary at the minimum point $\tilde{\mathbf{w}} = \tilde{\mathbf{w}}^*$, viz.

$$\delta I^L(\tilde{\mathbf{w}}^*, \tilde{\mathbf{w}}) = 0 \quad \text{for every } \tilde{\mathbf{w}} \in \mathcal{W}. \tag{2.25}$$

This is equivalent to saying that (cf. the equivalence between (1.35) and (2.13)) $\tilde{\mathbf{w}}^*$ is an eigenmode which satisfies the homogeneous equations

$$\begin{aligned} (C^L_{ijkl} w_{l,k})_{,i} &= 0 \ \text{ in } V, & [C^L_{ijkl}\, w_{l,k}] n_i &= 0 \ \text{ on } S_D, \\ C^L_{ijkl} w_{l,k} \nu_i &= 0 \text{ on } S_T, & w_j &= 0 \text{ on } S_u. \end{aligned} \tag{2.26}$$

On substituting (2.16) we may interpret (2.26) as the equilibrium equations for $\dot{N}^L_{ij}$ under dead loading. This means that the solution in velocities for the *comparison* solid, if exists, is nonunique for any actual loading $\dot{\bar{b}}_j$ , $\dot{\bar{T}}_j$ . For, on account of linearity of the constitutive rate equations (2.16) for the comparison material, the eigenmode $\tilde{\mathbf{w}}^*$ with an arbitrary scalar multiplier $\gamma$ can be added to a solution $\tilde{\mathbf{v}}$ to the non-homogeneous problem (1.35), with $\dot{N}_{ij}$ replaced by $\dot{N}^L_{ij}$ , to generate another solution $\tilde{\mathbf{v}} + \gamma\tilde{\mathbf{w}}$ [5].

Now, as indicated above, the moduli $C^L_{ijkl}$ within the body can be equal to the moduli from a specific constitutive cone for the actual nonlinear law. For certain incremental loadings (but not always), among the solutions $\tilde{\mathbf{v}} + \gamma\tilde{\mathbf{w}}$ for the comparison solid there can be those which involve only strain rates from that cone. It is evident that those fields $\tilde{\mathbf{v}} + \gamma\tilde{\mathbf{w}}$ are solutions not only for the comparison material but also for the actual solid.

---

2) It is worth pointing out that (2.24) need not always hold at the critical instant, even if the comparison moduli vary smoothly along the deformation path.

In those circumstances, a solution to the actual first-order rate boundary value problem is nonunique [23]; the critical instant under consideration is hence called a point of bifurcation in velocities. If $C^L_{ijkl}$ satisfy (2.17) then this is the *primary* bifurcation in velocities. On the basis of Hill's theory, that primary bifurcation point was extensively studied analytically or numerically in the literature, starting from [24]. Note that infinitely many solutions in velocities can exist, corresponding to a single eigenmode $\tilde{\mathbf{w}}^*$ and to an interval of the multiplier $\gamma$. This does not mean, however, that infinitely many post-bifurcation branches are initiated in that way at the single equilibrium state under consideration. For, higher-order conditions of continuing equilibrium usually eliminate most of the first-order solutions, for instance, all of them except three: the fundamental one and the two associated with the limits of the interval of $\gamma$ within which the involved strain-rates are directed into the constitutive cone where $C^L_{ijkl}$ apply [25] [26]. The critical point can also correspond not to a "true" bifurcation of the deformation path but rather to a so-called limit point; cf. the discussion of eigenstates.

If $\tilde{C}^L_{ijkl}$ do not coincide with the actual moduli field then the critical instant precedes in general the primary bifurcation point which therefore cannot be determined by examining the incrementally linear comparison material only.

**Example: Column buckling**

Consider a perfect column (Fig. 2) subject to an increasing compressive load $P(\theta)$. The column is assumed not to buckle in the elastic range but otherwise to be sufficiently

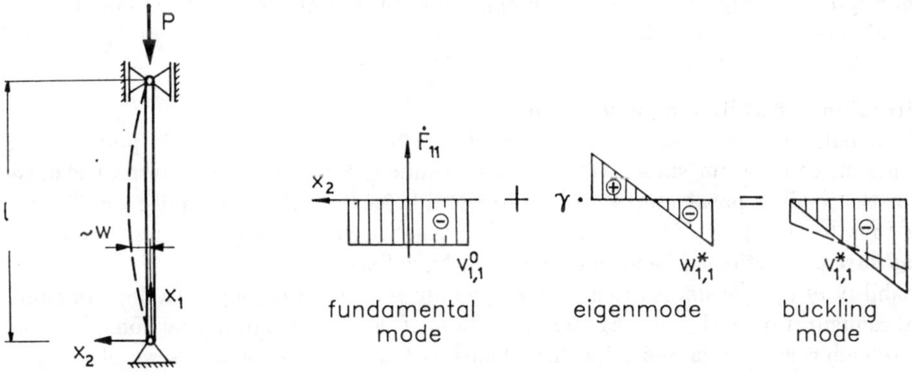


Fig. 2, Mode superposition at the primary bifurcation point for a column under increasing compressive load.

slender to adopt the simplest assumptions of the beam theory (for a more exact analysis, see [24]). Consequently, on using (1.15) with $F_{ij} = \delta_{ij}$ and $\sigma_{ij} = \sigma_1\delta_{1i}\delta_{1j}$ and substituting (2.21), the integrand in (2.18) in the curent configuration taken as reference is approximated by

$$\begin{aligned} C^L_{ijkl} w_{j,i} w_{l,k} &\approx L^p (w_{1,1})^2 + \sigma_1 (w_{2,1})^2 \\ &= L^p x_2^2 (w'')^2 + \sigma_1 (w')^2, \end{aligned} \tag{2.27}$$

where $L^p$ is the current uniaxial tangent modulus, $w(x_1)$ is the deflection mode of the column axis and a prime denotes differentiation by $x_1$. Volume integration of (2.27) yields the sufficiency condition for uniqueness in the form

$$I^L(\tilde{\mathbf{w}}) = \int_0^l (L^p I_c (w'')^2 - P(w')^2)\, dx_1 > 0 \qquad \text{for every } w(x_1), \tag{2.28}$$

where $I_c = \int_A x_2^2\, da$ and $A$ is the column cross-section. For a uniform column with the boundary conditions as indicated in Fig. 2, and at a tangent modulus *continuously varying* with $\theta$, the standard method of the calculus of variation yields the critical load and eigenmode in the classical form

$$P_T = \frac{\pi^2 L^p I_c}{l^2}, \qquad w^*(x_1) = \sin \frac{\pi x_1}{l}. \tag{2.29}$$

This critical load $P_T$ (the tangent modulus load) corresponds to the primary bifurcation point discussed above. The eigenmode with a multiplier from some interval can be superimposed, at *increasing* $P$, on the fundamental mode of uniform compression to produce another solution which describes incipient buckling without abrupt unloading *at* the instant of bifurcation (Fig. 2). Higher-order conditions of continuing equilibrium imply that the actual value of the multiplier corresponds to the neutral loading at an outer point of some cross-section (cf. [25]).

### 2.5 Directional stability of equilibrium

On a path of quasi-static deformation of the body, consider a single point which represents an equilibrium state. We will now examine stability of that state at a fixed value of the loading parameter $\lambda$. Accordingly, admissible velocities vanish on $S_u$ and belong thus to the space $\mathcal{W}$. We assume the standard case of *dead* loading where the *nominal* surface tractions $T_j$ and body forces $b_j$ have fixed values.

Stability *of equilibrium* is essentially a dynamic concept. Roughly speaking, an equilibrium configuration of the body is said to be stable if the distance from that configuration in any dynamic motion caused by a disturbance can be made as small as we please if a measure of the disturbance itself is sufficiently small. That definition has to be made precise by specifying the class of disturbances under consideration and the measures of the distance from the examined equilibrium state and of the disturbance strength.

Path-dependence of an inelastic material creates difficulties in obtaining a condition sufficient for stability. There are also mathematical problems which are met when the continuum problem is to be examined rigorously. Therefore, the reasonable first step is to study a particular kind of stability when only *direct* paths of departure from the equilibrium configuration are taken into account. This means that a perturbed motion is *a priori* assumed to be such that variations of the "direction" of the velocity field along the path are negligible. Stability in that restricted sense is called here *directional stability.*

Along an admissible *direct* path starting from the given equilibrium state, the work of deformation in the body can be expressed as

$$W = \int_V N_{ij} \Delta u_{j,i} \, d\xi + \frac{1}{2} \int_V \Delta N_{ij} \Delta u_{j,i} \, d\xi + o((\Delta\theta)^2), \tag{2.30}$$

where $N_{ij}$ stands for the initial nominal stresses at the equilibrium state, $\Delta\theta$ is a small increment of a time-like parameter $\theta$ which measures the "length" of the path, $\Delta u_j$ are current displacements reached along the path and $\Delta N_{ij}$ are the respective stress increments. Since $\Delta u_j$ vanish on $S_u$ , from (1.31) we obtain that the former integral in (2.30) is exactly equal to the work $W^{load}$ done by the dead loads. Hence, the work difference reads

$$W - W^{load} = \frac{1}{2}(\Delta\theta)^2 \int_V \dot{N}_{ij} \dot{u}_{j,i} \, d\xi + o((\Delta\theta)^2) \tag{2.31}$$

where the rates are taken with respect to $\theta$ at the equilibrium state. It follows that if

$$\int_V \dot{N}_{ij}(\nabla \mathbf{w}) w_{j,i} \, d\xi > 0 \quad \text{for every } \tilde{\mathbf{w}} \in \mathcal{W}, \ \tilde{\mathbf{w}} \neq \tilde{\mathbf{0}} \tag{2.32}$$

then any movement from the equilibrium configuration in any direction requires some additional energy to be supplied to the system from external sources. *In that sense* (2.32) *is a condition sufficient for directional stability of equilibrium* [4]. On substituting (1.11) the condition (2.32) can be rewritten in a more compact form as

$$I(\tilde{\mathbf{w}}) > 0 \quad \text{for every } \tilde{\mathbf{w}} \in \mathcal{W}, \ \tilde{\mathbf{w}} \neq \tilde{\mathbf{0}}, \tag{2.33}$$

where

$$I(\tilde{\mathbf{w}}) \equiv \int_V U(\nabla \mathbf{w}) \, d\xi, \qquad \tilde{\mathbf{w}} \in \mathcal{W}. \tag{2.34}$$

A slightly stronger condition will be introduced later.

It is pointed out that (2.33) cannot be regarded as being sufficient for stability of equilibrium in a dynamic sense for arbitrarily circuitous paths unless further assumptions are introduced [4] [5] [27] [28] [53].

Compare now the obtained criteria for uniqueness and for stability of equilibrium. It is evident that if $\bar{v}_j = 0$ on $S_u$ then (2.33) is a consequence of (2.4) since then the zero velocity field is admissible as a particular member of the pair considered in (2.4). If $\bar{v}_j \neq 0$ then the connection is slightly more complicated. Let $\Delta\tilde{\mathbf{v}} = \gamma\tilde{\mathbf{w}}$ with $\tilde{\mathbf{w}}$ and one member $\tilde{\mathbf{v}}$ fixed and $\gamma$ being a positive number. If $\gamma$ increases unboundedly then the integral in (2.4) divided by $\gamma^2$ tends to the integral in (2.33), by continuity and homogeneity of the relationship $\dot{N}_{ij}(\dot{F}_{lk})$. If (2.4) holds for all pairs then the limit value is never negative; however, in certain cases it may be equal to zero. With the exception of such singular points, (2.33) is a consequence of (2.4). The converse need *not* be true, so that in general the criteria for uniqueness and for stability of equilibrium are distinct and *bifurcation may take place before the loss of stability of equilibrium* [29] [4] [27].

## 2.6 Primary eigenstate

Suppose that the constitutive rate equations (1.10) admit a potential (1.16) and consider the critical instant when the condition (2.33) fails for the first time along a given deformation path. Similarly as is Section 2.4 above, suppose that at the critical instant the functional (2.34) is just positive semi-definite on $\mathcal{W}$, in the sense that

$$\begin{cases} I(\tilde{\mathbf{w}}) \geq 0 & \text{for every } \tilde{\mathbf{w}} \in \mathcal{W}, \\ I(\tilde{\mathbf{w}}^*) = 0 & \text{for some } \tilde{\mathbf{w}}^* \in \mathcal{W}, \ \tilde{\mathbf{w}}^* \neq \tilde{0}. \end{cases} \tag{2.35}$$

This implies that $I$ is stationary at its minimum point $\tilde{\mathbf{w}} = \tilde{\mathbf{w}}^*$, viz.

$$\delta I(\tilde{\mathbf{w}}^*, \tilde{\mathbf{w}}) = 0 \quad \text{for every } \tilde{\mathbf{w}} \in \mathcal{W}. \tag{2.36}$$

Under dead loading and for $\bar{v}_j = 0$ on $S_u$, the functional $J$ defined in (2.12) coincides with the functional $I$ so that (2.36) coincides then with (2.13). It follows that $\tilde{\mathbf{w}}^*$ satisfying (2.35) (with any positive multiplier) is a non-trivial solution to the *homogeneous* (but generally nonlinear) first-order rate boundary value problem, obtained e.g. by setting $\dot{\lambda} = 0$. The body configuration at which such a nontrivial solution exists is called *an eigenstate*, and the condition (2.35) defines *the primary eigenstate* [27] [19]. It may be remarked that the method of determining the primary bifurcation point discussed in Section 2.4 reduces in effect to finding the primary eigenstate for the incrementally linear *comparison* material.

**Exercise**

Examine conditions for the eigenstate at uniform uniaxial tension under prescribed nominal surface tractions.

**Example: Column buckling (continued)**

For the perfectly straight column discussed at the end of Section 2.4, the integrand in the stability condition (2.33) takes the form analogous to (2.27) but with one important distinction: at the zone where $\nabla\mathbf{w}$ corresponds to unloading (i.e. where $w_{1,1} < 0$), the tangent (plastic) modulus must be replaced by an elastic modulus. The critical load corresponding to the primary eigenstate (2.35), i.e. to the possibility of incipient buckling at *fixed* load, is called the reduced modulus load and denoted by $P_R$. Since the elastic modulus is greater that $L^p$, we have $P_R > P_T$. The task of determining the value of $P_R$ is not undertaken here.

In the *secondary* post-bifurcation process of buckling, or for an imperfect column, the primary eigenstate is attained when the *controlled* load $P$ reaches its maximum (limit) value along the loading path. The equilibrium states beyond that stage can be regarded as unstable; this question will be examined later on in more detail.

**Illustrative diagram**

The basic results of Hill's theory quoted in this chapter are illustrated in Fig. 3 on a diagram. The critical values of the loading parameter are denoted by $\lambda_T$ and $\lambda_R$, in analogy to the critical loads $P_T$ and $P_R$ in the example of column buckling, but the picture is now quite general and pertains to many other particular problems. It must be noted, however, that some modifications may be necessary in certain cases. For instance, bifurcation at the tangent modulus load may be "smooth" and not in velocities as indicated in the figure,

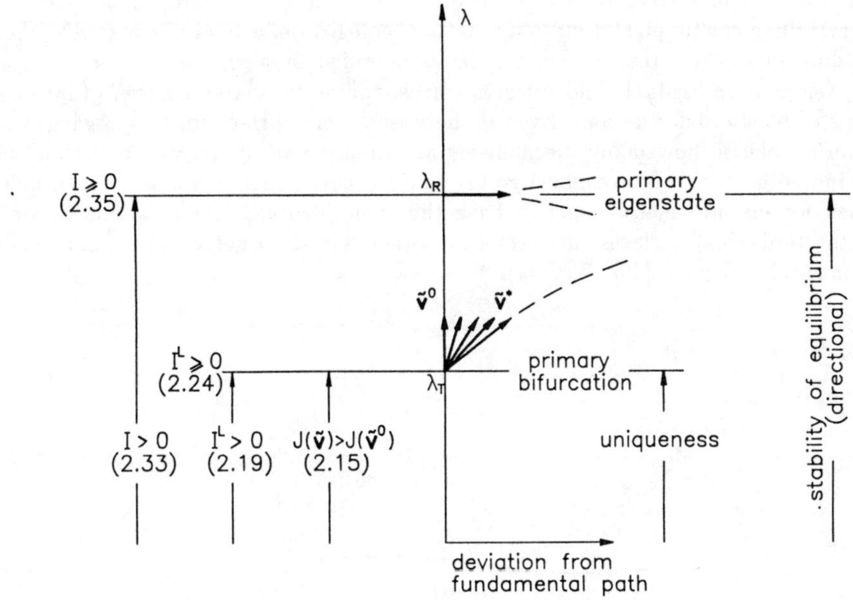


Fig. 3, Schematic illustration of the basic results of Hill's theory. It is assumed that the relative convexity property (2.17) holds and $C^L_{ijkl}$ coincide with the current tangent moduli.

the values of $\lambda_T$ and $\lambda_R$ may coincide with each other or else they may be "jumped over" when the tangent modulus along the deformation path suffers a discontinuous drop, etc. Such cases will be briefly discussed in the other Chapters.

## 3. LINEARIZATION OF BIFURCATION PROBLEM AT A YIELD-SURFACE VERTEX

### 3.1 Structure of constitutive laws for polycrystals

The bifurcation problem in velocities is inherently nonlinear for an incrementally nonlinear constitutive law. However, we have seen that in certain circumstances the *primary* bifurcation point can be found, or a lower bound to that point can be determined, from a linearized problem formulated for the comparison material. That approach is justified if the relative convexity property (2.17) has been proved for a specific material model. Within the class of classical elastic-plastic solids obeying the normality flow rule relative to a smooth yield surface the property holds automatically, and the linearization method is thus based on a firm theoretical foundation. Nevertheless, it has turned out that the obtained bifurcation stresses are frequently unrealistically high for structural metals.

The discrepancy with experiment can, at least qualitatively, be attributed to the effect of formation of a yield-surface vertex. Consider a representative heterogeneous element of a polycrystalline elastic-plastic material. If all rheological effects are disregarded then an elastic domain in the stress or strain space is bounded by a yield surface defined as an inner envelope of individual yield surfaces corresponding to a large number of internal mechanisms of plastic deformation. Even if the virgin yield surface may be regarded as smooth, during plastic flow many mechanisms are simultaneously activated so that the respective individual yield surfaces must go through the current loading point. Since there is no reason for distinct mechanisms to have the same loading/unloading macroscopic criterion, the individual surfaces intersect each other and a corner on the global yield surface is inevitably formed [30] (Fig. 4a). [3)]

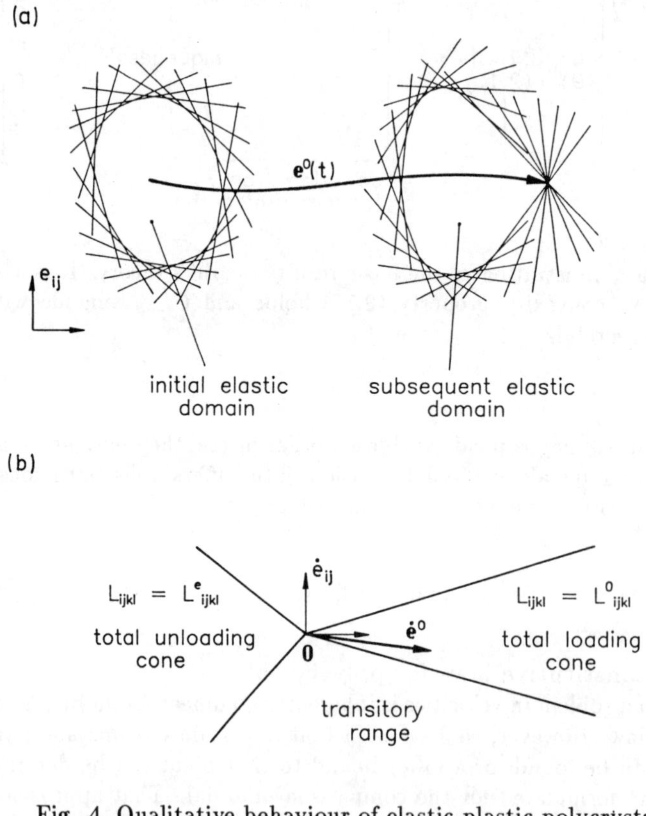


Fig. 4, Qualitative behaviour of elastic-plastic polycrystals:
(a) formation of a vertex on the theoretical yield surface at the current loading point,
(b) structure of the incremental constitutive law.

[3)] Such theoretically sharp corners need not be easy to be detected experimentally [31].

This is connected with a more complex structure of the incremental constitutive law than the classical one, as shown schematically in Fig. 4b. The strain-rate space (in fact, the set of strain-rate directions) is divided into three constitutive domains: the cone of fully active loading, the cone of total unloading and the intermediate range of partial unloading (*op.cit*); the former cone may degenerate to a single ray. Of course, the term "fully active" or "total" loading concerns only those mechanisms of plastic flow which are stressed to their current yield point. It is essential that the direction of the plastic component of the strain rate is *not* fixed but depends on the strain-rate direction, also within the total loading cone. This is associated with significant reduction of the incremental stiffness of the material against a change of the straining direction in comparison with the classical elastic-plastic model [32]. As a first approximation, the incremental constitutive relationship in the total loading cone may be assumed integrable along a deformation path which at each instant is directed inside the total loading cone, which justifies the use of the moduli from the deformation theory of plasticity within that cone. Specific models of that type have been developed [33].

In view of the indicated complexity of the whole incremental constitutive law for structural metals, the linearization of the bifurcation problem by using only the "total loading" moduli becomes especially attractive. However, (2.17) has the status of a mathematical condition which may or may not be satisfied if these moduli are substituted in place of $C^L_{ijkl}$ . Fulfillment of (2.17) depends on the actual constitutive relationship in the range of partial unloading whose specification is necessarily connected with introducing some arbitrary assumptions. The usual reference to Sewell's result [9] that the total loading moduli in the class of models described by (1.22)÷(1.28) do satisfy (2.17) is therefore not quite satisfactory, especially due to the assumption (1.26) adopted for large N. We show below how the linearization of the bifurcation problem for velocities can be justified by appealing to another constitutive inequality recently derived from micromechanical considerations. The essence of that approach is that qualitative properties of the material are specified *at a micro-level*, without specifying the macroscopic constitutive law (1.10) itself. Before presenting that approach, we discuss below a modification of Hill's theory in the case when a fundamental solution in velocities is known, which is the common case in practice.

### 3.2 Tangent comparison solid

The theory developed by Hill does not require any solution in velocities to be known in advance. In practice, a fundamental solution is usually known or even trivial. As mentioned by Hill [6], the uniqueness condition (2.3) can be weakened if one solution in velocities, say $\tilde{\mathbf{v}}^0$, is known. For, we can then take $\tilde{\mathbf{v}}^0$ as a fixed member of the pair of possible solutions in (2.2). Denote the respective rates of the nominal stress and deformation gradient by $\dot{N}^0_{ij}$ and $\dot{F}^0_{ij}$ . It is then clear that under controlled nominal loads the solution $\tilde{\mathbf{v}}^0$ is unique when

$$\int_V (\dot{N}_{ij} - \dot{N}^0_{ij})(\dot{F}_{ji} - \dot{F}^0_{ji})\, d\xi > 0 \quad \text{for every } \tilde{\mathbf{v}} \in \mathcal{V},\ \tilde{\mathbf{v}} \neq \tilde{\mathbf{v}}^0 . \tag{3.1}$$

In comparison with (2.4) the distinction is that now only *one* field is variable; the condition (3.1) is thus generally weaker than (2.4) and more easy to handle.

At a certain stage of the fundamental deformation process, in analogy to (2.18), consider the following quadratic functional

$$I^0(\tilde{\mathbf{w}}) = \frac{1}{2}\int_V C^0_{ijkl} w_{j,i} w_{l,k}\, d\xi\,, \qquad \tilde{\mathbf{w}} \in \mathcal{W}\,, \tag{3.2}$$

based on the actual *tangent moduli*

$$C^0_{ijkl} = C_{ijkl}(\dot{\mathbf{F}}^0, \mathbf{H})\,, \qquad \dot{F}^0_{ij} = v^0_{i,j}\,, \tag{3.3}$$

which correspond to the fundamental *forward* increment of the deformation. The integral in (3.2) is well defined if the tangent moduli themselves are well defined almost everywhere in $V$, i.e. except possibly in a region of zero volume only (on the elastic-plastic interface, for instance). This is always assumed below whenever the functional $I^0$ appears in considerations.

As we have seen in preceding Chapter, the inequality

$$I^0(\tilde{\mathbf{w}}) > 0 \quad \text{for every } \tilde{\mathbf{w}} \in \mathcal{W},\ \tilde{\mathbf{w}} \neq \tilde{\mathbf{0}} \tag{3.4}$$

excludes bifurcation in velocities if (2.17) is satisfied within the body for $\tilde{C}^L_{ijkl} = \tilde{C}^0_{ijkl}$ . If the fundamental solution $\tilde{\mathbf{v}}^0$ is known then the property (2.17) can be replaced by a less restrictive requirement. Namely, suppose that the following inequality is satisfied [34]

$$(\dot{N}_{ij}(\dot{\mathbf{F}}) - \dot{N}^0_{ij})(\dot{F}_{ji} - \dot{F}^0_{ji}) \geq C^0_{ijkl}(\dot{F}_{ji} - \dot{F}^0_{ji})(\dot{F}_{lk} - \dot{F}^0_{lk}) \quad \text{for every } \dot{\mathbf{F}} \tag{3.5}$$

which can be equivalently rewritten as

$$(\dot{N}_{ij}\,(\dot{\mathbf{F}}) - C^0_{ijkl}\dot{F}_{lk})(\dot{F}_{ji} - \dot{F}^0_{ji}) \geq 0 \quad \text{for every } \dot{\mathbf{F}}\,. \tag{3.6}$$

In comparison with (2.17), one pair of the stress- and deformation-rates is now fixed, so that (3.5) is indeed a weaker restriction then (2.17) with $C^L_{ijkl} = C^0_{ijkl}$ . Nevertheless, (3.5) *ensures that the first-order solution* $\tilde{\mathbf{v}}^0$ *is unique if* (3.4) *holds.* For, substitution of (3.5) into (3.4) yields the sufficiency condition (3.1). The incrementally linear comparison solid with the instantaneous moduli $C^0_{ijkl}$ satisfying (3.5) will be called the *tangent comparison solid.* Henceforth up to the end of this Section we assume that the incremental constitutive law admits a potential (1.16) so that $C^0_{ijkl} = C^0_{klij}$ .

That straightforward modification of Hill's theory brings definite advantages. In Section 3.3 it will be shown that (3.5) is not only a mathematical condition less restrictive than (2.17) but can also be derived from micromechanical considerations. The difference between the two constitutive inequalities is illustrated graphically in Fig. 5, where $U^0$ is the quadratic potential for the tangent comparison solid, defined by $U^0(\dot{\mathbf{F}}) = \frac{1}{2}C^0_{ijkl}\dot{F}_{ji}\dot{F}_{lk}$.

It is evident that (3.5) implies

$$U(\dot{\mathbf{F}}) \geq U^0(\dot{\mathbf{F}}) \quad \text{for every } \dot{\mathbf{F}}\,, \tag{3.7}$$

e.g. when $|\dot{\mathbf{F}}^0|/|\dot{\mathbf{F}}|$ approaches 0 . If (3.7) holds almost everywhere in the body then (3.4) implies fulfillment of the condition (2.33) of directional stability of equilibrium.

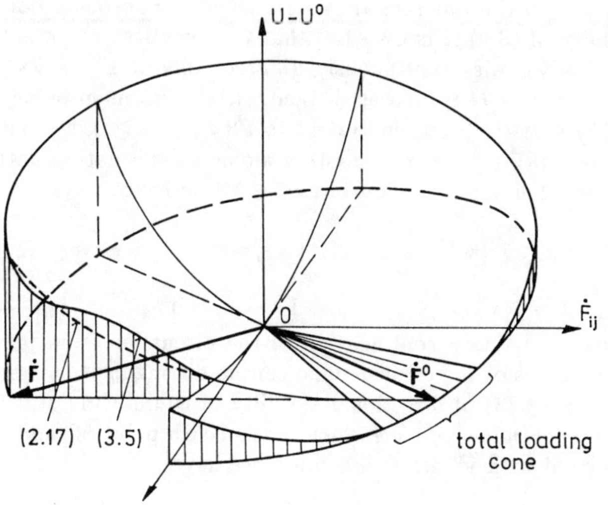


Fig. 5, Difference between the relative convexity property (2.17) and its weakened form (3.5). The former requires that $U - U^0$ is a convex function along any ray in $\dot{\mathbf{F}}$-space while the latter requires $U - U^0$ to be only non-decreasing along a ray with the origin at $\dot{\mathbf{F}}^0$.

A fundamental distinction between (3.1) and (2.4) is that (3.1) does *not* ensure convexity of the functional $J(\tilde{\mathbf{v}})$ defined by (2.12). It turns out, however, that (3.1) is still *sufficient for the minimum principle* (2.15) *to be valid* [35]. In particular, (3.4) with (3.5) ensure that the unique solution $\tilde{\mathbf{v}}^0$ assigns to $J$ an absolute minimum value in $\mathcal{V}$. To demonstrate this directly, one can use the identity

$$J(\tilde{\mathbf{v}}) - J(\tilde{\mathbf{v}}^0) = I^0(\tilde{\mathbf{v}} - \tilde{\mathbf{v}}^0) + \int_V (U(\dot{\mathbf{F}}) - U^0(\dot{\mathbf{F}}))\, d\xi\,, \qquad \tilde{\mathbf{v}} \in \mathcal{V} \tag{3.8}$$

valid under the assumptions that $\tilde{\mathbf{v}}^0$ is a solution and $C^0_{ijkl} = C^0_{klij}$ . Substitution of (3.4) and (3.7) into (3.8) yields (2.15), although $J$ need not be convex.

Comparison with the standard theorem of the calculus of variations suggests that the functional $I^0$ has the following interpretation

$$I^0(\tilde{\mathbf{w}}) = \delta^2 J(\tilde{\mathbf{v}}^0, \tilde{\mathbf{w}}) \tag{3.9}$$

where

$$\delta^2 J(\tilde{\mathbf{v}}^0, \tilde{\mathbf{w}}) \equiv \frac{1}{2} \lim_{\gamma \to 0} (\delta J(\tilde{\mathbf{v}}^0 + \gamma \tilde{\mathbf{w}}, \tilde{\mathbf{w}}) - \delta J(\tilde{\mathbf{v}}^0, \tilde{\mathbf{w}}))/\gamma \tag{3.10}$$

is the second weak variation of $J$ at $\tilde{\mathbf{v}}^0$ in the direction $\tilde{\mathbf{w}}$. In fact, although the usual regularity assumptions of the theorem are not satisfied, a rigorous proof of (3.9) is available [36]. For validity of (3.9) it is essential that the constitutive potential U is (almost everywhere in the body) twice continuously differentiable at $\dot{\mathbf{F}}^0 = \nabla \mathbf{v}^0$, as it has been assumed above by requiring $I^0$ to be well defined. It is worth mentioning that if (3.9) and (3.7) hold then $J(\tilde{\mathbf{v}}) - J(\tilde{\mathbf{v}}^0) \geq 0$ is equivalent to $I^0(\tilde{\mathbf{v}} - \tilde{\mathbf{v}}^0) \geq 0$, by (3.8).

If the tangent moduli $C^0_{ijkl}$ vary smoothly along a deformation path then a critical stage can be reached when for some nonzero $\tilde{\mathbf{w}}^* \in \mathcal{W}$ we have

$$I^0(\tilde{\mathbf{w}}^*) = 0 \leq I^0(\tilde{\mathbf{w}}) \quad \text{and} \quad \delta I^0(\tilde{\mathbf{w}}^*, \tilde{\mathbf{w}}) = 0 \quad \text{for every } \tilde{\mathbf{w}} \in \mathcal{W} \tag{3.11}$$

as a particular case of (2.24) and (2.25) when $\frac{1}{2}I^L = I^0$. Then $\tilde{\mathbf{w}}^*$ is an eigenmode for the tangent comparison solid. As a continuation of the discussion from preceding Chapter, the following typical cases of the critical stage can be distinguished.

(i) The condition (2.33) of directional stability of equilibrium still holds, and there exists a constant $c$ *independent of place* such that at each point of the body at which the moduli $C^0_{ijkl}$ associated with $\tilde{\mathbf{v}}^0$ are well defined, we have

$$C_{ijkl}(\dot{\mathbf{F}}) = C^0_{ijkl} \quad \text{if } |\dot{\mathbf{F}} - \dot{\mathbf{F}}^0| < c\,. \tag{3.12}$$

Then the velocity field $\tilde{\mathbf{v}}^* = \tilde{\mathbf{v}}^0 + \gamma\tilde{\mathbf{w}}^*$ also corresponds to the moduli field $C^0_{ijkl}$ provided $|\gamma|$ is sufficiently small. We have thus the situation discussed in Section 2.4: the primary bifurcation point has been reached, and $\tilde{\mathbf{v}}^*(\gamma)$ represents an infinite family of velocity solutions to the actual first-order rate boundary value problem. It is clear that $J(\tilde{\mathbf{v}}^*) = J(\tilde{\mathbf{v}}^0)$, as can be seen e.g. from (2.13) or from (3.8).

(ii) The condition (2.33) holds but no constant $c$ satisfying (3.12) can be found. The latter case is obvious if the incremental constitutive law is thoroughly nonlinear but is possible also for the classical elastic-plastic solids. Then in general no bifurcation in velocities can take place at the critical instant but rather a smooth bifurcation can be inferred. More precisely, if (1.30) is valid almost everywhere within the body then the eigenmode $\tilde{\mathbf{w}}^*$ with an *arbitrary* multiplier can be added to the fundamental solution in quasi-static *accelerations* to generate another solution to the (linear) second-order rate boundary value problem [14]. The actual order of path bifurcation follows from an analysis of the post-bifurcation expansion and may be different from two [22].

(iii) The condition (3.11) is met simultaneously with the condition (2.35) for the primary eigenstate discussed in Section 2.6. This take place if the "direction" of the eigenmode $\tilde{\mathbf{w}}^*$ corresponds to the tangent moduli field, in the sense that $C_{ijkl}(\nabla\mathbf{w}^*, \mathbf{H}) = C^0_{ijkl}$ almost everywhere in the body. In typical circumstances, that case occurs along a smooth path when the loading function $\lambda(\theta)$ reaches an analytic extremum point (called also a limit point if $\lambda$ is a multiplier of controlled loads). For, at that point $\dot{\lambda}$ vanishes so that the fundamental solution $\tilde{\mathbf{v}}^0$, if exists and is finite for the "time" scale defined by a choice of $\theta$, represents itself an eigenmode $\tilde{\mathbf{w}}^*$ satisfying both (3.11) and (2.36). This conclusion remains valid also for the configuration-dependent loading discussed in Chapter 5.

### 3.3 A constitutive inequality and uniqueness criterion

Suppose that the constitutive relationship (1.10) is intended to describe adequately the incremental elastoplastic behaviour of a polycrystalline metal at the macro-level. As discussed in Section 3.1, the assumption of a smooth yield surface is an oversimplification, at least in bifurcation problems. Suppose thus that (1.10) represents a nonlinear relationship at a yield surface vertex. We may assume that time-independent constitutive relations at the level of a single grain have the known structure (1.22)÷(1.25) which results from the piecewise linearity and normality postulates, commonly accepted for ductile metal crystals deformed by multislip [8]. The symmetry conditions (1.27) are also assumed but fulfillment of (1.26) is not necessary here so that nonuniqueness of the constitutive response *at the micro-level* is not excluded. Under these restrictions on qualitative properties of the material at the micro-level and with the help of the averaging theorems due to Hill [37], it has been shown in [38] that the macroscopic constitutive relationship (1.10) should satisfy the following constitutive inequality

$$\int \dot{F}^*_{ji}\, dN_{ij} \geq \int \dot{N}^*_{ij}\, dF_{ji}\,; \tag{3.13}$$

$\mathbf{F}$ and $\mathbf{N}$ can be replaced by any other pair of work-conjugate measures of strain and stress. The inequality (3.13) is to hold for all segments of every piecewise-smooth deformation path while the starred rates are virtual and may vary arbitrarily along the path. It is understood that $\dot{\mathbf{N}}^*$ is related to $\dot{\mathbf{F}}^*$, similarly as $d\mathbf{N}$ to $d\mathbf{F}$, by the currently valid incremental constitutive law which varies along the path but is unaffected in any way by variations of the starred rates (so that the starred (virtual) and unstarred (actual) rates in (3.13) cannot be interchanged). Validity of (3.13) has recently been examined in [39] for a class of materials obeying the maximum dissipation principle and extended, in particular, to an elastic material with a number of interacting planar Griffith's microcracks. Interpretation of (3.13) in terms of the second-order work of deformation is given in [40].

**Exercise**

Verify validity of (3.13) for the classical elastoplastic solids obeying the normality flow rule relative to a smooth yield surface.

The following two consequences [38] of (3.13) are of special interest for the theory of bifurcation and stability. The first is that (3.13) *implies existence of the constitutive potential* (1.16) (under the *right-hand* continuity restriction on the tangent moduli). The second (immediate) consequence of (3.13) is that along a *regular* deformation path with the actual rates $\dot{\mathbf{F}}^0$ and $\dot{\mathbf{N}}^0$ , (3.13) implies (and is implied by)

$$\dot{N}^0_{ij}\dot{F}^*_{ji} - \dot{N}^*_{ij}\dot{F}^0_{ji} \geq 0 \quad \text{for every } \dot{\mathbf{F}}^*,\ \dot{N}^*_{ij} = \dot{N}_{ij}(\dot{\mathbf{F}}^*,\mathsf{H})\,. \tag{3.14}$$

A deformation path (for a material element) is called regular if it consists only of regular points at which the current constitutive relationship $\dot{N}_{ij}(\dot{\mathbf{F}})$ along the path does not change discontinuously. [4] As discussed in [38], (3.14) may be regarded as a mathematical

[4] In the sense that the virtual stress-rate $\dot{N}^*_{ij}$ constitutively related to any *fixed* $\dot{\mathbf{F}}^*$ varies continuously along the path.

formulation of the physical condition that the actual rates $\dot{\mathbf{F}}^0$ and $\dot{\mathbf{N}}^0$ do not correspond to abrupt unloading at the micro-level. The set of $\dot{\mathbf{F}}^0$ satisfying (3.14) at the current state may thus be identified with the non-unloading range; this is either the closure of the total loading cone discussed above or degenerates to a single ray in the strain-rate space. From (3.13) it can be deduced that the non-unloading range should always vary with the deformation in such a manner that *it contains the actual rate of strain* (or stress) *at almost every instant*, except at singular points. At a singular point, typically at a sufficiently "sharp" corner on the strain path, the forward rate of strain falls *momentarily* outside the non-unloading range, and (3.14) does not hold at that point. Such points are excluded on a regular path. This does not mean that along a regular path there is no unloading, but rather that the unloading at the micro-level takes place gradually (i.e. smoothly at the macro-level).

The left-hand expression in (3.14) is of the form of Hill's "bilinear invariant" [37]. This results in invariance of the constitutive inequality (3.14) and (3.13) under transformation to another work-conjugate pair of stress and strain measures, possibly under a change of the reference configuration. For instance, (3.14) is equivalent to

$$\overset{\circ}{\tau}{}^0_{ij}\dot{\varepsilon}_{ij} - \overset{\circ}{\tau}_{ij}\dot{\varepsilon}^0_{ij} \geq 0 \quad \text{for every } \dot{\varepsilon}_{ij}\,. \tag{3.15}$$

Significance of (3.14) for the bifurcation theory results from the following lemma proved in [38]: If *the potential* (1.16) *exists and the tangent moduli* $C^0_{ijkl}$ *are well defined then* (3.14) *is equivalent to* (3.5). This gives a clear physical meaning to the inequality (3.5) as a condition that $\dot{\mathbf{F}}^0$ does not induce an abrupt unloading at the micro-level. Since existence of the potential (1.16) is a consequence of (3.13), the condition (3.5) which justifies the use of (3.4) as the uniqueness criterion turns out to be a consequence of (3.13) provided only that $\dot{\mathbf{F}}^0$, $\dot{\mathbf{N}}^0$ correspond to a regular deformation path and that the tangent moduli $C^0_{ijkl}$ are well defined.

A process of deformation of a finite body may be called *regular* if the set of material points at which the current constitutive relationship $\dot{N}_{ij}(\dot{\mathbf{F}})$ changes discontinuously or the tangent moduli are not well defined is of zero volume at every instant. The uniqueness criterion can then be formulated as follows [38]:

*In a regular deformation process under controlled nominal loads, fulfillment of* (3.4) *excludes bifurcation in velocities provided the constitutive inequality* (3.13) *holds.*

This uniqueness criterion provides another justification for the common procedure of excluding bifurcation in velocities on the basis of (3.4) only. The previous justification based on (2.17) required the material model to be specified since (2.17) had the status of a mathematical condition which needed verification. The essence of the above result is that the actual incremental constitutive law need not be fully specified to apply the criterion (3.4) since (3.13) may be assumed *a priori* as a constitutive inequality derived from micromechanical considerations. As discussed in Section 3.1, the incremental constitutive law intended to model adequately the real behaviour of structural polycrystalline metals may be difficult to be determined experimentally.

In the uniqueness range defined by the conditions in the above criterion, both the minimum property (2.15) and the condition (2.33) of directional stability of equilibrium are satisfied. This follows from (3.5) or (3.7) which, as indicated above, in a regular process must hold almost everywhere in the body for (3.13) to be valid.

A method of verification of (3.15) for phenomenological models of elastic-plastic behaviour at a yield-surface vertex has been proposed in [38].

## 4. NONUNIQUENESS AND INSTABILITY IN POST-CRITICAL RANGE

### 4.1 Post-critical inequalities

We have discussed conditions sufficient for uniqueness of a solution in velocities and for directional stability of equilibrium. Then, the critical stage of the deformation has been examined when one of those conditions fails for the first time. Circumstances have been established in which the solution in velocities is nonunique at the critical instant, either at varying loading (the primary bifurcation point) or at constant loading (the primary eigenstate).

The question now arises: what can be said about uniqueness or stability *beyond* the critical point of primary bifurcation or primary eigenstate, respectively? This question is essential for complete understanding of the bifurcation phenomenon, nevertheless, it is difficult to find in the literature any definite answer or even a discussion on that point. Following [36], we will now consider this problem (and also later on in Chapter 6) under the assumption that the constitutive rate equations admit the potential (1.16).

It is possible that the condition (3.4) is again satisfied along the *secondary* post-bifurcation branch. For instance, in elastic-plastic solids with a smooth yield surface, local elastic unloading starts on that branch immediately beyond the bifurcation point so that the elastoplastic tangent moduli in (3.2) are replaced gradually by "stiffer" elastic moduli [25] [26]. The situation is similar when the effect of formation of a yield-surface vertex is taken into account, with the difference that the "stiffer" moduli are initially taken from the range of *partial* unloading before total elastic unloading occurs anywhere. Now, if (3.13) holds and the secondary post-bifurcation segment of the deformation process is regular then fulfillment of (3.4) excludes bifurcation in velocities as stated in Section 3.3, implying also the minimum property (2.15) and the condition (2.33) of directional stability of equilibrium. In other words, in those circumstances there is no qualitative difference between the fundamental pre-bifurcation path and the secondary post-bifurcation path as far as incremental uniqueness and stability are concerned.

The situation is generally different on the *fundamental* post-bifurcation branch. If the quadratic form $I^0(\tilde{\mathbf{w}})$ is positive definite before the critical point and positive semi-definite at the critical point then, if the deformation path if *smoothly* continued, we will usually have

$$I^0(\tilde{\mathbf{w}}) < 0 \quad \text{for some } \tilde{\mathbf{w}} \in \mathcal{W} \tag{4.1}$$

beyond that point. Then (3.1) fails for $\tilde{\mathbf{v}}^* = \tilde{\mathbf{v}}^0 + \gamma\tilde{\mathbf{w}}$ with $|\gamma|$ sufficiently small, and uniqueness of the solution $\tilde{\mathbf{v}}^0$ is not guaranteed. However, it is not evident whether uniqueness itself is lost since (3.1) is merely sufficient for uniqueness in velocities.

Since nonnegativeness of the second variation is necessary for a minimum, from (3.9) we obtain that (4.1) implies

$$J(\tilde{\mathbf{v}}) < J(\tilde{\mathbf{v}}^0) \quad \text{for some } \tilde{\mathbf{v}} \in \mathcal{V}\,, \tag{4.2}$$

so that the minimum principle (2.15) necessarily fails in the range of (4.1).

On the other hand, for incrementally nonlinear solids (4.1) does not contradict (2.33) so that, as pointed out in Chapter 2, stability of equilibrium can be preserved also on the fundamental post-bifurcation branch. Hence, in typical circumstances (but not always), along that branch a deformation range can be expected where (4.1) and (4.2) are satisfied simultaneously with (2.33). Only at a later stage, usually beyond the primary eigenstate, we will have

$$I(\tilde{\mathbf{w}}) < 0 \quad \text{for some } \tilde{\mathbf{w}} \in \mathcal{W}\,. \tag{4.3}$$

The condition (4.3) has been introduced by Hill and Sewell [27] as a *part* of a set of sufficiency conditions for instability of equilibrium in a dynamic sense. Although (4.3) means that a spontaneous departure from the equilibrium state is *energetically* possible (cf. the derivation of (2.33)), it is not evident whether such a departure can actually take place under arbitrarily small disturbances.

The following two questions:

- Is (4.3) *sufficient* for instability of equilibrium in a dynamic sense?
- Is (4.1) *sufficient* for non-uniqueness of the fundamental solution $\tilde{\mathbf{v}}^0$?

are addressed below under an additional assumption that the continuum problem has been spatially *discretized.* This means that up to the end of Chapter 4, admissible variations of velocities (or of displacements) are restricted to the form

$$\mathbf{w}(\boldsymbol{\xi}) = \sum_{\alpha=1}^{M} w_\alpha\, \boldsymbol{\phi}_\alpha(\boldsymbol{\xi})\,, \tag{4.4}$$

where $\boldsymbol{\phi}_\alpha$ are given continuous and piecewise smooth "shape functions" vanishing on $S_u$ and $w_\alpha$ are arbitrary scalar coefficients. Their number $M$ is finite so that the fields (4.4) form a finite-dimensional space, for simplicity denoted still by $\mathcal{W}$ as in the continuum problem. The definition of a solution, (1.33) with (1.32), is accordingly changed following the standard discretization procedure. It will later become clear that the above assumption has been introduced for mathematical reasons; the conclusions will pertain to a number of continuous problems as well.

## 4.2 Instability of equilibrium

Throughout this section the loading parameter $\lambda$ is regarded as fixed, $t$ denotes a natural time, and inertia forces are taken into account. Let $\Delta\mathbf{N}(t) = \mathbf{N}(t) - \mathbf{N}(0)$ denote a stress increment along a deformation path starting from an equilibrium state. By subtracting the virtual work expressions in the initial and current instants, the (discretized) equations of motion can be written down in the form

$$\int_V (\Delta N_{ij} w_{j,i} + \bar{\rho} a_j w_j)\, d\xi = 0 \quad \text{for every } \tilde{\mathbf{w}} \in \mathcal{W}\,, \tag{4.5}$$

where $\bar{\rho}$ is the material density in the reference configuration and $\mathbf{a} = \ddot{\mathbf{u}}$ denote accelerations (time derivatives are now taken with respect to $t$).

We will examine the possibility of departure from an equilibrium configuration $\tilde{\mathbf{u}}^0$ along a *direct* path of deformation which, by definition, can be approximated by a straight path

$$\tilde{u}_j(t) = \tilde{u}_j^0 + B(t)\tilde{w}_j\,, \quad B(0) = 0\,, \quad \dot{B} \geq 0\,, \tag{4.6}$$

where $\tilde{\mathbf{w}} \in \mathcal{W}$ is fixed. We introduce a constitutive approximation that along straight (direct) paths and for small $B$ the rates in (1.10) can be replaced by small increments of $N_{ij}$ and $F_{ij}$ with a negligible error, so that

$$\Delta N_{ij}(t) = B(t)\dot{N}_{ij}(\nabla \mathbf{w}) \tag{4.7}$$

where $\dot{N}_{ij}(\cdot)$ denotes the constitutive function at the considered equilibrium state. Of course, the approximation (4.7) would be unacceptable for arbitrarily circuitous paths.

Since $\mathcal{W}$ is now finite-dimensional and $U$ depends continuously on $\dot{\mathbf{F}}$, the functional (2.34) is continuous in $\mathcal{W}$ and, when constrained to a level set of the function

$$\Phi(\tilde{\mathbf{w}}) \equiv \frac{1}{2}\int_V \bar{\rho} w_j w_j \, d\xi\,, \tag{4.8}$$

attains a minimum at some $\tilde{\mathbf{w}}^* \in \mathcal{W}$. By the method of Lagrangian multipliers, there is a number $\mu$ such that

$$\delta I(\tilde{\mathbf{w}}^*, \tilde{\mathbf{w}}) = \mu \delta \Phi(\tilde{\mathbf{w}}^*, \tilde{\mathbf{w}}) \quad \text{for every } \tilde{\mathbf{w}} \in \mathcal{W}\,. \tag{4.9}$$

At the equilibrium state and under the assumption that (1.16) holds, this reads

$$\int_V \dot{N}_{ij}(\nabla \mathbf{w}^*) w_{j,i} \, d\xi = \mu \int_V \bar{\rho} w_j^* w_j \, d\xi \quad \text{for every } \tilde{\mathbf{w}} \in \mathcal{W}\,. \tag{4.10}$$

On multiplying both sides of (4.10) by $B$ and using (4.7) we obtain that the stresses along the deformation path (4.6) satisfy (within the approximations involved in (4.7)) the discretized equations of motion (4.5) provided in (4.6) we take $\tilde{w}_j = \tilde{w}_j^*$ and $B(t)$ such that

$$\ddot{B} = -\mu B\,. \tag{4.11}$$

Clearly, the sign of $\mu$ decides whether there is a tendency to decrease or to increase the speed of departure from equilibrium.

By homogeneity of degree two of $I$ and $\Phi$, substitution of $\tilde{\mathbf{w}} = \tilde{\mathbf{w}}^*$ into (4.9) yields $I(\tilde{\mathbf{w}}^*) = \mu \Phi(\tilde{\mathbf{w}}^*)$ and

$$\mu = \min_{\substack{\tilde{\mathbf{w}} \in \mathcal{W} \\ \tilde{\mathbf{w}} \neq \tilde{\mathbf{0}}}} \frac{I(\tilde{\mathbf{w}})}{\Phi(\tilde{\mathbf{w}})}\,. \tag{4.12}$$

Suppose now that (4.3) holds; then it is clear that $\mu < 0$. As an admissible function $B(t)$ in (4.6) satisfying (4.11) we can thus take

$$B(t) = (\varepsilon/\kappa)\sinh(\kappa t)\,, \quad \kappa = (-\mu)^{1/2}\,, \quad \varepsilon > 0 \tag{4.13}$$

which corresponds to a free inertial motion starting from the equilibrium state at $t = 0$ with initial velocities $\varepsilon w_j^*$. Evidently, a given small finite distance from the initial equilibrium configuration is exceeded for $t$ sufficiently large no matter how small $\varepsilon$ is. We have thus proved the following theorem on instability of equilibrium (in the first approximation) [36]:

*If the discretized system is considered then* (4.3) *implies instability of equilibrium in the dynamic sense for vanishingly small initial disturbances, under the approximation* (4.7).

The method used here to demonstrate dynamic instability of equilibrium of an inelastic time-independent system is an extension of the well known "kinetic" method for linear elastic systems where $\mu$ is defined as a square of the lowest natural frequency of vibrations. Of course, for inelastic solids $\mu$ defined by (4.12) has no longer this special interpretation.

The obtained result indicates that the condition (2.33), usually treated as a condition sufficient for stability of equilibrium, may be interpreted (with a sign $\geq$) as a condition necessary for stability of equilibrium in the dynamic sense. The most essential assumptions needed to demonstrate validity of that conclusion appear to be the following: conservativeness of the loading and existence of the potential (1.16). The approximations (4.4) and (4.7) have been primarily used to simplify the mathematical proof.

### 4.3 Continuous range of bifurcation points

Consider now a process of quasi-static deformation at varying $\lambda$. Since U is a continuously differentiable function of $\dot{\mathbf{F}}$, $J$ is differentiable in the class $\mathcal{V}$ which has been assumed now to be finite dimensional. A minimizer of $J$ in $\mathcal{V}$ must satisfy (2.13) and is thus automatically a (discretized) solution. In general, $J$ may be unbounded from below so that an absolute minimum need not be attained. However, this cannot happen if the equilibrium state at which the discretized rate-problem is formulated is directionally stable in the sense of (2.33). In fact, one can prove the following existence theorem [36]: *If* (2.33) *holds then the discretized rate-problem has a solution which assigns to the functional $J$ its absolute minimum value in $\mathcal{V}$.*

Details of the proof are omitted here; they can be found in [36], and also in [41] in a different version. It is recalled that the above theorem is valid for a broad class of incrementally nonlinear constitutive laws expressible in the potential form (1.16); an essential assumption is that the incremental loading is conservative (not necessarily configuration-insensitive as assumed above).

In the uniqueness range when (3.1) holds, the minimum principle (2.15) is valid and the solution $\tilde{\mathbf{v}}^0$ coincides with the solution guaranteed by the above existence theorem. But if (4.2) holds then the guaranteed solution assigns to $J$ a lower value than $\tilde{\mathbf{v}}^0$ and must thus be distinct from $\tilde{\mathbf{v}}^0$ . In turn, if $\tilde{\mathbf{v}}^0$ does minimize $J$ in $\mathcal{V}$ but the minimum is attained also at some other $\tilde{\mathbf{v}}^* \in \mathcal{V}$ then $\tilde{\mathbf{v}}^*$ represents another first-order solution distinct from $\tilde{\mathbf{v}}^0$. Hence, *for uniqueness of a solution $\tilde{\mathbf{v}}^0$ to the discretized rate-problem formulated at a (directionally) stable equilibrium state it is* necessary *that $\tilde{\mathbf{v}}^0$ satisfies the minimum principle* (2.15).

If the tangent moduli $C^0_{ijkl}$ corresponding to $\tilde{\mathbf{v}}^0$ are well defined almost everywhere in $V$ then (3.9) is valid. Since nonnegativeness of the second variation is necessary for a minimum, a necessary condition for $\tilde{\mathbf{v}}^0$ to be a minimizer of $J$ in $\mathcal{V}$ is that

$$I^0(\tilde{\mathbf{w}}) = \frac{1}{2} \sum_{\alpha,\beta=1}^{M} K^0_{\alpha\beta} w_\alpha w_\beta \geq 0 \quad \text{for every } w_\alpha\,, \tag{4.14}$$

where

$$K^0_{\alpha\beta} = \int_V C^0_{ijkl} \phi_{\alpha j,i} \phi_{\beta l,k} \, d\xi \tag{4.15}$$

is the tangent stiffness matrix associated with the solution $\tilde{\mathbf{v}}^0$. As stated above, on account of incremental nonlinearity of the constitutive law, (2.33) can be satisfied along a deformation path also when (4.14) fails, that is, when (4.1) holds. We arrive thus at the following conclusion [36] [41]:

*For a discretized system, there is a bifurcation in velocities at every point on a fundamental solution path along which* (4.1) *holds simultaneously with* (2.33).

A fundamental distinction can thus be observed between the spectrum of bifurcation points for a typical elastic structure and for the incrementally nonlinear discretized system. In the former case the tangent stiffness matrix (independent of the actual incremental solution) must be singular at a bifurcation point implying that such points form a *discrete* set along a deformation path (of course, provided that trivial bifurcations related e.g. to a rigid-body rotation about a symmetry axis have been eliminated). In the latter case, the above conclusion means that *the bifurcation points form in general a continuous range along a deformation path*; of course, the actual tangent stiffness matrix $K^0_{\alpha\beta}$ (dependent on the actual incremental solution $\tilde{\mathbf{v}}^0$ ) need *not* be singular in that range. This result, known for the Shanley column [29] for many years, at the present generality seems to be proved only recently as indicated above. Along a smooth deformation path, the continuous range of bifurcations in velocities can be entered at the primary bifurcation point at which (3.4) fails and (3.11) is met (so that $K^0_{\alpha\beta}$ *is* momentarily singular), and terminated at the primary eigenstate at which (2.33) fails and (2.35) is met. This conclusion is *not* necessarily limited to discretized systems; it can remain valid also for many continuous systems under appropriate regularity assumptions which, however, need not always be satisfied. If the primary bifurcation is induced by a discontinuous drop of the tangent moduli then the continuous range of bifurcations can be entered even if the tangent stiffness matrix $K^0_{\alpha\beta}$ is not singular at *any* instant.

The previously discussed bifurcation at an eigenstate for the incrementally linear comparison material turns out thus to be merely a special kind of bifurcation in incrementally nonlinear solids. Contrary to that special case, in the range (4.1) the secondary solution $\tilde{\mathbf{v}}^*$ which minimizes $J(\mathbf{v})$ in $\mathcal{V}$ must correspond to a moduli field different from the fundamental moduli field $\tilde{C}^0_{ijkl}$ . For, if the moduli fields were the same then straightforward transformations with the help of symmetry of the moduli would yield $J(\tilde{\mathbf{v}}^*) - J(\mathbf{v}^0) = I^0(\tilde{\mathbf{v}}^* - \tilde{\mathbf{v}}^0) = 0$ which would contradict (4.2).

**Illustrative diagram**

The results presented in this section are illustrated schematically in a diagram in Fig. 6. Comparison with Fig. 3 shows how the present conclusions complement Hill's theory which has not dealt with the problems of non-uniqueness and instability in a post-critical range. A picture similar to that in Fig. 6 was sometimes in the literature assumed as granted on the basis of an analysis of simplest systems as the Shanley column, but a general proof was lacking. The proof given in [36] can be extended to a class of continuous systems under certain requirements of mathematical nature; cf. the discussion in Chapter 6. We will see later that a deformation path in the range of (4.2) (typically, beyond the primary bifurcation point) can be regarded as unstable, in the sense of instability of a quasi-static deformation *process*.

The question: what happens if the critical value $\lambda_T$ of the loading parameter $\lambda$ suffers a discontinuous drop along the deformation path due to a sudden decrease of the tangent

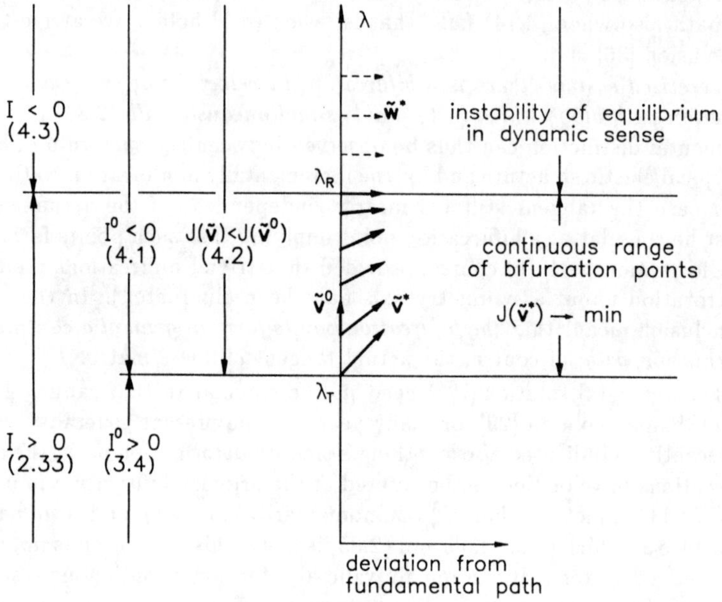


Fig. 6, Schematic illustration of the results of Chapter 4 for a discretized problem.

modulus, so that the range $\lambda > \lambda_T$ is entered at some other critical value $\lambda_D > \lambda_T$, can now be given a general answer, at least for discretized systems. If $\lambda_T < \lambda_D < \lambda_R$ then a bifurcation in velocities takes place at the critical value $\lambda_D$ provided $\lambda$ is further increasing. If $\lambda_D > \lambda_R$ then the equilibrium configuration at the critical instant is unstable (at least in the first approximation) so that a dynamic departure from that configuration becomes possible at constant $\lambda$.

## 5. ENERGY INTERPRETATIONS IN THE BIFURCATION THEORY

### 5.1 The energy functional

The work of deformation in the body can be written down as

$$W = \int_V \int N_{ij}\, dF_{ji}\, d\xi \,, \tag{5.1}$$

where the stresses are determined pointwise by integration of the constitutive rate equations (1.10) along the deformation path. A potential energy of the loading device which applies nominal surface tractions $\overline{T}_j$ and nominal body forces $\bar{b}_j$ independently of the body configuration can be expressed as

$$\Omega = \Omega(\tilde{\mathbf{u}}, \lambda) = -\int_V \bar{b}_j u_j \, d\xi - \int_{S_T} \overline{T}_j \, u_j \, da \,. \tag{5.2}$$

Introduce *the energy functional* [42] [34]

$$E = W + \Omega \tag{5.3}$$

defined for any kinematically admissible deformation process at varying or fixed $\lambda$. In general, $E$ is a functional of the deformation history due to path-dependence of $W$. An increment of the value of $E$ can be interpreted as the amount of energy which has to be supplied from external sources to the mechanical system consisting of the deformed body *and* the loading device in order to produce quasi-statically a deformation increment, generally with the help of additional perturbing forces. It is emphasized that an increment of the value of ($-\Omega$) is generally *not* equal to the work done by the controlled loads unless $\bar{b}_j$ and $\overline{T}_j$ are constant in time.

In the cases of an elastic support or fluid-pressure loading, the nominal surface tractions $T_j$ on $S_T$ are not only functions of $\xi_k$ and $\lambda$ but depend also on the actual displacements or their surface gradients. For one- or two- -dimensional idealizations of the solid body, generalized body forces dependent in a similar way on the displacement field are induced by lateral surface tractions; other examples can also be given. Suppose thus (cf. [17] [6]) that the incremental loading consists not only of a controllable part (distinguished by a bar), as it was assumed above, but also of a deformation-sensitive part, that is

$$\dot{T}_j = \dot{\overline{T}}_j + f_j(\tilde{\mathbf{v}}) \,, \quad \dot{b}_j = \dot{\bar{b}}_j + g_j(\tilde{\mathbf{v}}) \,, \tag{5.4}$$

where $f_j(\tilde{\mathbf{v}})$ and $g_j(\tilde{\mathbf{v}})$ are linear homogeneous expressions in the velocity $v_l$ and its gradient $v_{l,k}$ at the considered material point. Coefficients in the expressions may depend piecewise continuously on $\xi_k$ and sufficiently smoothly on $\lambda$ and $\tilde{\mathbf{u}}$. The energy functional (5.3) can still be defined provided the loading is *conservative* in an overall sense (cf. [17] [43]). This means that for a *fixed* value of the loading parameter $\lambda$, the total work done by the body forces $b_j = b_j(\boldsymbol{\xi}, \lambda, \tilde{\mathbf{u}})$ and surface tractions $T_j = T_j(\boldsymbol{\xi}, \lambda, \tilde{\mathbf{u}})$ in any virtual motion compatible with the kinematic constraints and leading from a configuration $\tilde{\mathbf{u}}$ to any sufficiently close configuration $\tilde{\mathbf{u}}^*$ is assumed to be *path-independent*, viz.

$$\int_V \int_{\mathbf{u}}^{\mathbf{u}^*} b_j du_j \, d\xi + \int_{S_T} \int_{\mathbf{u}}^{\mathbf{u}^*} T_j du_j \, da = \Omega(\tilde{\mathbf{u}}, \lambda) - \Omega(\tilde{\mathbf{u}}^*, \lambda) \,, \quad \lambda = \text{const.} \tag{5.5}$$

The functional $\Omega(\tilde{\mathbf{u}}, \lambda)$ is defined to within an additive function of $\lambda$ which may be chosen arbitrarily. For a physically appropriate choice of that function, $\Omega$ can be identified with the potential energy of the loading device and substituted into (5.3).

At a given body configuration $\tilde{\mathbf{u}}$ at a certain $\lambda$, introduce the bilinear functional $Q$ of velocities (or their variations) and the respective quadratic functional $R$, defined by

$$Q(\tilde{\mathbf{v}}, \tilde{\mathbf{v}}^*) = \int_V g_j(\tilde{\mathbf{v}}) v_j^* \, d\xi + \int_{S_T} f_j(\tilde{\mathbf{v}}) v_j^* \, da \,, \quad R(\tilde{\mathbf{v}}) = -\frac{1}{2} Q(\tilde{\mathbf{v}}, \tilde{\mathbf{v}}) \,. \tag{5.6}$$

It can be shown [36] that (5.5) implies

$$Q(\tilde{\mathbf{v}},\tilde{\mathbf{v}}^*) = Q(\tilde{\mathbf{v}}^*,\tilde{\mathbf{v}}) \quad \text{for every } \tilde{\mathbf{v}},\tilde{\mathbf{v}}^* \in \mathcal{V}\cup\mathcal{W}\,; \tag{5.7}$$

this is an extension of Hill's [17] "self-adjointness" condition for surface loading to the loading (5.4).

By the trapezoid rule of quadrature we obtain, instead of (5.2), the following expression for the difference between the values of $\Omega$ in two body configurations close to each other, correct to second-order:

$$\Omega(\tilde{\mathbf{u}}+\tilde{\mathbf{w}},\lambda) - \Omega(\tilde{\mathbf{u}},\lambda) = -\int_V b_j w_j \, d\xi - \int_{S_T} T_j w_j \, da + R(\tilde{\mathbf{w}}) + \ldots, \tag{5.8}$$

where $b_j$ , $T_j$ and $R$ are evaluated at $(\tilde{\mathbf{u}},\lambda)$.

From (5.4), (5.6) and (5.7) we obtain the equality

$$\int_V \dot{b}_j w_j \, d\xi + \int_{S_T} \dot{T}_j w_j \, da = \int_V \dot{\bar{b}}_j w_j \, d\xi + \int_{S_T} \dot{\bar{T}}_j w_j \, da - \delta R(\tilde{\mathbf{v}},\tilde{\mathbf{w}}) \tag{5.9}$$

valid for every $\tilde{\mathbf{v}} \in \mathcal{V}$ and $\tilde{\mathbf{w}} \in \mathcal{W}$. By substituting (5.9) into (1.33) it can be shown that the conditions and theorems from the bifurcation theory formulated for controlled nominal loads remain valid for the configuration-dependent conservative loading provided the expressions for $J$, $\frac{1}{2}I^L$, $I^0$, $I$ are modified simply by adding the quadratic functional $R$ [17]. Details are omitted here.

**Exercise**

Derive an explicit expression for $R(\tilde{\mathbf{w}})$ in the cases of elastic foundation and fluid-pressure loading.

### 5.2 Basic identity

Consider first the case of controlled nominal loads when (5.2) applies. The first-order rate of the energy functional $E$ at a given state of the body is expressed as the affine (linear but inhomogeneous) functional of a velocity field

$$\dot{E}(\tilde{\mathbf{v}}) = \int_V N_{ij} v_{j,i} \, d\xi - \int_V \bar{b}_j v_j \, d\xi - \int_{S_T} \bar{T}_j v_j \, da - \int_V \dot{\bar{b}}_j u_j \, d\xi - \int_{S_T} \dot{\bar{T}}_j u_j \, da\,. \tag{5.10}$$

The first term equals to $\dot{W}$ and the remaining contribute to $\dot{\Omega}$. Since the value of the last two terms in (5.10) is prescribed independently of $\tilde{\mathbf{v}}$, it can be seen by reference to the virtual power principle (1.31) that the considered state is in equilibrium if and only if the value of $\dot{E}$ is independent of $\tilde{\mathbf{v}} \in \mathcal{V}$. The equilibrium conditions can thus be written down as

$$\dot{E}(\tilde{\mathbf{v}}) = \text{ const. in } \mathcal{V}\,. \tag{5.11}$$

Consider now the second-order rate of $E$ evaluated at an equilibrium state. In calculating the second time derivative of the deformation work $W$ one should take into account the possibility of velocity gradient discontinuities which move relative to the material. This can be done by using the transport theorem and kinematic compatibility conditions [5]. It turns out that velocity-gradient discontinuities introduce no extra term in the expression for $\ddot{W}$ *evaluated at an equilibrium state*, which reads [6]

$$\ddot{W} = \int_V (\dot{N}_{ij}\, v_{j,i} + \dot{\bar{b}}_j \dot{v}_j)\, d\xi + \int_{S_u} N_{ij}\nu_i \dot{v}_j\, da + \int_{S_T} \dot{\bar{T}}_j \dot{v}_j\, da\,. \tag{5.12}$$

To show validity of (5.12) at the absence of such discontinuities, it suffices to differentiate the first term in (5.10) by $\theta$, use the Green theorem and substitute the equilibrium conditions for $N_{ij}$ .

Calculation of $\ddot{\Omega}$ in the case of configuration-independent loading is straightforward if there are no moving discontinuities in $\bar{b}_j$ and $\bar{T}_j$ [7]. By adding the expression for $\ddot{\Omega}$ to that in (5.12), the integrals of $\dot{\bar{b}}_j \dot{v}_j$ over $V$ and of $\dot{\bar{T}}_j \dot{v}_j$ over $S_T$ cancel and we obtain

$$\begin{aligned}\ddot{E} = &\int_V (\dot{N}_{ij} v_{j,i} - 2\dot{\bar{b}}_j v_j)\, d\xi - 2\int_{S_T} \dot{\bar{T}}_j v_j\, da \\ &+ \int_{S_u} N_{ij}\nu_i \ddot{u}_j\, da - \int_V \ddot{\bar{b}}_j u_j\, d\xi - \int_{S_T} \ddot{\bar{T}}_j u_j\, da\,.\end{aligned} \tag{5.13}$$

The last three terms in (5.13) have prescribed values so that $\ddot{E}$ at a given equilibrium state is a functional of velocities only. Comparison of (5.13) and (2.12) yields finally the following identity [42] [34]

$$\frac{1}{2}\ddot{E}(\tilde{\mathbf{v}}^*) - \frac{1}{2}\ddot{E}(\tilde{\mathbf{v}}) = J(\tilde{\mathbf{v}}^*) - J(\tilde{\mathbf{v}}) \quad \text{for every } \tilde{\mathbf{v}}, \tilde{\mathbf{v}}^* \in \mathcal{V}\,. \tag{5.14}$$

Note that the potentiality condition (1.16) is not necessary for validity of (5.14).

Validity of (5.14) can be extended to the case of configuration-dependent conservative loading (5.4) provided the functional $J$ is appropriately modified. Time differentiation of (5.8) with $\dot{u}_j = v_j$ and $\dot{w}_j = v_j^* - v_j$ , followed by substitution of (5.4), (5.6) and (5.7) shows (cf. [34]) that the contribution of the configuration-sensitive part of the incremental loading to the left-hand side of (5.14) is equal to

$$-Q(\tilde{\mathbf{v}}, \tilde{\mathbf{v}}^* - \tilde{\mathbf{v}}) + R(\tilde{\mathbf{v}}^* - \tilde{\mathbf{v}}) = R(\tilde{\mathbf{v}}^*) - R(\tilde{\mathbf{v}})\,. \tag{5.15}$$

Hence, (5.14) remains valid if the velocity functional $J$ is defined, instead of (2.12), by

$$J(\tilde{\mathbf{v}}) = \int_V (U(\nabla\mathbf{v}) - \dot{\bar{b}}_j v_j)\, d\xi - \int_{S_T} \dot{\bar{T}}_j v_j\, da + R(\tilde{\mathbf{v}})\,. \tag{2.12$'$}$$

---

[5] There is a misprint in the derivation given in [34]: in eq. (18) there should be $\Delta\ddot{W}$ instead of $\ddot{W}$.

[6] It can be remarked that $\dot{v}_j$ on $S$ in (5.12) need not be equal to the respective limit value on the boundary of the acceleration field $\dot{v}_j$ in $V$.

[7] If such discontinuities are present then certain extra terms appear in the expression for $\ddot{\Omega}$. However, those terms are independent of the actual continuation of deformation process and have thus no influence on the final result (5.14).

This is exactly the modification needed in the bifurcation theory, as mentioned above.

### 5.3 Energy interpretations

Henceforth we shall always assume that the constitutive rate equations admit the potential (1.16). Then the first-order rate equations of continuing equilibrium can be given the variational formulation (2.13), with $J$ defined either by (2.12) for controlled nominal loads or by (2.12′) for the configuration-dependent conservative loading discussed above. Now, by (5.14), the stationarity principle (2.13) has the following energy interpretation:

$$\delta\ddot{E}(\tilde{\mathbf{v}},\tilde{\mathbf{w}})=0 \quad \text{for every } \tilde{\mathbf{w}}\in\mathcal{W}\,. \tag{5.16}$$

On account of (5.11), this may be expressed in words as follows: any solution in velocities corresponds to a stationary value of the increment of $E$ evaluated with accuracy to second order terms.

In the uniqueness range where the minimum property (2.15) holds, substitution of (5.14) shows that the unique solution $\mathbf{v}^0$ strictly minimizes the value of $\ddot{E}$ among all kinematically admissible velocity fields, that is,

$$\ddot{E}(\tilde{\mathbf{v}})>\ddot{E}(\tilde{\mathbf{v}}^0) \quad \text{for every } \tilde{\mathbf{v}}\in\mathcal{V},\ \tilde{\mathbf{v}}\neq\tilde{\mathbf{v}}^0\,. \tag{5.17}$$

From (3.9) and (5.14) we obtain that the quadratic functional (3.2) based on the tangent moduli, if it is well defined, has the following energy interpretation

$$I^0(\tilde{\mathbf{w}})=\frac{1}{2}\delta^2\ddot{E}(\tilde{\mathbf{v}}^0,\tilde{\mathbf{w}})\,. \tag{5.18}$$

This remains valid in the case of the configuration-dependent conservative loading if the quadratic functional $R(\tilde{\mathbf{w}})$ is added to the expression for $I^0$.

If the constitutive inequality (3.13) holds and $\tilde{\mathbf{v}}^0$ corresponds to a regular deformation process then the inequality

$$\delta^2\ddot{E}(\tilde{\mathbf{v}}^0,\tilde{\mathbf{w}})>0 \quad \text{for every } \tilde{\mathbf{w}}\in\mathcal{W},\ \tilde{\mathbf{w}}\neq\mathbf{0}\,, \tag{5.19}$$

as being equivalent to (3.4), is sufficient for uniqueness of $\tilde{\mathbf{v}}^0$ and also for (5.17). In other words, in those circumstances, if the actual solution $\mathbf{v}^0$ corresponds to a proper relative minimum of $\ddot{E}(\tilde{\mathbf{v}})$ then existence of other stationarity points of $\ddot{E}(\mathbf{v})$ is excluded and the minimum turns out to be absolute in $\mathcal{V}$.

Consider now the instant of primary bifurcation in velocities when (3.11) is met while (3.12) is valid, still under the assumption that (3.5) is satisfied. As already discussed, the bifurcation solutions $\tilde{\mathbf{v}}^*=\tilde{\mathbf{v}}^0+\gamma\tilde{\mathbf{w}}^*$ correspond then to the same moduli field $\tilde{C}^0_{ijkl}$ and assign to the functional $J$ the same value as the fundamental solution $\tilde{\mathbf{v}}^0$. It is clear from (3.8) and (3.7) that no field from $\mathcal{V}$ can correspond to a lower value of $J$. Hence, by (5.14), at the critical point under consideration we have

$$\ddot{E}(\tilde{\mathbf{v}}^*)=\ddot{E}(\tilde{\mathbf{v}}^0)\leq\ddot{E}(\tilde{\mathbf{v}}) \quad \text{for every } \tilde{\mathbf{v}}\in\mathcal{V}. \tag{5.20}$$

Conversely, any kinematically admissible velocity field $\tilde{\mathbf{v}}^*$ satisfying (5.20) assigns to the functional $J$ a stationary value and represents thus another solution to the first-order problem. We have thus the following energy interpretation of the typical primary bifurcation in velocities: the minimum value of the second-order increment of the energy functional (5.3) is attained not solely for the fundamental velocity solution but also for some other deformation mode [44] [34].

Beyond that critical instant, the following condition

$$\delta^2 \ddot{E}(\tilde{\mathbf{v}}^0, \tilde{\mathbf{w}}) < 0 \quad \text{for some } \tilde{\mathbf{w}} \in \mathcal{W}, \tag{5.21}$$

equivalent to (4.1), is usually fulfilled on the regular *fundamental* path (but this is not a rule on a *secondary* post-bifurcation path). In that range the fundamental solution in velocities can no longer minimize $\ddot{E}$ (since non-negativeness of the second variation is necessary for a minimum) and *may* be nonunique. In fact, it is nonunique if a minimum of $\ddot{E}$ is actually attained in $\mathcal{V}$; according to the existence theorem from Section 4.3, this is ensured for discretized systems so long as the equilibrium is directionally stable, i.e. if (5.22) holds.

The condition (2.33) of directional stability of equilibrium, contrary to those in the bifurcation theory, has been derived from the energy balance and requires thus no additional energy interpretation. However, for the sake of comparison, we express it explicitly in terms of the energy functional $E$. If $\lambda$ is fixed then an increment of $E$ along any kinematically admissible path is equal to the deformation work minus the work of the given loads and coincides thus with the expression in (2.31). We thus have $I(\tilde{\mathbf{w}}) = \frac{1}{2}\ddot{E}(\tilde{\mathbf{w}})$ for $\tilde{\mathbf{w}} \in \mathcal{W}$, and the condition (2.33) can be equivalently rewritten as

$$\ddot{E}(\tilde{\mathbf{w}}) > 0 \quad \text{at } \lambda = \text{ const.} \quad \text{for every } \tilde{\mathbf{w}} \in \mathcal{W}, \ \tilde{\mathbf{w}} \neq \tilde{\mathbf{0}}. \tag{5.22}$$

Obviously, this equivalence remains valid for the configuration-dependent conservative loading provided $R(\tilde{\mathbf{w}})$ is added to the expression (2.34) for $I(\tilde{\mathbf{w}})$.

**Illustrative diagram**

The schematic diagram in Fig. 7 shows a summary of the results presented in the preceding Chapters which are now interpreted in terms of the second time derivative of the energy functional $E$ defined by (5.3). A discussion of the inequalities (6.1) and (6.3) as conditions sufficient for instability of a deformation process and of an equilibrium state, respectively, will be given in Chapter 6. Note the role played in the diagram by the assumed constitutive inequality (3.13) which along a regular deformation path is equivalent to (3.5) and implies equivalence between (5.21) and (6.1).

A modification of the diagram is needed if at the critical instant $\lambda = \lambda_T$ at which (3.11) holds by assumption, the minimum in (5.17) is still strict, i.e. the condition (5.20) assumed in the diagram is not fulfilled. This implies that bifurcation *at* $\lambda_T$ cannot be in velocities but rather can occur "smoothly", in a higher-order rate of displacements. As already mentioned, this is a rule when the incremental response of the material is "thoroughly" nonlinear but is possible also in the classical elastic-plastic solids. Another possible modification has been indicated in the discussion of Fig. 6; further possibilities will be considered in Chapter 6.

No matter what happens at $\lambda_T$, the path branching *between* $\lambda_T$ and $\lambda_R$ indicated in the diagram can occur through a finite difference in velocities. For discretized systems and for a certain class of continuous systems, branching can be identified with bifurcation, but

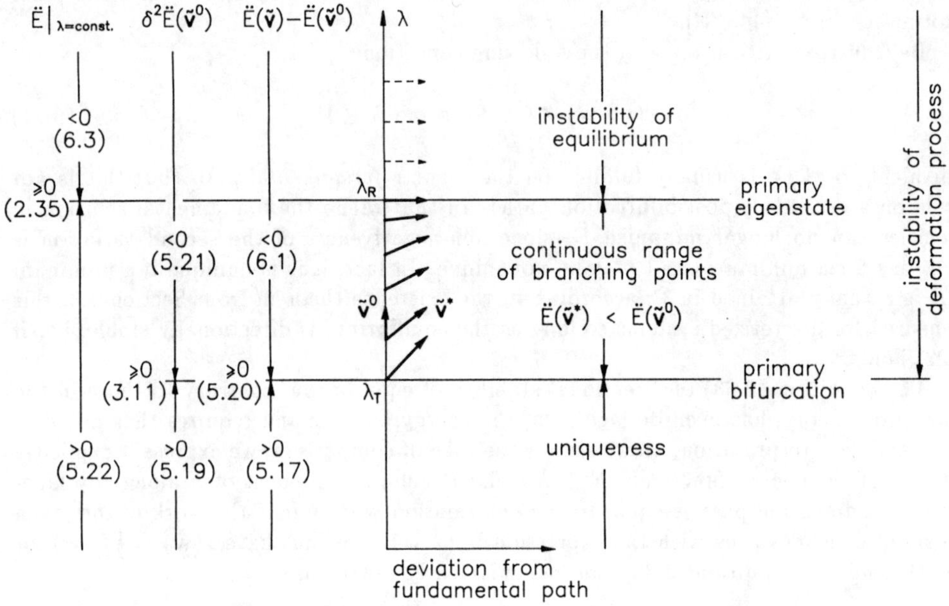


Fig. 7, Energy interpretation of the conditions for bifurcation and instability. It is assumed that the constitutive inequality (3.13) holds and that the deformation process is sufficiently regular (and spatially discretized when necessary).

as discussed below, for a general continuous system the branching may take place under application of arbitrarily small perturbing forces.

## 6. UNIFIED APPROACH TO BIFURCATION AND INSTABILITY PROBLEMS

### 6.1 **Various forms of instability**

In Hill's theory outlined in Chapter 2 the term "stability" concerns exclusively an equilibrium state at a fixed value of the loading parameter. This is sufficient for examining conditions under which dynamic "snap-through" phenomena may appear. On the other hand, in the engineering language it is common to speak about "instabilities" also when discussing such phenomena as buckling at increasing load or necking in bars or sheets at increasing elongation. The basic tool in analysis of such "geometric" instabilities is not the theory of stability of equilibrium but the bifurcation theory, complemented by studies of post-bifurcation behaviour and imperfection sensitivity (cf. [25] [45] [26] [46] [47] [48] [49] and the references quoted therein). Initiation of shear band formation (and also of localized necking in thin sheets) is studied by using a similar approach. However, in the latter case the theory was developed to some extent independently (see [50] and the references therein) since the bifurcation theory formulated for a finite body with well-defined boundary conditions did not apply without modifications to such "material" instabilities (especially in inhomogeneous bodies). Of still another kind are local instabilities at the body surface which manifest themselves in the form of surface undulations (cf. [51]).

The question arises whether all those observable instabilities of "dynamic", "geometric", "local" or "material" kind, apparently quite distinct from each other, can be investigated in a unified way by using a *single* criterion for predicting their appearance. We will show below, following [34] [36], that an affirmative answer can be given under the general assumptions already adopted in presentation of the bifurcation theory, by extending the energy criterion of instability of equilibrium to quasi-static deformation *processes*. The essence of the unified approach lies in finding a simple and general condition which could help to solve, at least partially, the following problem: what deformation processes *cannot* be practically realized in a physical system.

### 6.2 Instability of a quasi-static deformation process

Consider a deformation process which represents a theoretical quasi-static solution to an initial - boundary value problem formulated for an inelastic continuous body. Real material properties and boundary conditions are assumed to be adequately modeled. Now, the observable "instabilities" of various kind mentioned above clearly demonstrate that apparently quite correct solutions at sufficiently advanced deformations can be no longer in accord with experimental observations. This means that certain deformation processes, for instance, of macroscopically uniform deformations, are not practically realizable in a post-critical range. One of possible explanations is that this is related to an instability of the deformation *process* (or *path*, in an equivalent terminology) beyond a certain critical stage.

Traditionally, stability of a quasi-static deformation path is tacitly identified with stability of equilibrium states traversed by the path. However, for path-dependent and incrementally nonlinear solids distinction must be made between these two kinds of instability. Stability of equilibrium means that the effect of disturbing influences becomes negligible when the strength of a disturbance is vanishingly small while the value of the loading parameter is kept fixed. Now, if that final effect is very small but nonzero and if a quasi-static loading program is continued, there is no guarantee that the distance between the *subsequent* configurations along the perturbed and fundamental paths at the same loading levels will remain small; note that continuous dependence of this distance on initial data is necessarily broken e.g. at a bifurcation point. This observation alone indicates that stability of an equilibrium state (at constant loading) and of a deformation process (at varying loading) are essentially different concepts. The distinction is even more clear when the effect of application of infinitesimal perturbing forces is studied. Depending on whether such forces are applied at constant loading or at varying loading, different branches of the incrementally nonlinear constitutive law can be activated. In the former case the straining "direction" is defined by the disturbance itself while in the latter it is primarily determined by the fundamental velocity field; for elastic-plastic solids the "stiffness" in the latter case can be much smaller.

The problem arises how to define stability of a deformation process. A natural approach is to identify the process with a quasi-static motion and to apply concepts of the general theory of stability of motion. The sense of the term "stability" lies, roughly speaking, in resistance to small disturbing influences (called also *disturbances*) which should be carefully distinguished from geometric or material *imperfections* : the former perturb the motion of a given material system while the latter change the system itself. There is a vast literature on various theories of stability of motion, in particular, on Liapounov's theory where *initial* disturbances are considered. In reality, any material body is subject to perpetual disturbing influences *during* a deformation process; in the general theory of stability of motion such influences are called *persistent disturbances* (cf. e.g. [52]). They can be represented, for instance, by small perturbing forces acting in addition to the prescribed loads. The idea of stability for persistent disturbances is not less natural than that for initial disturbances; moreover, in the case of path-dependent solids (where material properties at a perturbed configuration of the body are dependent on the way how this configuration has been reached) it is more easily applicable.

We shall adopt the following concept of instability of a deformation process for persistent disturbances: If application of infinitesimal perturbing forces in any finite interval of time can cause finite deviations from a theoretical, unperturbed deformation process then such a process is regarded as being unstable and, consequently, practically unrealizable in a physical system.

This is still not a precise definition of instability since the measures of the distance and of perturbing forces have not been specified. While it is not difficult to formulate some mathematically rigorous definition of stability or instability, the problem is to ensure that it is physically sound and that it yields practically applicable instability criteria from which the onset of observable instabilities could be found. Therefore, we use another approach: we will first select a certain qualitative property of post-critical solutions as a *possible candidate* for the general condition sufficient for instability of a deformation process, and then examine its connection with the concept of instability for persistent disturbances mentioned above.

### 6.3 The energy criterion of instability of a process

The fundamental assumptions are the same as in discussing the energy interpretations in the bifurcation theory:

- the constitutive rate equations (1.10) admit the potential (1.16),
- the incremental loading (5.4) is conservative in the overall sense (5.5).

As we have seen in the previous chapter, along a typical regular deformation path the value of $\ddot{E}(\tilde{\mathbf{v}})$ in $\mathcal{V}$ is minimized by the fundamental solution $\tilde{\mathbf{v}}^0$ up to a certain critical stage and not beyond that stage. This motivates introduction of the following definition: *A quasi-static deformation process is said to be unstable in the energy sense at some stage of the deformation if the respective velocity solution* $\tilde{\mathbf{v}}^0$ *does not minimize the value of* $\ddot{E}$ *in* $\mathcal{V}$, *that is, if*

$$\ddot{E}(\tilde{\mathbf{v}}) < \ddot{E}(\tilde{\mathbf{v}}^0) \quad \text{for some } \tilde{\mathbf{v}} \in \mathcal{V}\,. \tag{6.1}$$

Equivalently, a necessary condition for stability of the fundamental process in the energy sense is that at every stage

$$\ddot{E}(\tilde{\mathbf{v}}) \geq \ddot{E}(\tilde{\mathbf{v}}^0) \quad \text{for every } \tilde{\mathbf{v}} \in \mathcal{V}\,. \tag{6.2}$$

This criterion was proposed [42][34] as an intuitive criterion of instability of a plastic deformation process, and its justification has been given more recently [36]. Since at an equilibrium state the value of $\dot{E}(\tilde{\mathbf{v}})$ is independent of $\tilde{\mathbf{v}}$, the condition (6.1) (or (6.2)) can be written down in an equivalent incremental form as $\Delta E < \Delta E^0$ (or $\Delta E \geq \Delta E^0$), where $\Delta E^0$ and $\Delta E \equiv (\Delta\theta)\dot{E} + \frac{1}{2}(\Delta\theta)^2\ddot{E}$ are the increments of $E$ evaluated *to second order* along the fundamental path and along any kinematically admissible branching path, respectively.

We shall discuss below the question whether fulfillment of (6.1) does correspond to instability of the deformation process against application of vanishingly small perturbing forces. Two cases are to be distinguished, depending on whether the equilibrium state is directionally stable or not.

**Instability of equilibrium**

In the special cases when $\lambda=$ const. or when $U$ is quadratic, (6.1) reduces to

$$\ddot{E}(\tilde{\mathbf{w}}) < 0 \quad \text{at } \lambda = \text{ const.} \quad \text{for every } \tilde{\mathbf{w}} \in \mathcal{W}, \tag{6.3}$$

which is the energy condition of instability of *equilibrium*, equivalent to (4.3). In general, we only have the implication [34]:

$$(6.3) \Rightarrow (6.1)\,. \tag{6.4}$$

This is in agreement with the observation that instability of equilibrium is a narrower concept than instability of a quasi-static deformation process.

To prove the implication (6.4), consider the quantity

$$\frac{1}{\gamma^2}(J(\tilde{\mathbf{v}}^0 + \gamma\tilde{\mathbf{w}}) - J(\tilde{\mathbf{v}}^0)) \tag{6.5}$$

and put $\gamma \to \infty$ while $\tilde{\mathbf{w}} \in \mathcal{W}$ is held fixed. In the case of controlled nominal loads, from (2.12) it is clear that only the volume integral of $U(\nabla\mathbf{v} + \gamma\nabla\mathbf{w})/\gamma^2$ can contribute to a non-zero limit value of (6.5). On account of homogeneity and continuity of $U(\dot{\mathbf{F}})$, the pointwise limit value of this integrand at each point of differentiability of the field $\tilde{\mathbf{w}}$ is equal to $U(\nabla\mathbf{w})$. The limit value of the expression (6.5) as $\gamma \to \infty$ is thus equal to $I(\tilde{\mathbf{w}})$. It follows that if (4.3) holds for some $\tilde{\mathbf{w}}$ then (4.2) is satisfied for $\tilde{\mathbf{v}} = \tilde{\mathbf{v}}^0 + \gamma\tilde{\mathbf{w}}$ for $\gamma$ sufficiently large. Since (4.3) and (4.2) are equivalent to (6.3) and (6.1), respectively, the implication (6.4) has been proved. Evidently, the implication remains valid for the configuration-sensitive incremental loading (5.4) since adding to $J(\tilde{\mathbf{v}})$ and $I(\tilde{\mathbf{v}})$ of the quadratic term $R(\tilde{\mathbf{v}})$ and $R(\tilde{\mathbf{w}})$, respectively, has no influence on the above argument.

We have shown in Section 4.2 that fulfillment of (4.3), and hence of (6.3), implies instability of the current equilibrium state in a dynamic sense for vanishingly small disturbances of *initial* velocities, at least in the first approximation and for a discretized system. The argument can easily be modified to extend its validity to the case of persistent disturbances represented by vanishingly small perturbing forces. Let the function $B(t)$ be defined not by (4.13) but by

$$B(t) = \begin{cases} (\varepsilon/\kappa)(\cosh(\kappa t) - 1) & \text{if } \; 0 \leq t \leq \tau, \\ (\varepsilon/\kappa)(\cosh(\kappa t) - \cosh(\kappa(t-\tau))) & \text{if } \; t \geq \tau, \end{cases} \tag{6.6}$$

for some $\tau > 0$. This corresponds to imposing a system of perturbing forces in an initial time interval $[0, \tau)$ at unperturbed initial velocities at $t = 0$. For $t < \tau$ we have $\ddot{B} = \kappa^2 B + \varepsilon\kappa$, and by an analogous argument as in Section 4.2 we obtain that the perturbing forces corresponding to the path (4.6) with $\tilde{\mathbf{w}} = \tilde{\mathbf{w}}^*$ are constant in time in the interval $[0, \tau)$ and can be made as small as we please if $\varepsilon$ is taken to be sufficiently small. For $t \geq \tau$ the function (6.6) corresponds to an inertial motion free of perturbing forces since (4.11) is satisfied. As before, a finite distance from equilibrium is reached at $t$ sufficiently large no matter how small $\varepsilon$ and $\tau$ are. Hence, under the same approximations as previously, we arrive at the following conclusion [36]: *If* (6.3) *holds then a finite distance from the equilibrium configuration can be reached at* $\lambda = const.$ *in a dynamic motion caused by arbitrarily small perturbing forces acting in an arbitrarily short time interval.*

The above demonstration of an instability of equilibrium associated with (6.3) remains valid for the continuum problem provided the absolute minimum of $I(\tilde{\mathbf{w}})$ on the hypersurface $\Phi(\tilde{\mathbf{w}}) = \text{const.}$ in $\mathcal{W}$ is actually attained. The latter requirement need not always be satisfied if $\mathcal{W}$ is infinite-dimensional (the functional $I(\tilde{\mathbf{w}})$ may be unbounded from below on that hypersurface). However, from a physical point of view it looks unlikely that stability of equilibrium in a dynamic sense can be lost just due to discretization of the system (of course, discretization could have a *stabilizing* effect due to elimination of certain instability modes). The approximation (4.7) seems also to have no essential destabilizing effect since the deviation from the equilibrium configuration, albeit finite, has been assumed small (the assumption of a direct path of departure from equilibrium might have again a *stabilizing* effect, e.g. due to elimination of cycling instability modes - cf. [53]).

**Instability of deformation process**

Consider now a fundamental process of quasi-static deformation at varying $\lambda$ and assume that the traversed equilibrium states are directionally stable, in the sense of (5.22). Denote the actual velocity solution by $\tilde{\mathbf{v}}^0$ and consider a range of the deformation where (6.1) is satisfied.

Let us discuss first a discretized system. At any equilibrium state the conditions (5.22) and (6.1) are equivalent to (2.33) and (4.2), respectively. Hence, from the conclusions of Section 4.3 it follows that at any instant in the fundamental process under consideration there is another first-order solution $\tilde{\mathbf{v}}^* \neq \tilde{\mathbf{v}}^0$ which minimizes $\ddot{E}(\tilde{\mathbf{v}})$ in $\mathcal{V}$. We may assume that for the same loading program there exists a bifurcating branch of quasi-static deformation at *varying* $\lambda$ initiated with the displacement-rate field $\tilde{\mathbf{v}}^*$ (for at least one minimizer if the minimum is not strict). The bifurcating branch deviates from the fundamental one on a finite distance after some increment of $\theta$, while the bifurcation is possible at *every* point of the fundamental path. This is nothing else as a special case of the instability of a process for persistent disturbances which are represented here by *zero* body forces and are thought merely as an agency which selects the actual solution in velocities. If the velocity solutions at a bifurcation point have comparable chances to determine the actual continuation of the deformation process then any segment of the fundamental deformation path, no matter how short, has zero probability to be followed due to existence of infinitely many alternatives. Moreover, the secondary continuations of the deformation process are energetically preferable, according to the minimum property of $\tilde{\mathbf{v}}^*$ and the interpretation of an increment of $E$. From a physical point of view, this must be interpreted as an instability of the fundamental deformation process. The instability has fundamentally different character than the instability of equilibrium discussed above: now the fundamental *path* can be left *quasi-statically* under *varying* $\lambda$, while previously a departure from an equilibrium *state* took place *dynamically* at *fixed* $\lambda$. However, in *both* cases the fundamental path may be treated as unrealizable in a physical system.

The above demonstration of instability of the deformation process, associated with (6.1), remains valid for the continuum problem provided the absolute minimum of $J(\tilde{\mathbf{v}})$ in $\mathcal{V}$ is actually attained. This need not be the case when the functional space $\mathcal{W}$ is infinite-dimensional (even if $\mathcal{W}$ were enlarged to be a complete space). The demonstration of instability then breaks down since existence of a continuous range of bifurcations cannot be definitely concluded. Following [36] it will be indicated below that in a general case the instability can rather be interpreted as *sensitivity of the incremental solution to vanishingly small perturbing forces*, in accord with the concept of instability for persistent disturbances.

Measures of the distance between two configurations of a continuous body are not equivalent to each other. Taking into account that rigid body displacements at fixed $\lambda$ have been excluded, we will use the measure

$$\| \tilde{\mathbf{w}} \| \equiv \left( \int_V w_{j,i} w_{j,i} \, d\xi \right)^{1/2} \tag{6.7}$$

which is a norm on the linear space $\mathcal{W}$.

For the discretized problem the condition (2.33) ensured that the distance from equilibrium attainable on a straight (direct) path vanished with a vanishing input of the energy supplied by a disturbance. The same condition in an infinite-dimensional space does not ensure this rigorously since $I(\tilde{\mathbf{w}})$ can approach zero in a non-trivial way without achieving it. It is thus reasonable to assume the condition of directional stability of equilibrium in a slightly stronger form than (5.22), viz.

$$\ddot{E}(\tilde{\mathbf{w}}) \geq \alpha \|\tilde{\mathbf{w}}\|^2 \quad \text{at } \lambda = \text{ const.} \qquad \text{for every } \tilde{\mathbf{w}} \in \mathcal{W}, \tag{6.8}$$

where $\alpha$ is a positive parameter, independent of place, which may be arbitrarily small. For a discretized problem, (6.8) is equivalent to (5.22), and for the continuum problem, the equivalence can be expected to be only violated at certain critical instants.

If (6.8) holds then it can be proved that

$$\ddot{E}(\tilde{\mathbf{v}}) \to +\infty \qquad \text{when} \qquad \|\tilde{\mathbf{v}}\| \to \infty, \ \tilde{\mathbf{v}} \in \mathcal{V} \tag{6.9}$$

and that $\ddot{E}(\tilde{\mathbf{v}})$ is bounded from below in $\mathcal{V}$. Since an absolute minimum of $\ddot{E}(\tilde{\mathbf{v}})$ in $\mathcal{V}$ need not in general be attained, we modify the argument from Section 4.3 and examine a minimizing sequence $\{\tilde{\mathbf{v}}_n\} \in \mathcal{V}$ , $n = 1, 2, \ldots$, rather than a minimizer itself. For instance, the minimizing sequence can be determined by the Ritz method and the velocity fields $\tilde{\mathbf{v}}_n$ can be solutions to a sequence of discretized rate-problems.

It can be shown (a detailed proof given in [36] is not brief and is omitted here) that the minimizing sequence has the following property: the velocity fields $\tilde{\mathbf{v}}_n$ are solutions to the perturbed first-order rate boundary value problems, such that for any fixed material subdomain $G \subseteq V$ we have

$$|\dot{\mathbf{P}}^*(G, \tilde{\mathbf{v}}_n)| \leq \varepsilon_n \to 0 \quad \text{as } n \to \infty, \tag{6.10}$$

where $\dot{\mathbf{P}}^*(G, \tilde{\mathbf{v}}_n)$ is the rate of the resultant perturbing force acting on $G$. If (6.8) holds simultaneously with (6.1) then the distance between velocity fields $\tilde{\mathbf{v}}_n$ from the minimizing sequence and the fundamental solution $\tilde{\mathbf{v}}^0$ , when measured in the sense of (6.7), remains finite as the infimum of $J(\tilde{\mathbf{v}})$ is approached at $n \to \infty$, viz.

$$C_2 > \|\tilde{\mathbf{v}}_n - \tilde{\mathbf{v}}^0\| > C_1 \quad \text{for } n > K \tag{6.11}$$

for some positive constants $C_1$ , $C_2$ and $K$ . Hence, we arrive at the following instability theorem [36]:

*If* (6.8) *holds and the fundamental process is unstable in the energy sense* (6.1) *then finite deviations from the fundamental velocity field, measured by* (6.7), *can be caused by vanishingly small perturbations of the equations of continuing equilibrium, in the sense of* (6.10). [8)]

Let (6.8) and (6.1) hold along the fundamental path of quasi-static deformation at varying $\lambda$ in a certain interval of $\theta$. As indicated above, at each equilibrium state of the body within that interval there exists a sequence of perturbed velocity fields $\tilde{\mathbf{v}}_n$ satisfying (6.11) while the rates of perturbing forces tend to zero in the sense of (6.10). Consider a branching path initiated with a displacement rate field $\tilde{\mathbf{v}}_n$ ; we assume that if the *rates* of perturbing forces are kept fixed then variations of the velocity field along any such path in a small time interval can be neglected. According to (6.11), the measure (6.7) of the distance between body configurations in the fundamental and perturbed processes grows in time with a *finite* rate and achieves a given small but finite value after a small increment of $\theta$. This distance can be reached at perturbing forces as small as we please if $n$ is taken sufficiently large, with the initial instant of their application being arbitrarily chosen from the considered time interval.

This is the final interpretation of the instability of a quasi-static deformation process implied by (6.1) when each of the traversed equilibrium states is (directionally) stable. This corresponds to the instability of a process for arbitrarily small *persistent* disturbances when the distance is defined by (6.7) and the perturbing forces are measured in the sense of (6.10). However, it must be noted that the transition from the above theorem formulated in terms of the *rates* to the final statement expressed in *increments,* although it may seem intuitively obvious, is not mathematically rigorous in view of the additional assumption made. In turn, from a physical point of view the conclusion about instability is strengthened by the fact that the initiation of a branching path with a velocity field $\tilde{\mathbf{v}}_n$ for $n$ sufficiently large is energetically preferable to further continuation of the fundamental process. This follows from the interpretation of $\ddot{E}(\tilde{\mathbf{v}})$ and from the inequality $\ddot{E}(\tilde{\mathbf{v}}_n) < \ddot{E}(\mathbf{v}^0)$ which is implied by (6.1) for $n$ sufficiently large.

**Example: The Shanley column**

The meaning and range of application of the instability condition (6.1) have already been illustrated on a schematic diagram in Fig. 7 for a general system. We will examine now the classical Shanley model of an elastic-plastic column with a central two-flange

---

8) In an equivalent formulation, if (6.8) and (6.1) hold simultaneously then there is $\varepsilon > 0$ such that for every $\delta > 0$, however small, there is $\tilde{\mathbf{v}}_\delta \in \mathcal{V}$ satisfying $|\dot{\mathbf{P}}^*(G, \tilde{\mathbf{v}}_\delta)| < \delta$ for every fixed $G \subseteq V$ and such that $\|\tilde{\mathbf{v}}_\delta - \tilde{\mathbf{v}}^0\| \geq \varepsilon$.

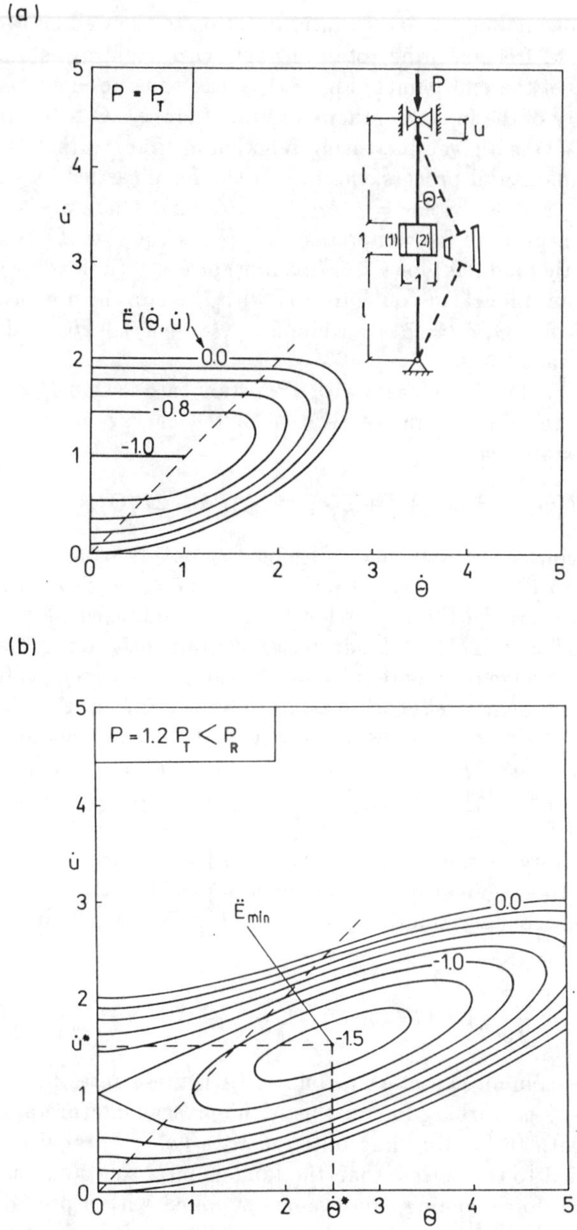


Fig. 8, Representation of the functional $\ddot{E}(\tilde{\mathbf{v}})$ for the Shanley column (a) at the tangent modulus load, (b) between the tangent and reduced modulus loads. Numerical values correspond to $\dot{P}/2L^p = 1$, $\ddot{P} = 0$, $L^p/L^e = 0.5$, $L^p = 0.5$. Contours of $\ddot{E} > 0$ are not shown.

hinge [23] and give, following [36], an energy interpretation to the well known results. The model has two degrees of freedom: the rotation angle $\Theta$ of rigid arms and the relative vertical displacement $u$ of the end points (Fig. 8a); $\tilde{\mathbf{v}}$ can thus be identified with $(\dot{\Theta}, \dot{u})$. We will examine stability of the fundamental deformation process $\Theta \equiv 0$ when the vertical compressive load $P \equiv \lambda(t)$ is a given increasing function of time, so that $\Omega = -Pu$. At a certain stage of the fundamental process, the rate of the force carried by an elastoplastic flange $(K)$, $K = 1$ or 2, is given by $\dot{P}_K = L(\dot{e}_K)\dot{e}_K$ (no sum), where $\dot{e}_K$ is the shortening rate of the flange (taken positive for compression). $L(\dot{e})$ is equal to $L^p$ (current tangent modulus) or to $L^e$ (elastic modulus) for $\dot{e} > 0$ (loading) or $\dot{e} < 0$ (unloading), respectively, with $L^e > L^p > 0$. At an unbuckled configuration with the current dimensions as shown in Fig. 8a we have $\dot{e}_1 = \dot{u} + \dot{\Theta}$, $\dot{e}_2 = \dot{u} - \dot{\Theta}$ while $\ddot{u} = \frac{1}{2}(\ddot{e}_1 + \ddot{e}_2) + 2l\dot{\Theta}^2$ (all variables are treated as nondimensional).

Assume for simplicity that $P$ increases with a constant rate so that $\ddot{P} = 0$. Elementary calculations show that the second time derivative of the energy functional (5.3) at an unbuckled equilibrium state reads

$$\ddot{E}(\dot{\Theta}, \dot{u}) = L(\dot{e}_1)\dot{e}_1^2 + L(\dot{e}_2)\dot{e}_2^2 - 2\dot{P}\dot{u} - 2Pl\dot{\Theta}^2. \tag{6.12}$$

The graph of $\ddot{E}(\dot{\Theta}, \dot{u})$ consists of four quadrics joined smoothly along the lines $\dot{\Theta} = \pm\dot{u}$. The uniqueness range (5.19) corresponds to $P < P_T$, where $P_T = L^p/l$ is the tangent modulus load, and in this range the expression (6.12) is minimized by the fundamental continuation $\dot{\Theta}^0 = 0$, $\dot{u}^0 = \dot{P}/2L^p$. If $L^p$ decreases continuously along the fundamental path then the uniqueness range terminates at $P = P_T$; this is the primary bifurcation point at which an absolute minimum of $\ddot{E}(\dot{\Theta}, \dot{u})$ is attained at any $\tilde{\mathbf{v}}$ from the "fan" of velocity solutions: $\dot{u} = \dot{u}^0$, $|\dot{\Theta}| \leq \dot{u}^0$ (cf. Fig. 8a). However, if $L^p$ and thus also $P_T$ decrease discontinuously then the range $P > P_T$ need not be preceded by $P = P_T$. Within the range $P_T < P < P_R$, where $P_R = 2L^eL^p/l(L^e + L^p)$ is the reduced modulus load, the fundamental solution $(\dot{\Theta}^0, \dot{u}^0)$ becomes a saddle point of $\ddot{E}$ so that (6.1) is met while the condition (5.22) of stability of equilibrium is still satisfied. In accord with the theorem from Section 4.3, along this segment of the fundamental path the absolute minimum value of $\ddot{E}$ is attained at the secondary solution points (cf. Fig. 8b) which are easily found to be

$$\dot{\Theta}^* = \pm\frac{\dot{P}}{2L^p}z, \qquad \dot{u}^* = \frac{\dot{P}}{2L^p}(1 + (P/P_T - 1)z), \qquad \text{where} \quad z \equiv \frac{1 - L^p/L^e}{2(1 - P/P_R)}. \tag{6.13}$$

As $P$ is increasing, the column can start to buckle with these rates at *any* point along the segment, without any perturbing forces: this is the present interpretation of the path instability associated with (6.1). Buckling initiated with $(\dot{u}^*, \dot{\Theta}^*)$ requires incrementally less energy to be supplied to the *system* than the fundamental solution, i.e. $\Delta E^* < \Delta E^0$. If $P(t)$ is a gravitational force coming from a mass supplied with a prescribed rate then $\Delta E^* - \Delta E^0$ can be identified with the decrease of the work needed to transport the mass; note that $\dot{u}^* > \dot{u}^0$.

For $P > P_R$ the condition (6.1) is satisfied simultaneously with (6.3), and $\ddot{E}$ is unbounded from below. In this range, instability of the fundamental process is interpreted as instability of the traversed equilibrium states at fixed $P$ in the dynamic sense discussed above.

## 6.4 Applications of the energy criterion

### 6.4.1 Unified treatment of various critical points

Once the general criterion (6.1) for instability of a deformation process has been formulated and justified, it can be applied in a consistent manner to a variety of instability problems mentioned in Section 6.1. The range of unstable deformations where (6.1) holds can be determined and rejected as apparently unrealizable in a physical system. The initial point of that range represents the critical point on the deformation path, henceforth referred to as E , beyond which some symptoms of instability can be expected. The typical critical points from Hill's theory (i.e. the primary bifurcation point and the primary eigenstate) as well as the critical point associated with shear band bifurcation can be generated in that way from the single condition (6.1) just as *particular cases* of the onset of path instability in the energy sense. An additional advantage of the present energy approach, besides of a unified treatment of various kinds of instability, is that it provides also information about the post-critical behaviour of the system, namely, indicates which post-critical paths cannot be followed in practice and why. Usually, a numerical analysis of development of initial imperfections is adopted to obtain that kind of additional information, and in simplest cases the predictions are the same (cf. e.g. [26][48]). However, such agreement cannot always be expected since numerical results may be strongly influenced by the form and magnitude of the assumed imperfections, and certain instabilities may remain undetected in the course of numerical computations. Of course, the initial imperfection approach has other advantages as e.g. assessment of the influence of geometrical inaccuracies on the buckling loads.

Suppose that the fundamental deformation process is regular in the sense defined in Section 3.3, and that the micromechanically based constitutive inequality (3.13) holds. This means that the actual actual velocity solution $\tilde{\mathbf{v}}^0$ corresponds to the rates $\dot{F}^0_{ij}$ , $\dot{N}^0_{ij}$ of deformation and stress which satisfy the inequality (3.5). (It is recalled that (3.5) is a weaker restriction than the relative convexity property (2.17) with $C^L_{ijkl} = C_{ijkl}(\mathbf{F}^0)$). Then (6.1) is *equivalent* to (4.1) (cf. Fig. 7), as follows from (5.14), (3.9), (3.8) and (3.7). For a discretized problem, this can be reformulated as follows: the fundamental process becomes unstable in the energy sense (6.1) exactly when the tangent stiffness matrix (4.15) associated with $\tilde{\mathbf{v}}^0$ becomes indefinite. A number of possibilities now arises depending on the properties of the functionals $I^0(\tilde{\mathbf{w}}), J(\tilde{\mathbf{v}})$ and $I(\tilde{\mathbf{w}})$, where $\tilde{\mathbf{w}} \in \mathcal{W}$ and $\tilde{\mathbf{v}} \in \mathcal{V}$, at the critical instant E. Equivalently, one can discuss properties of $\delta^2\ddot{E}(\tilde{\mathbf{v}}^0,\tilde{\mathbf{w}}), \ddot{E}(\tilde{\mathbf{v}})$ and $\ddot{E}(\tilde{\mathbf{w}})$; in the discussion below we use the triple $I^0(\tilde{\mathbf{w}})$, $\ddot{E}(\tilde{\mathbf{v}})$ and $I(\tilde{\mathbf{w}})$. The list of possibilities considered below is not intended to be complete.

#### (i) E as the primary bifurcation point

Suppose that at E the condition (5.20) is satisfied simultaneously with (2.33). Then, as already explained in the preceding section , E coincides with the "usual" primary bifurcation point (in velocities) found from the linearized criterion (3.11). The energy criterion provides additional information that the fundamental post-bifurcation path becomes unstable in the sense discussed in the preceding Section and may thus be rejected as a physically unrealistic solution. Stability of the secondary post-bifurcation branches can also be investigated by using (6.1) provided that $\tilde{\mathbf{v}}^0$ changes its previous meaning and denotes the velocity solution along the actually examined branch. If a stable secondary branch exists then it can replace the fundamental one beyond $\mathsf{E}$, which in most typical circumstances can correspond to an observable "geometric" instability.

A smooth bifurcation is found at $\mathsf{E}$ when the condition (3.11) is satisfied but (5.17) still holds. Then bifurcation in velocities (which would require fulfillment of (5.20)) is impossible but the eigenmode $\tilde{\mathbf{w}}^*$ can be added to the fundamental solution in quasi-static *accelerations* to generate another second-order solution, as discussed in Section 3.2.

$\mathsf{E}$ can correspond also to a primary bifurcation point at which (4.1) is met and which is *not* an eigenstate for the tangent comparison material. This can happen if at $\mathsf{E}$ the tangent moduli $C^0_{ijkl}$ suffer a jump related to an instantaneous drop of the incremental "stiffness" of the material. Although $C^0_{ijkl}$ are discontinuous at $\mathsf{E}$ with respect to $\theta$, they can be still well defined at $\mathsf{E}$ since they relate the *forward* rates of strain and stress. Suppose that at such $\mathsf{E}$ there exists a global minimizer $\tilde{\mathbf{v}}^*$ of $\ddot{E}(\tilde{\mathbf{v}})$ in $\mathcal{V}$, that is,

$$\ddot{E}(\tilde{\mathbf{v}}^0) > \ddot{E}(\tilde{\mathbf{v}}^*) \le \ddot{E}(\tilde{\mathbf{v}}) \quad \text{for every } \tilde{\mathbf{v}} \in \mathcal{V}\,. \tag{6.14}$$

By the stationarity principle (5.16), $\tilde{\mathbf{v}}^*$ is a solution so that at $\mathsf{E}$ we have bifurcation in velocities. Existence of that bifurcation cannot be stated by consideration of the incrementally linear comparison solid only. For discretized problems, fulfillment of (6.14) is ensured when (4.1) is satisfied simultaneously with (2.33), as follows from (5.14) and the theorem from Section 3.3.

**Example: Bifurcation under plane strain tension at a non-singular tangent stiffness matrix**

The example is taken from the paper [41]. A rectangular specimen is subject to

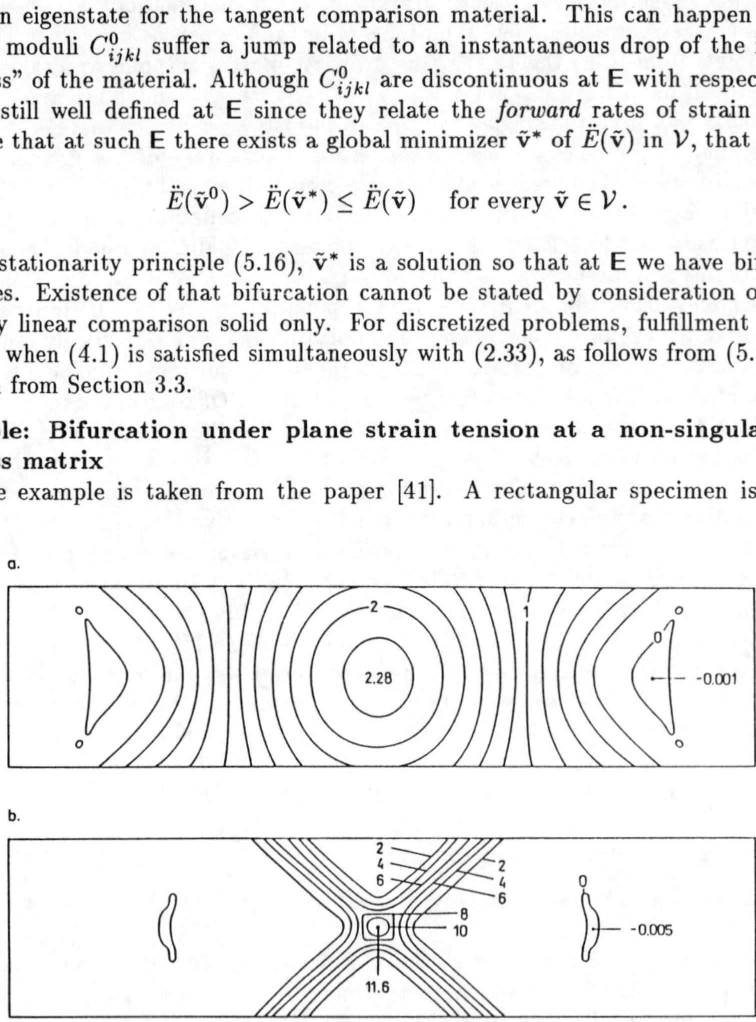


Fig. 9, Distribution of maximum principal strain rate in bifurcation solutions at plane strain tension of a rectangular specimen for (a) $L^p = 0.57\tau$, (b) $L^p = 0.14\tau$, where $\tau$ and $L^p$ are the current equivalent Kirchhoff stress and hardening modulus, respectively.

plane strain tension; the material model is from the class proposed by Christoffersen and Hutchinson [33] which describes an incrementally nonlinear response at a yield-surface vertex. The hardening rate is assumed to be initially sufficiently high to ensure uniqueness of a homogeneous solution. Bifurcation is induced by a subsequent discontinuous drop of the current equivalent hardening modulus $L^p$ to a value such that the current loading parameter $\lambda$, identified e.g. with elongation of the specimen, falls inside the range between $\lambda_T$ and $\lambda_R$ (cf. Fig. 7). The bifurcation solutions visualized in Fig. 9 have been found by minimizing the velocity functional (2.12), which is equivalent to minimizing $\ddot{E}(\tilde{\mathbf{v}})$, with the help of the finite element method. The velocity field corresponding to Fig. 9a describes a diffuse necking mode, while in the case (b) of a more pronounced drop of the material stiffness, a shear band mode is clearly visible. The actual tangent stiffness matrix is indefinite for the fundamental homogeneous solution and positive definite for the secondary solutions. Of course, due to inherent nonlinearity of the problem, the magnitude of the bifurcation mode cannot be freely scaled down, contrary to the linearized bifurcation problem discussed in Section 2.4.

**(ii) E as the primary eigenstate**

Suppose that at E the condition (6.2) is satisfied but (2.33) fails. Since (6.2) excludes (4.1), we must have (2.35) so that the primary eigenstate is met. Typically, the

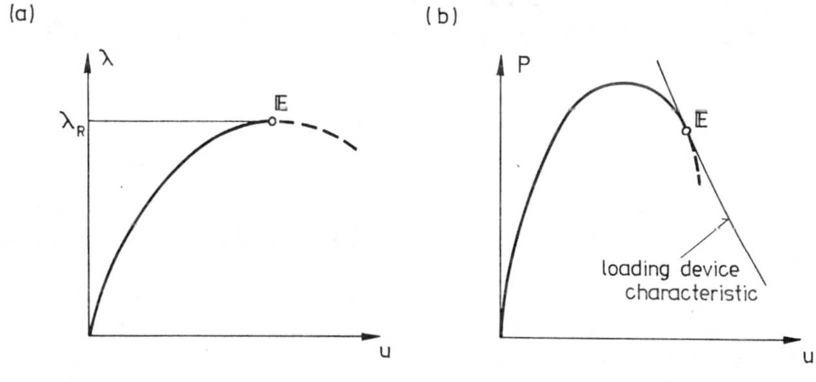


Fig. 10, (a) Generalized limit point and (b) its interpretation for an elastic loading device.

fundamental post-critical path becomes then unstable in the dynamic sense associated with fulfillment of (6.3). Depending on the problem, there may or may not be another post-critical branch along which (6.2) would still be satisfied. For instance, if the point E is met as a limit point then its appearance can be related to an observable "dynamic" instability (snap-through). The limit point can be understood in a generalized sense of an analytic extremum of a controlled loading parameter $\lambda$, and not necessarily of a load multiplier (Fig. 10). For instance, for a tensile specimen tested in a loading device of finite elastic stiffness, the limit point is met when the load-displacement curve becomes tangent to the (possibly nonlinear) characteristic of the loading device.

(iii) **E as the onset of material instability**

"Local" or "material" instabilities can be investigated following the same general approach as above, without the need to formulate any auxiliary problem (as is usually done in the "local" bifurcation approach by discussing a rate-problem for an infinite homogeneous continuum [50]). For, it suffices to consider local conditions whose fulfillment implies (6.1). We do not discuss that problem in much detail here, and restrict ourselves to certain introductory remarks. From the identity (5.14) and the well-known Legendre-Hadamard necessary condition for a minimum of an integral functional, it follows that a *necessary* condition for (6.2) is that at every regular point in the body (i.e. at which the velocity gradient $v^0_{j,i}$ and the respective tangent moduli $C^0_{ijkl}$ are well defined and continuous with respect to $\xi_i$) the tangent moduli are at least *semi-strongly elliptic,* viz.

$$C^0_{ijkl} n_i n_k g_j g_l \geq 0 \quad \text{for all } n_i, g_j \,. \tag{6.15}$$

Therefore, if the tangent moduli in a deformation process satisfy somewhere in the body the inequality

$$C^0_{ijkl} n_i n_k g_j g_l < 0 \quad \text{for some } n_i, g_j \tag{6.16}$$

then the process is unstable in the energy sense (6.1). It can be seen that the condition of strong ellipticity loss results in a natural way from the energy criterion of instability of a deformation *process.* (Clearly, the condition analogous to (6.16) can be applied to two-dimensional continua intended, for instance, to model a thin metal sheet.) The associated critical condition is (cf. (3.11))

$$\det(C^0_{ijkl}\, n_i n_k) = 0 \tag{6.17}$$

which is commonly interpreted as a critical condition for shear band bifurcation [50]. Observe that here (6.15) results in a natural way as a particular condition for the critical instant E.

In turn, by appealing to a theorem due to Graves [54] we obtain that a necessary condition for the functional $I(\tilde{\mathbf{w}})$ to reach an absolute minimum at $\tilde{\mathbf{w}} = \tilde{\mathbf{0}}$, or equivalently, to take only nonnegative values, is that

$$U(\dot{\mathbf{F}}) \geq 0 \quad \text{for all } \dot{F}_{ji} = g_j n_i \tag{6.18}$$

at every material point. It follows that if somewhere in the body we have

$$U(\dot{\mathbf{F}}) < 0 \quad \text{for some } \dot{F}_{ji} = g_j n_i \tag{6.19}$$

then the equilibrium state is unstable in the energy sense (6.3). Since we have assumed that the constitutive inequality (3.5) holds, (6.19) cannot take place without (6.16), on account of (3.7) which is a consequence of (3.5). The critical condition associated with (6.18) is (cf. (2.36))

$$\dot{N}_{ij}(\dot{\mathbf{F}})\, n_i = 0 \quad \text{for some } \dot{F}_{ji} = g_j n_i \tag{6.20}$$

which corresponds to the possibility of shear band bifurcation when the material outside the band is non-deforming [55].

The conditions (6.16) and (6.19) may be regarded as conditions sufficient for local (or material) instability [34]; it is pointed out that these conditions have been derived for a possibly inhomogeneous, finite body under arbitrary boundary conditions. The distinction between the two conditions is that (6.16) implies an instability of the *process* of uniform straining of a material element while (6.19) implies an instability of *equilibrium* of the element. Transition from the range (6.15) (with strict inequality for nonzero $n_i$ , $g_j$ ) to (6.16) without violating (6.18) is related to initiation of quasi-static localization of deformation into shear bands. In turn, in the range (6.19) the shear band formation may take place as a *dynamic* process representing a kind of "internal snap-through".

**Remark**

If (3.5) does not hold then bifurcations can take place before (6.1) is met. This indicates that while the energy approach and the bifurcation approach are closely connected to each other, they are not equivalent. It can be mentioned that if (3.5) does not hold then (6.15) and (6.18) are merely consequences of a more general condition necessary for material stability [34] which is not discussed here.

### 6.4.2. Upper bound technique for determining the onset of path instability

The condition (6.1) provides a basis for a simple upper-bound technique for determining the critical point beyond which the deformation path corresponding to the velocity field $\tilde{\mathbf{v}}^0$ is expected to loose its practical significance. Namely, it suffices to find just one kinematically admissible velocity field satisfying (6.1) to demonstrate instability of the path. This task is further simplified if the implication (4.1) $\Rightarrow$(4.2) $\Leftrightarrow$ (6.1) is taken into account. It follows that we can consider a certain *restricted* class of possible instability modes of required type from $\mathcal{W}$ and then by direct substitution determine the range where (4.1) is satisfied for some field from that class, implying (6.1). To do this, we do not need to solve any equations; of course, the better the choice of the examined class of modes the better is the estimation of the critical point. In particular, *all* admissible modes can be taken into account; then the "most dangerous" instability modes are found as eigenfields for the linearized homogeneous problem for the comparison material. The critical point is then found when the lowest eigenvalue reaches zero, which is nothing else than the usual procedure for determining the primary bifurcation point from (2.25) or an equivalent condition. The upper bound technique with a restricted class of modes is much simpler while the predictions for critical stresses or strains may be quite close to exact bifurcation values. It can be mentioned that the technique works also for stress-strain curves with a dicontinuous tangent, while the bifurcation approach based on the incrementally linear comparison material in general breaks then down.

**Example: Upper bound to the onset of necking in 3D-specimens subject to uniaxial tension**

Consider a specimen whose cross-section orthogonal to the tensile $x_1$-axis in the current configuration is independent of $x_1$ but is otherwise arbitrary (Fig. 11a). The usual assumptions of shear-free end conditions and the prescribed elongation rate are adopted. The material is assumed to be homogeneous and orthotropic relative to $x_i$-axes, but detailed

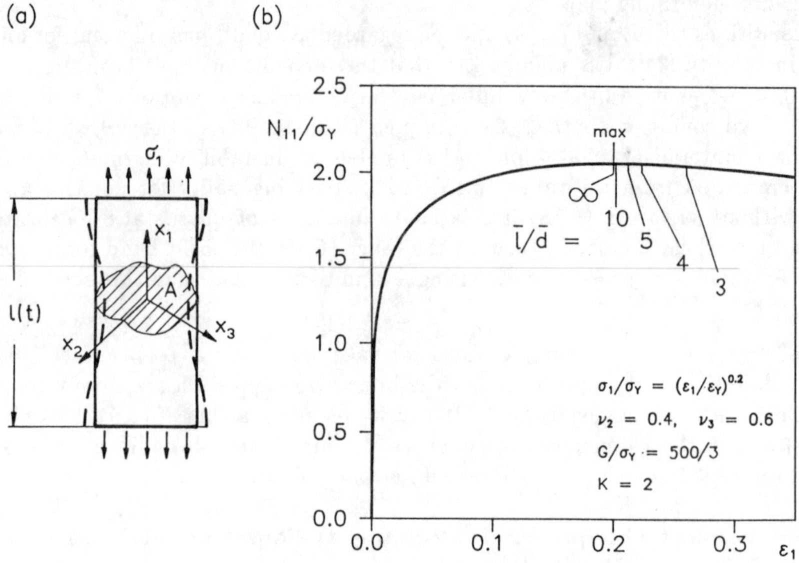


Fig. 11, Upper bound to the onset of necking in uniaxial tension; (a) the specimen, (b) results for an orthotropic, initially cylindrical specimen of different initial length/diameter ($\bar{l}/\bar{d}$) ratios on a plot of normalized nominal stress vs logarithmic strain.

specification of its incremental constitutive law (possibly highly nonlinear) is unknown. Only the uniaxial stress-strain curve and the cross-section evolution in the *fundamental, uniform* deformation process are treated as known, so that the current uniaxial tangent modulus $L^p = \dot{\tau}_1^0/\dot{\varepsilon}_1^0$ and the ratios $\nu_2 = -\dot{\varepsilon}_2^0/\dot{\varepsilon}_1^0$, $\nu_3 = -\dot{\varepsilon}_3^0/\dot{\varepsilon}_1^0$ are given.

In the upper bound technique, a kinematically admissible mode of instability can be assumed arbitrarily. Let the velocity field $\tilde{\mathbf{v}}$, to be substituted in (6.1) as the necking mode, be defined as $\tilde{\mathbf{v}} = \tilde{\mathbf{v}}^0 + \tilde{\mathbf{w}}$, where

$$\begin{aligned} w_1 &= \phi_1\phi_2'\phi_3', \\ w_2 &= -\nu_2\phi_1'\phi_2\phi_3', \qquad\qquad \phi_r = \phi_r(x_r), \quad \phi_r' = d\phi_r/dx_r, \\ w_3 &= -\nu_3\phi_1'\phi_2'\phi_3. \end{aligned} \tag{6.21}$$

As explained above, the task of determining the range where (6.1) holds can be simplified by substituting the field $\tilde{\mathbf{w}}$ into (4.1) (with the current configuration as reference). The mode (6.21) has been chosen such that

$$w_{3,3} : w_{2,2} : w_{1,1} = \dot{\varepsilon}_3^0 : \dot{\varepsilon}_2^0 : \dot{\varepsilon}_1^0, \tag{6.22}$$

which eliminates from the integrand in (4.1) all tangent moduli of the type $L^0_{rrss}$ by replacing them by the single uniaxial modulus $L^p$. A further natural assumption

$$\phi_1 = \sin\frac{K\pi x_1}{l}, \qquad \phi_2 = \sin\frac{K\pi x_2\sqrt{\nu_2}}{l}, \qquad \phi_3 = \sin\frac{K\pi x_3\sqrt{\nu_3}}{l}, \qquad K = 1,2,\ldots, \tag{6.23}$$

compatible with the kinematical constraints $w_1 = 0$ at $x_1 = 0$ and $x_1 = l$, yields $\dot\varepsilon_{12} = \dot\varepsilon_{13} = 0$. The only remaining unknown modulus $L^0_{2323}$ is assumed to be not greater than the elastic shear modulus $G$.

On using the above specifications and (1.15), after straightforward transformation we obtain that (4.1) is satisfied if

$$L^p < \sigma_1 - \frac{\nu_2\Phi_2 + \nu_3\Phi_3}{\Psi}\sigma_1 - \frac{4\nu_2\nu_3\Xi}{\Psi}G, \tag{6.24}$$

where

$$\begin{aligned}
\Phi_2 &= \int\!\!\int_A \sin^2\frac{K\pi x_2\sqrt{\nu_2}}{l}\cos^2\frac{K\pi x_3\sqrt{\nu_3}}{l}\,dx_2\,dx_3,\\
\Phi_3 &= \int\!\!\int_A \sin^2\frac{K\pi x_3\sqrt{\nu_3}}{l}\cos^2\frac{K\pi x_2\sqrt{\nu_2}}{l}\,dx_2\,dx_3,\\
\Psi &= \int\!\!\int_A \cos^2\frac{K\pi x_2\sqrt{\nu_2}}{l}\cos^2\frac{K\pi x_3\sqrt{\nu_3}}{l}\,dx_2\,dx_3,\\
\Xi &= \int\!\!\int_A \sin^2\frac{K\pi x_2\sqrt{\nu_2}}{l}\sin^2\frac{K\pi x_3\sqrt{\nu_3}}{l}\,dx_2\,dx_3.
\end{aligned} \tag{6.25}$$

In the range (6.24) the uniform deformation *process* is thus necessarily unstable, in the sense discussed in Section 6.3; in view of the assumed form of the instability mode, necking is the expected symptom of instability. Note that equilibrium can still be stable if the loading device is sufficiently stiff.

In Fig. 11b the critical points beyond which (6.24) holds are illustrated for a specimen of an initially circular but currently elliptic cross-section (on account of assumed orthotropy) and for a power hardening uniaxial stress-strain curve. It can be seen that instability is predicted somewhat beyond the maximum load point at which $L^p = \sigma_1$. The last two correction terms in (6.24) tend to zero for infinitely slender specimens, in agreement with the classical Considère criterion. In comparison with the similar results obtained earlier by using the bifurcation approach (cf. [47] for a review), the present conclusion has been obtained for a wider class of materials and for arbitrary cross-sections.

### 6.4.3 A computational algorithm for automatic choice of the post-bifurcation branch

On the basis of (6.2) as a necessary condition of stability of a deformation process, a computational method has been proposed in [41] for crossing bifurcation points with automatic branch switching. Post-bifurcation branches which are unstable according to (6.1), the fundamental branch in particular, are automatically rejected in the algorithm. The algorithm has been implemented into a finite element code developed by K.Thermann; the results in Fig. 9 illustrate applicability of the method also in the case when the stress-strain curve is not smooth. Further details can be found in [41].

REFERENCES


1. Hill, R: On the problem of uniqueness in the theory of a rigid-plastic solid, *J. Mech. Phys. Solids,* **5** (1957), 153-161, 302-307.
2. Hill, R: On uniqueness and stability in the theory of finite elastic strain, *J. Mech. Phys. Solids,* **5** (1957) 229-241.
3. Hill, R: Stability of rigid-plastic solids, *J. Mech. Phys. Solids,* **6** (1957), 1-8.
4. Hill, R: A general theory of uniqueness and stability in elastic-plastic solids, *J. Mech. Phys. Solids,* **6** (1958), 236-249.
5. Hill, R.: Some basic principles in the mechanics of solids without a natural time. *J. Mech. Phys. Solids,* **7** (1959), 209-225.
6. Hill, R.: Aspects of invariance in solids mechanics, *Advances in Applied Mechanics, Vol.* **18**, Acad. Press, New York 1978, 1-75.
7. Hill, R: On the classical constitutive laws for elastic/plastic solids, *Recent Progress in Applied Mechanics, The Folke Odkvist Volume* (Ed B. Broberg *at. al.*), Almqvist & Wiksell, Stockholm 1967, 241-249.
8. Hill, R. and Rice, J.R.: Constitutive analysis of elastic-plastic crystals at arbitrary strain, *J. Mech. Phys. Solids,* **20** (1972), 401-413.
9. Sewell, M.J.: A survey of plastic buckling, in: *Stability* (Ed. H.H.E. Leipholz), Univ. of Waterloo Press, Ontario 1972, 85-197.
10. Koiter, W.T.: Stress-strain relations, uniqueness and variational theorems for elastic-plastic materials with a singular yield surface, *Quart. Appl. Math.,* **11** (1953), 350-353.
11. Mandel, J.: Generalisation de la théorie de plasticité de W.T. Koiter, *Int. J. Solids Structures,* **1** (1965), 273-295.
12. Hill, R: Generalized constitutive relations for incremental deformation of metal crystals by multislip, *J. Mech. Phys. Solids,* **14** (1966), 95-102.
13. Klushnikov, V.D.: *Stability of Elasic-Plastic Systems* (in Russian), Nauka, Moscow 1980.
14. Petryk, H. and Thermann, K.: Second-order bifurcation in elastic-plastic solids, *J. Mech. Phys. Solids,* **33** (1985), 577-593.
15. Petryk, H. and Mróz, Z.: Time derivatives of integrals and functionals defined on varying volume and surface domains, *Arch. Mech.,* **38** (1986), 697-724.
16. Thomas, T.Y.: *Plastic Flow and Fracture in Solids*, Acad. Press, New York 1961.
17. Hill, R: Uniqueness criteria and extremum principles in self-adjoint problems of continuum mechanics, *J. Mech. Phys. Solids,* **10** (1962), 185-194.
18. Nguyen, Q.S.: Stabilité et bifurcation des systèmes dissipatifs standarts à comportement indépendant du temps physique, *C. R. Acad. Sci. Paris, t.* **310**, *Serie II,* (1990), 1375-1380.
19. Hill, R: Eigenmodal deformations in elastic/plastic continua, *J. Mech. Phys. Solids,* **15** (1967), 371-386.
20. Raniecki, B: Uniqueness criteria in solids with non-associated plastic flow laws at finite deformations, *Bull. Acad. Polon. Sci., Sér. sci. techn.,* **27** (1979), 391-399.
21. Raniecki, B. and Bruhns, O.T.: Bounds to bifurcation stresses in solids with non-associated plastic flow law at finite strain, *J. Mech. Phys. Solids,* **29** (1981), 153-172.
22. Triantafyllidis, N.: On the bifurcation and postbifurcation analysis of elastic-plastic

solids under general prebifurcation conditions, *J. Mech. Phys. Solids,* **31** (1983), 499-510.
23. Hill, R: Bifurcation and uniqueness in non-linear mechanics of continua, in: *Problems of Continuum Mechanics, N. I. Muskhelishvili Anniversary Volume*, SIAM, Philadelphia 1961, 155-164.
24. Hill, R. and Sewell, M.J.: A general theory of inelastic column failure - I, *J. Mech. Phys. Solids,* **8** (1960), 105-111.
25. Hutchinson, J.W.: Post-bifurcation behavior in the plastic range, *J. Mech. Phys. Solids,* **21** (1973), 163-190.
26. Hutchinson, J.W.: Plastic buckling; *Advances in Applied Mechanics, Vol.* **14**, Acad. Press, New York 1974, 67-144.
27. R. Hill and M.J. Sewell, A general theory of inelastic column failure - II, *J. Mech. Phys. Solids,* **8** (1960), 112-118.
28. Nguyen, Q.S. and Radenkovic, D.: Stability of equilibrium in elastic plastic solids, *Lecture Notes in Mathematics, Vol.* **503**, Springer, Berlin 1975, 403-414.
29. Shanley, F.R.: Inelastic column theory, *J. Aero. Sci.,* **14** (1947), 261-267.
30. Hill, R.: The essential structure of constitutive laws for metal composites and polycrystals, *J. Mech. Phys. Solids* **15** (1967), 79-95.
31. T.H. Lin, Physical theory of plasticity, *Advances in Applied Mechanics, Vol.* **11**, Acad. Press, New York 1971, 255-311.
32. Hutchinson, J.W.: Elastic-plastic behaviour of polycrystalline metals and composites. *Proc. Roy. Soc. Lond.,* **A 319** (1970), 247-272.
33. Christoffersen, J.; Hutchinson, W: A class of phenomenological corner theories of plasticity. *J. Mech. Phys. Solids,* **27** (1979), 465-487.
34. Petryk, H.: On energy criteria of plastic instability, in: *Plastic Instability, Proc. Considère Memorial,* Ecole Nat. Ponts Chauss., Paris 1985, 215-226.
35. Petryk, H., On the theory of bifurcation in elastic-plastic solids with a yield-surface vertex, in: *Inelastic Solids and Structures, Antoni Sawczuk Memorial Volume* (Ed. M.Kleiber and J.A.König), Pineridge Press, Swansea 1990, 131-143.
36. Petryk, H.: The energy criteria of instability in time-independent inelastic solids, *Arch. Mech.,* **43** (1991), No 4 (in press).
37. Hill, R: On constitutive macro-variables for heterogeneous solids at finite strain, *Proc. Roy. Soc. Lond.,* **A 326** (1972), 131-147.
38. Petryk, H.: On constitutive inequalities and bifurcation in elastic-plastic solids with a yield-surface vertex, *J. Mech. Phys. Solids,* **37** (1989), 265-291.
39. Nguyen, Q.S. and Petryk, H.: A constitutive inequality for time-independent dissipative solids, *C. R. Acad. Sci. Paris, t.* **312**, *Serie II,* (1991), 7-12.
40. Petryk, H.: On the second-order work in plasticity, *Arch. Mech.,* **43** (1991) No 2/3 (in press).
41. Petryk, H. and Thermann, K.: On discretized plasticity problems with bifurcations, *Int. J. Solids Structures,* (in press).
42. Petryk, H., A consistent energy approach to defining stability of plastic deformation processes, in: *Stability in the Mechanics of Continua, Proc. IUTAM Symp. Nümbrecht* 1981 (Ed. F.H. Schroeder), Springer, Berlin - Heidelberg 1982, 262-272.

43. Sewell, M.J.: On configuration-dependent loading, *Arch. Rat. Mech. Anal.*, **23** (1967), 327-351.
44. Petryk, H., On the onset of instability in elastic-plastic solids, in: *Plasticity Today: Modelling, Methods and Applications* (Ed. A. Sawczuk and G. Bianchi), Elsevier, London 1985, 429-447.
45. Hutchinson, J.W.: Imperfection sensitivity in the plastic range, *J. Mech. Phys. Solids*, **21** (1973), 191-204.
46. Storåkers, B.: On uniqueness and stability of elastic-plastic deformation, *Arch. Mech.*, **27** (1975), 821-839.
47. Miles, J.P.: On necking phenomena and bifurcation solutions, *Arch. Mech.*, **32** (1980), 909-931.
48. Needleman, A. and Tvergaard, V.: Aspects of plastic postbuckling behavior, in: *Mechanics of Solids, The Rodney Hill 60-th Anniversary Volume* (Ed. H.G. Hopkins and M.J. Sewell), Pergamon Press, Oxford 1982, 453-498.
49. Tvergaard, V.: On bifurcation and stability under elastic-plastic deformation, *Plasticity Today: Modelling, Methods and Applications* (Ed. A. Sawczuk and G. Bianchi), Elsevier, London 1985, 377-398.
50. Rice, J.R.: The localization of plastic deformation, in: *Theoretical and Applied Mechanics* (Ed. W.T. Koiter), North- Holland, Amsterdam 1977, 207-220.
51. Hutchinson J.W. and Tvergaard, V.: Surface instabilities on statically strained plastic solids, *Int. J. Mech. Sci.*, **22** (1980), 339-354.
52. N.N. Krasovskii, *Stability of Motion,* Stanford Univ. Press 1963.
53. Petryk, H., On stability and symmetry conditions in time-independent plasticity, *Arch. Mech.*, **37** (1985), 503-520.
54. Graves, L.M.: The Weierstrass condition for multiple integral variation problems, *Duke Math. J.*, **5** (1939), 656-660.
55. Kolymbas, D.: Bifurcation analysis for sand samples with a non-linear constitutive equation, *Ing.-Archiv*, **50** (1981), 131-140.

# GENERIC SINGULARITIES FOR PLASTIC INSTABILITY PROBLEMS

**A. Léger**
**E.D.F.-D.E.R., Clamart, France**

**M. Potier-Ferry**
**University of Metz, Metz, France**

## ABSTRACT

Plastic instability is discussed from the point of view of generic bifurcation theory. We try to get an exhaustive classification of the singularities that can occur for the so-called dissipative systems. This goal is achieved for systems with two degrees of freedom. For continuous systems, we analyse the post-bifurcation behavior of representative beam buckling problems with variable cross-sections.

## INTRODUCTION

Despite of very old studies [8] [28], beam buckling was not really understood before the work of SHANLEY [24], who established that the buckling criterion is in fact a bifurcation criterion. For complex structures, a quite general criterion has been stated by R. HILL [11] in the form of a uniqueness condition for the velocity problem. The first comprehensive analysis of the post-bifurcation behavior is due to J.W. HUTCHINSON [12] [13] [14]. Generally, the post-buckling behavior of an elastic-plastic solid involves the growth of an unloading zone which spreads out locally around a point neutrally loaded at the critical load. Asymptotic methods have been proposed to account for this free boundary problem which leads to a deflection-load relation that involves an expansion into fractional powers [12] [14] [27]. But almost all these studies deal with bifurcation from a homogeneous prebifurcation state, the only exceptions being the papers [3] [17].

By comparison, the theoretical framework for elastic stability problem is quite clear [2] [15] [23]. First, the stability and bifurcation condition is given by the energy criterion or the criterion of the second variation. Second, one is able to compute the bifurcation branches and their sensitivity to imperfections. Next one can characterize the bifurcation points and singular points that are generic, i.e. that are robust with respect to perturbations of the system, and one can classify these singularities according to their robustness and to the symmetry of the problem [23]. The latter point of view is, in fact, the one of the so-called catastrophe theory or generic bifurcation theory [4] [9].

In this paper we present the plastic stability analyses versus this generic bifurcation theory. A complete classification of the singularities would be probably very intricate because such a classification involves already five "elementary catastrophes" for systems with two degrees of freedom. For continuous systems, new generic bifurcation points are discussed, that occur when the body is partially plastified in the prebifurcation state. For original statements and more complete discussions, we refer to [16] [17] [22].

## I - SINGULARITIES AND BIFURCATION POINTS FOR ELASTIC SYSTEMS

We recall here the main outlines of elastic stability theory for the sake of comparison and to explain clearly the point of view of generic bifurcation theory.

We refer to [4] [15] [26] for classical statements of elastic stability, to [4] [9] [10] for generic bifurcation and singularity theory and to [23] to a tentative synthesis.

In classical elastic stability theory, the possible values of the displacement **u** are sought as depending on a scalar loading parameter $\lambda$. The set of admissible displacements is denoted by $\mathcal{U}$ and is assumed to be a vectorial space, for the sake of simplicity. The equilibrium states are stationary points of the potential energy $E(\mathbf{u},\lambda)$ so that

$$(1\text{-}1) \qquad \delta E = D_u E\,(\mathbf{u}, \lambda)\,(\delta \mathbf{u}) = 0 \qquad \forall \delta \mathbf{u} \in \mathcal{U}.$$

The potential energy and therefore the equation (1-1) are assumed to be smooth functions of the displacement and of the load parameter, which is a fundamental assumption and a main difference with elasto-plasticity.

When there is no ambiguity, we shall write $D_u E(\delta \mathbf{u})$ and, in the same manner, $D_u^2 E(\delta \mathbf{u}_1, \delta \mathbf{u}_2)$ for the second variation, etc. An equilibrium state $\mathbf{u}_o$ for $\lambda = \lambda_o$ is said to be stable if the second variation of the potential energy nearby $u_o$ is strictly positive :

$$(1\text{-}2) \qquad \delta^2 E = D_u^2 E\,(\delta \mathbf{u}, \delta \mathbf{u}) > 0 \qquad \forall \delta \mathbf{u} \in \mathcal{U}, \qquad \delta \mathbf{u} \neq 0.$$

This criterion follows from the LEJEUNE-DIRICHLET theorem for discrete systems and has been established in some cases for continuous elastic systems [21]. So long as the criterion (1-2) holds, the inverse function theorem allows one to find a unique response curve $\mathbf{u}(\lambda)$ for any $(\mathbf{u}, \lambda)$ in a neighbourhood of $(\mathbf{u}_o, \lambda_o)$. Therefore, there is coincidence between points of possible local nonuniqueness - or singular points on the path $\mathbf{u}(\lambda)$ - and points where the stability can change. Such a state $(\mathbf{u}_o, \lambda_o)$ is characterized by :

$$(1\text{-}3) \qquad \exists\, \mathbf{U} \in \mathcal{U}, \qquad \mathbf{U} \neq 0: \qquad D_u^2 E\,(\mathbf{U}, \delta \mathbf{u}) = 0, \qquad \forall \delta \mathbf{u} \in \mathcal{U}.$$

The purpose of bifurcation theory is to find solution curves in the vicinity of neutral stability. It is well known that the singular points are generally extrema on the curve

$(\mathbf{u}, \lambda)$ and are called limit points [Figure 1(a)]. The solution branch can be expanded as in [23] Section 3-5 :

$$(1\text{-}4) \qquad \mathbf{u} = \mathbf{u}_o + \varepsilon \mathbf{U} + O(\varepsilon^2), \qquad \lambda = \lambda_o + \varepsilon^2 \lambda_2 + O(\varepsilon^3),$$

where

$$(1\text{-}5) \qquad \lambda_2 = - D_u^3 E\,(\mathbf{U}, \mathbf{U}, \mathbf{U})/2\partial_\lambda\, D_u\, E(\mathbf{U}),$$

provided that three generic conditions be satisfied : the eigenvector **U** is simple, the numerator and denominator in (1-5) are different from zero. In the terminology of catastrophe theory, the limit point (the fold of C.T.) is the only robust singularity. Non generic singularities are obtained if one of these three conditions is removed [23] but, generally, this cannot happen, except if the system has additional properties.

Most of the systems that are usually studied are symmetric with respect to a plane, an axis or a point. That is why modern statements of bifurcation theory account for symmetry properties [10] [23]. To modelize such a symmetry, one assumes that there exists a linear involutive operator R ($R^2$ = Id, R ≠ Id) that leaves the potential energy invariant :

$$(1\text{-}6) \qquad E\,(R\mathbf{u}, \lambda) = E\,(\mathbf{u}, \lambda), \qquad \forall \mathbf{u} \in \mathcal{U},$$

One can discuss the nature of possible singularities close to a state $(\mathbf{u}_o, \lambda_o)$ which satisfies (1-3), according to the linear mode **U**. When **U** is symmetric (R**U** = **U**), the previous analysis is not altered and the limit point is the only robust singularity. When the mode is antisymmetric (R**U** = - **U**), two branches intersect at $(\mathbf{u}_o, \lambda_o)$ : a branch of symmetric solutions $\mathbf{u}_s(\lambda)$ and a branch of nonsymmetric solutions that can be put in the form :

$$(1\text{-}7) \qquad \mathbf{u} = \mathbf{u}_s(\lambda) + \varepsilon \mathbf{U} + O(\varepsilon^2), \qquad \lambda = \lambda_o + \varepsilon^2 \lambda_2 + O(\varepsilon^4).$$

Then, in the case of symmetry breaking, the only robust singular point is the symmetric bifurcation point (the cusp of C.T.), that is pictured in Figure 1b. Both singularities, limit point and symmetric bifurcation, can exist in a single system, for instance, in the buckling of a lumped arch. Of course, there can be other catastrophes if there are additional symmetries.

We refer to [23] for proofs, for classification of non generic singularities and for a lot of elementary examples.

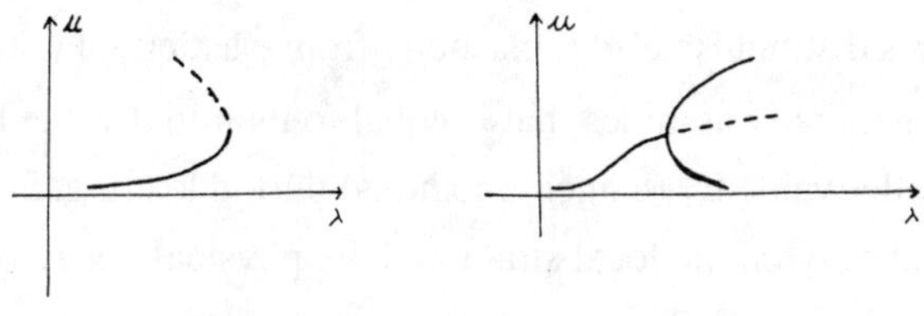


Figure 1. Singular points of elastic stability theory. Unstable states = dashed curves. (a) the limit point ; (b) the symmetric bifurcation (only the supercritical case).

## II - A CLASSIFICATION OF SINGULARITIES FOR DISCRETE IRREVERSIBLE SYSTEMS

### II-1 Some specific non-linearities

The framework of elastic stability cannot be applied to problems of plasticity because the constitutive laws are neither conservative, nor smooth. For instance, let us consider the simple uniaxial stress-strain law (Figure 2). The first characteristic of these laws is the existence of limits for the possible values of the stress, This is mathematically expressed by a threshold function, that depends also on the plastic strain $\alpha$ :

(2-1) $$\mathcal{F}(\sigma, \alpha) = \sigma - \sigma^{LIM}(\alpha) \leq 0$$

Before the limit is reached ($\mathcal{F} < 0$) ,the plastic strain is constant ($\alpha = 0$). When it is reached ($\mathcal{F} = 0$), the constitutive law is given by the classical alternative :

(2-2)
$$\text{or} \quad \dot{\alpha} > 0 \ \text{ and } \ \dot{\mathcal{F}} = 0 \qquad \text{(loading)}$$
$$\text{or} \quad \dot{\alpha} = 0 \ \text{ and } \ \dot{\mathcal{F}} \leq 0 \qquad \text{(unloading)}$$

The two latter conditions can be put in a more compact form, that is usually called a variational inequality [7] :

$$\dot{\alpha} \geq 0$$

(2-3) $\mathcal{F}(\beta - \dot{\alpha}) \leq 0$ for any $\beta \geq 0$.

Two main features distinguish elasto-plasticity from elasticity. First the slope $d\sigma/d\varepsilon$ is discontinuous. One generally assumes that is equal to its initial value E during unloading while it has a smaller value $E_T$ (tangent modulus) during loading. Second the stored energy does not depend only on the local strain and the physical behavior is dissipative. In that sense, plasticity is not similar to perfect contact problems, even if they lead to slope discontinuities and to variational inequalities formulation.

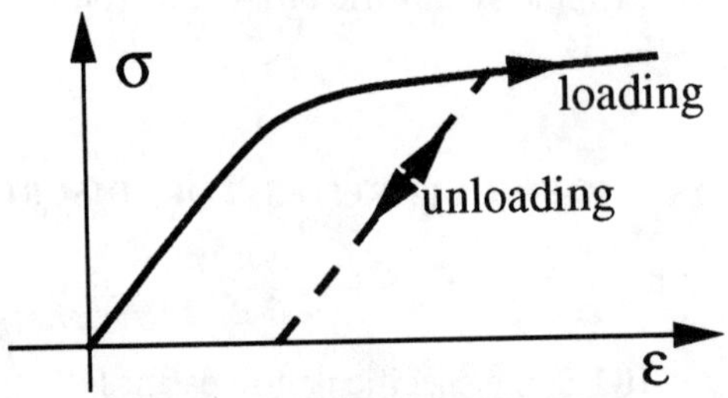


Figure 2. Uniaxial stress-strain law : slope discontinuity and dissipative behavior.

Nevertheless, friction laws can be put in a form that is similar to (2-1) (2-2). Let us limit ourselves to bidimensional problems and to persistent contacts along a line. The ratio of the tangential force on the normal forces is limited by the friction angle $\phi$ :

$$|T| / N \leq \text{tg}\, \phi$$

and the "flow rule" is given by the alternative (case $T > 0$)

(2-4)

or $\dot{\alpha} > 0$ and $T / N = \text{tg}\, \phi$ (slip)

or $\dot{\alpha} = 0$ and $|T| / N \leq \text{tg}\, \phi$ (no slip)

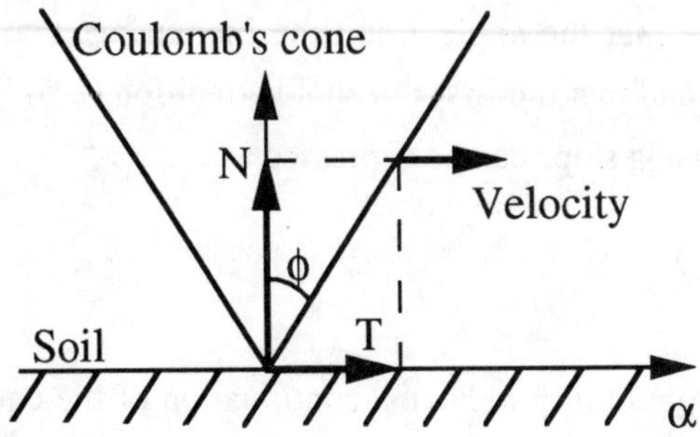


Figure 3. Sketch of Coulomb friction law.

II-2 Two generic singularities

The aim is to sketch a classification of the robust singularities that can be encountered for mechanical systems with bodies in the plastic range or with frictional contacts. The complete classification for systems with two degrees of freedom has been established by POTIER-FERRY [22]. Here we discuss a simple system with one degree of freedom to present the first two generic singularities.

Let us consider the equilibrium of a heavy material point, which lies on a rough soil. The heavy point can move along a curve $z = f(\alpha, \lambda)$, which lies in a vertical plane $O\alpha z$, the axis $Oz$ being upwards vertical. The shape of the soil varies with a parameter $\lambda$. For instance, $\lambda$ can be the angle of rotation of a rigid block. The threshold condition can be written only in terms of the curve f :

$$- \operatorname{tg} \phi \leq \frac{\partial f}{\partial \alpha} (\alpha, \lambda) \leq + \operatorname{tg} \phi \qquad (2\text{-}5)$$

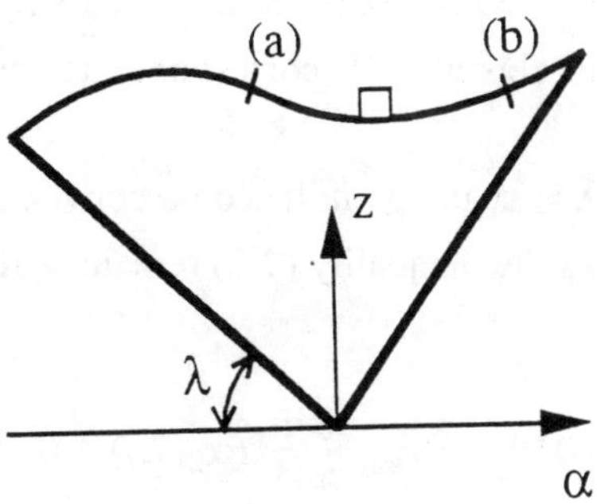


Figure 4. Friction system with one degree of freedom.

Now let us assume that the angle $\lambda$ is time increasing, that the slope $\partial f/\partial\alpha$ is negative and decreasing and that the first threshold condition (2-5) is reached at $\lambda = \lambda_1$. The condition of a decreasing slope can be expressed as :

$$\text{(2-6)} \qquad \frac{\partial^2 f}{\partial\alpha\partial\lambda}(\alpha, \lambda) < 0$$

Let us discuss the possibility to have a continuation of the equilibrium path $\alpha(\lambda)$ beyond $\lambda_1$. Because of condition (2-6) it is not possible to go back inside the Coulomb's cone. Hence the threshold condition (2-5) must hold along the continuation and the point moves in the positive direction :

$$\text{(2-7)} \qquad \frac{\partial f}{\partial\alpha}(\alpha(\lambda), \lambda) = -\operatorname{tg}\phi \qquad , \qquad \frac{d\alpha}{d\lambda} > 0$$

$$\text{(2-8)} \qquad \frac{\partial^2 f}{\partial\alpha^2}(\alpha(\lambda), \lambda)\frac{d\alpha}{d\lambda} + \frac{\partial^2 f}{\partial\alpha\partial\lambda}(\alpha(\lambda), \lambda) = 0$$

With account of (2-6) (2-7) (2-8), one sees that the continuation path exists only if the point lies on a convex part of the soil :

$$\text{(2-9)} \qquad \frac{\partial^2 f}{\partial\alpha^2}(\alpha, \lambda) > 0$$

If for $\lambda = \lambda_1$ the point lies on a concave part of the soil, there is no neighbour equilibrium state for $\lambda$ greater than $\lambda_1$. Then, there is an unstable progress of the point. This simple catastrophe is pictured in Figure 5-a. Such a singularity cannot occur for smooth systems, especially in elasticity. It corresponds to the limit load of plasticity theory.

On the contrary, if for $\lambda = \lambda_1$ the point lies on a convex part of the soil, there is a unique solution path so long as the inequality (2-7) remains valid. If on the path $\alpha(\lambda)$, it happens that

$$\text{(2-10)} \qquad \frac{\partial^2 f}{\partial\alpha^2}(\alpha_2, \lambda_2) = 0 \qquad , \qquad \frac{\partial^3 f}{\partial\alpha^3}(\alpha_2, \lambda_2) \neq 0$$

then the state $(\alpha_2, \lambda_2)$ is a limit point on the response curve as pictured in Figure 5-b. Indeed a result recalled in Section 2 can bc applied to the equation (2-7).

So there are already two elementary catastrophes for systems with one degree of freedom : first the unstable progress pictured in Figure 5-a, which means that the instability occurs at the threshold ; second the limit point of Figure 5-b, when the instability occurs after a dissipative process. It is clear that the same classification holds for larger systems with only one scalar threshold condition [22] Section 4. Unfortunately, the modelisation of dissipative bodies involves at least such one condition for each material point or each crack tip in fracture mechanics. So one can expect the classification of singularities is much more intricate for dissipative systems than for elastic systems.

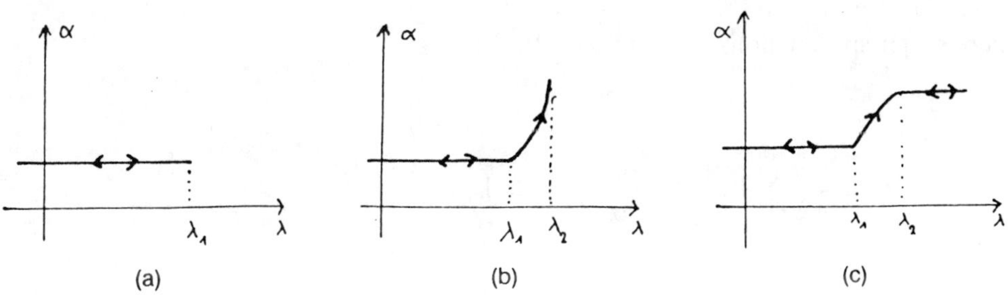


Figure 5. Singular points with only one degree of freedom. Two opposite arrows denote a reversible path, a single arrow a dissipative path. (a) unstable progress ; (b) limit point ; (c) a typical nonsingular path.

## II-3 A system with two degrees of freedom

A more representative example for beam, plate or shell buckling must have at least two degrees of freedom. Indeed such instability phenomena occur generally with symmetry breaking, which implies a sudden change of way for the equilibrium path. Significant examples are the famous discrete SHANLEY model and the interactive crack problems [19] that are discussed in other papers in this volume. Here we limit ourselves to a simple system with friction, that was introduced in [22].

Let us consider the equilibrium of two heavy material points, which can move along two parallel and rough rails. These rails lie in two vertical and parallel planes $O\alpha_1 z$

and $O\alpha_2z$, whose distance is denoted by l. They are modelized by two curves of respective equations $z = f(\alpha_1, \lambda)$ and $z = f(\alpha_2, \lambda)$. As in Section II-2, the shape of the rails varies with the parameter λ. Both points are fastened together by a right spring of rigidity k and of natural length $l_o$ (Figure 6). The motion of one point follows Coulomb friction law. After some lengthy computations, we get the derivatives of the ratios "tangent force/normal force" at a symmetric state :

$$(2\text{-}11) \qquad \begin{aligned} - \overline{\dot{(T_1/N_1)}} &= p\,\dot{\alpha}_1 + q\,\dot{\alpha}_2 - r \\ - \overline{\dot{(T_2/N_2)}} &= q\,\dot{\alpha}_1 + p\,\dot{\alpha}_2 - r \end{aligned}$$

where p, q, r are functions of λ and $\alpha_1 = \alpha_2 = \alpha$ :

$$p = \frac{\partial^2 f}{\partial \alpha^2} + \left(1 + \left(\frac{\partial f}{\partial \alpha}\right)^2\right) k\,(l - l_o) / mgl$$

$$(2\text{-}12) \qquad q = -\left(1 + \left(\frac{\partial f}{\partial \alpha}\right)^2\right) k\,(l - l_o) / mgl$$

$$r = -\frac{\partial^2 f}{\partial \alpha \partial \lambda} \quad , \qquad r \text{ assumed positive.}$$

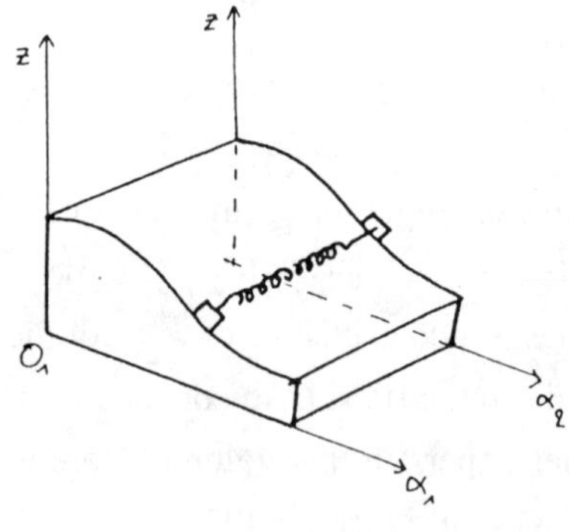


Figure 6. A simple friction model to study symmetry breaking.

Hence the behavior of the system close to a symmetric state is characterized by the three coefficients p, q, r, but mainly by the first two ones. Note that the two latter coefficients depend on the curvature of the rails and on the initial elongation of the spring.

So long as the friction threshold is not reached, the points do not move. Let us now look at the behavior of the system when the threshold is reached at a negative slope, as in Figure 6. In this case the set of admissible velocities is the following cone $C = \mathbb{R}_+ \times \mathbb{R}_+$. The velocities are solution of a nonlinear system of equations, that be put in the form of two alternatives as in (2-4) :

$$\text{(2-13)} \qquad \begin{cases} \text{or } p\,\dot{\alpha}_1 + q\,\dot{\alpha}_2 - r = 0 \quad , \quad \dot{\alpha}_1 > 0 \\ \text{or } \dot{\alpha}_1 = 0 \quad , \quad p\,\dot{\alpha}_1 + q\,\dot{\alpha}_2 - r \geq 0 \\ \text{and} \\ \text{or } q\,\dot{\alpha}_1 + p\,\dot{\alpha}_2 - r = 0 \quad , \quad \dot{\alpha}_2 > 0 \\ \text{or } \dot{\alpha}_2 = 0 \quad , \quad q\,\dot{\alpha}_1 + p\,\dot{\alpha}_2 - r \geq 0 \end{cases}$$

The system (2-13) can be put in the more compact form of a variational inequality

$$\text{(2-14)} \qquad \begin{cases} \dot{\alpha} \in C \\ (\delta\alpha - \dot{\alpha}).K.\dot{\alpha} - r\,(\delta\alpha_1 + \delta\alpha_2 - \dot{\alpha}_1 - \dot{\alpha}_2) \geq 0 \qquad \forall\, \delta\alpha \in C \end{cases}$$

where K is the simple matrix :

$$K = \begin{pmatrix} p & q \\ q & p \end{pmatrix}$$

We refer to NGUYEN paper and to [18] [19] for a stability criterion and to a non-bifurcation criterion. According to the first one, the equilibrium state is said to be stable if the quadratic form K is positive on the cone C. For instance, the symmetric state is stable if

$$\text{(2-15)} \qquad p > 0 \qquad \text{and} \qquad p + q > 0.$$

The non-bifurcation criterion is the positivity of the quadratic form on the space generated by the cone, i.e. in the same example :

(2-16) $p > |q| > 0.$

II-4 Five generic singularities for symmetric systems

Let us discuss the alternative equations (2-13). We underline that these equations govern the behavior of any symmetric "dissipative" system if only two thresholds are active, for instance for the discrete SHANLEY model or for a brittle body with two crack tips.

First we can get solutions such that the two points move forwards. In this case, the velocities solve a linear system that can be rewritten as :

$$(p - q)\,(\dot{\alpha}_1 - \dot{\alpha}_2) = 0 \quad , \qquad (p + q)\,(\dot{\alpha}_1 + \dot{\alpha}_2) = 2\,r$$

Except for $p = q$, this type of solution remains symmetric

(2-17) $\dot{\alpha}_1 = \dot{\alpha}_2 = r \,/\, (p + q)$

but it occurs only for positive values of $p + q$ :

(2-18) $p + q > 0$

Second we can find symmetry breaking solutions where one point (say the number one) moves forward while the second stops. This solution is :

(2-19) $\dot{\alpha}_2 = 0 \quad , \quad \dot{\alpha}_1 = r \,/\, p$

and it is subjected to the two following conditions :

(2-20) $q \geq p > 0$

Third we could have a solution such that the system stops completely, but this could occur only if $r \leq 0$, i.e. if the slope of the rail suddenly decreases, what we have excluded.

So any possible initial velocity is given by (2-17) or (2-19), but the corresponding inequality (2-18) or (2-20) has to be satisfied. When the friction threshold is just reached, the number of possible velocities depends on the place of the point (p,q) and more exactly on the two conditions (2-18) (2-20).

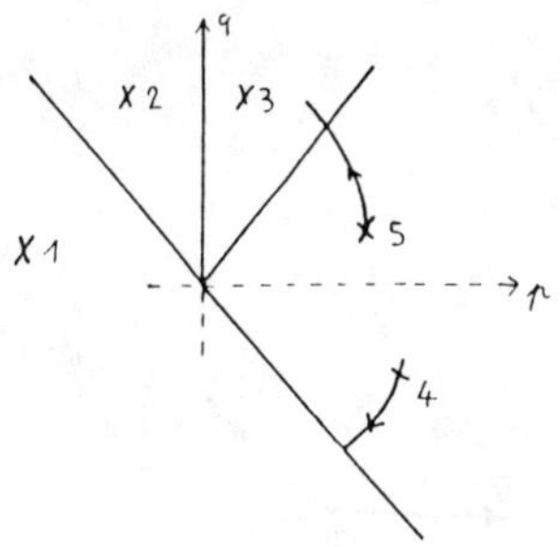


Figure 7. Five catastrophes can arise according to the position of $p_1$, $q_1$ (cases 1, 2, 3) or the path $p(\lambda)$, $q(\lambda)$ (cases 4, 5).

We can now discuss the shape of the solution curves $\alpha(\lambda)$ according to the parameters p and q. The constant path $\alpha = \alpha_1$ has no continuation for $\lambda$ greater than $\lambda_1$, if the value ($p_1$, $q_1$) of (p, q) at the state ($\alpha_1$, $\lambda_1$) are such that (point 1 of Figure 7) :

$$p_1 + q_1 < 0.$$

We find again the unstable progress pictured in Figure 5(a).

In the case (point 2 of Figure 7)

$$q_1 > - p_1 > 0$$

the path $\alpha = \alpha_1$ has a unique and symmetric extension, but these solutions do not satisfy (2-15) and hence they are unstable. Physically, this can be understood as an unstable progress but the picture is that of Figure 8.

If one has (point 3 of Figure 7)

$$q_1 > p_1 > 0$$

there is a symmetric continuation $\alpha_s(\lambda)$ of the trivial path. But from each point [$\alpha_s(\lambda)$, $\lambda$], there are also two bifurcating and symmetry-breaking solution paths, which are symmetric to one another. Each symmetric state is a bifurcation point beyond $\lambda_1$ [Figure

9(a)] and all those solutions are stable. Singularities of this type cannot occur for the classical systems, but they are well known in plasticity theory.

Let us now suppose that the parameters are in the angle

$$p_1 > | q_1 |.$$

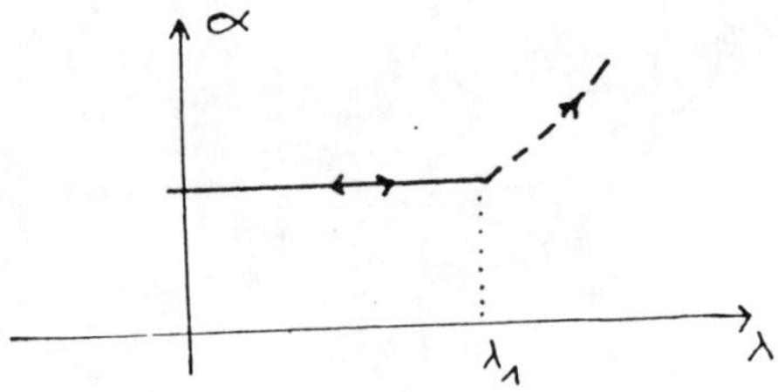


Figure 8. Unstable progress (second kind).

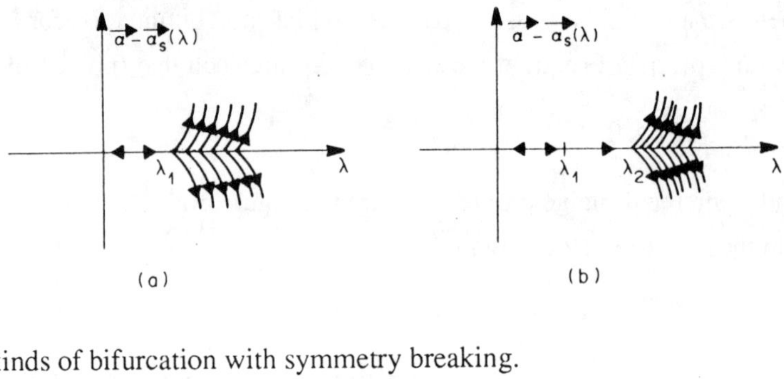


Figure 9. Two kinds of bifurcation with symmetry breaking.

Then, there is a unique and symmetric continuation $\alpha_s(\lambda)$ of the trivial path. The point $(\alpha_1, \lambda_1)$ is not singular. But we can discuss the stability and the bifurcations from those symmetric states because the previous arguments can be applied. In particular the parameters $p(\lambda)$ and $q(\lambda)$ are defined for $\lambda$ greater than $\lambda_1$. The continuation is unique until $p(\lambda)$ equals $\pm\, q(\lambda)$.

If $p(\lambda) + q(\lambda)$ goes to zero for $\lambda = \lambda_2$ (path 4 in Figure 7), $(\alpha_2, \lambda_2)$ is a limit point. All the solutions are symmetric.

If $p(\lambda) - q(\lambda)$ goes to zero for $\lambda = \lambda_2$ (path 5 in Figure 7), the symmetric path $\alpha_s(\lambda)$ continues beyond $\lambda_2$. But symmetry-breaking solutions bifurcate from each state $[\alpha_s(\lambda), \lambda]$ for $\lambda$ greater than $\lambda_2$. This is pictured in Figure 9(b). The schema is similar to one in Figure 9(a), but with a delay between the value $\lambda_1$ where the threshold is reached and the first bifurcation value $\lambda_2$.

There cannot exist an alternative robust singularities, because all the places in the (p, q) plane have been considered except $p_1 = \pm q_1$ or $p_1 = 0$. But those cases would lead to non generic singularities.

So we have found that five kinds of singularities can occur for symmetric systems with only two active thresholds. These five singularities can be encountered with the friction system, depending on the curvature of the soil and of the elongation of the spring. If the material points lie on a concave part of the rails, there is no continuing equilibrium path (unstable progress, Figure 5(a)). If the material points lie on a convex part of the rails, the behavior of the system is related to the elongation of the spring. If the spring is very compressed, there is a unique continuing equilibrium path, but it is unstable (unstable progress, Figure 8). A stable symmetry breaking occurs at the threshold if the spring moderately compressed (Figure 9(a)). If the spring is tensile or weakly compressed, there is no singularity at the threshold, but a unique stable and symmetric continuation path. The last two singularities can be obtained by following this path : one can get a symmetry breaking (Figure 9(b)) or a limit point (Figure 5(c)) if, during their progress, the material points arrive at an inflexion point of the soil.

## III - GENERIC BIFURCATIONS WITHIN STRUCTURAL PLASTICITY

At the end of section II, we have got two main ideas for the bifurcation analysis of dissipative systems. The first one is that the problem necessarily deals with a variational inequality, and the second is that we need a classification of the possible bifurcations. In the present section, we shall try by the same way to sketch a classification of the bifurcations that may occur generically within elastic-plasticity.

But, in order to have a better understanding of the qualitative phenonomenon, let us before revisit the Shanley criterion for the occurence of bending in an initially perfectly straight elastic-plastic compressed bearn. The main result, due to F.R. Shanley [24] is that the buckling load, that is precisely the value of the load at which bending begins, is characterized by the occurence of a neutrally loaded point. Actually, after a simple compression phase, a bifurcated solution appears under increasing load which general form is the sum of a compression part and a bending part. That means the initial stress rate along the first bifurcated branch is of the form :

$$\dot{\sigma}^{o}(x, y) = \dot{\sigma}^{f}(x) + C\,\Sigma(x, y). \tag{3.1}$$

x and y are respectively the longitudinal and the transversal coordinate of the beam, it is not assumed to have an uniform cross-section, the dot stands for a derivative with respect either to the load or to any time-like parameter monotonically connected to the load. The exponent f denotes the fundamental solution, namely the simple compression one, and $\Sigma$ indicates the shape of the bending part, that is the buckling mode.

Qualitatively, such a solution is possible with increasing load because $\dot{\sigma}^{f}$ balances $\dot{\lambda} > 0$, when $\Sigma$ is compatible only with $\dot{\lambda} = 0$. But the main particularity of this solution is that the amount of flexion, that is the constant C, is not given by the velocity problem itself but by a sign condition $\dot{\sigma}^{o}(x,y) \leq 0$ everywhere in the beam. Consequently, we see that a whole interval, depending only on the shape of the bending part, is allowed for the constant C, which ends are given by $\dot{\sigma}^{0} = 0$ somewhere in the beam.

But it is not possible for C to take values strictly inside this range. Indeed, as that would mean strict loading everywhere, the buckling would be, in such a way, computable within the framework of elastic stability, that is, under very general symmetry assumptions, would lead to the classical pitch-fork bifurcation as stated in section II, contrary to the condition of increasing load. Then C is necessarily at one end of the interval. But that means $\dot{\sigma}^{o}$ reaches zero somewhere in the domain. The corresponding point will naturally be designated as neutrally loaded point.

III-1 Basic assumptions and the velocity problem

We now establish a variational formulation of an incremental equilibrium elastic-plastic problem, without, at first, any structural assumptions. For reasons which will be clear latter, we choose two unknowns fields, namely both the displacement and the plastic strain, contrary to most of the classical formulations which often keep only the displacement [11], or merely only the plastic strain [19] [20].

The constitutive framework is assumed to be the generalized standart materials one, with only one yield surface. The extension to several yield surface will not be difficult. In addition, we assume that the free energy $\Psi$ is a function of both the displacement gradient $\nabla \mathbf{u}$ (remark that this dependance is compatible with geometric non linearities) and of a family of internal variables, including in particular the plastic strain, and globally denoted by $\alpha$ .

Then, an incremental form of the principle of virtual powers reads :

(3.2) $$\int_{\Omega} \mathbf{T} : \nabla \mathbf{u} \, d\Omega = < \dot{f}, v > ,$$

where f is the external loading, and $\mathbf{T}$ is a first Piola-Kirchhoff stress tensor given by :

$$\mathbf{T} - \rho \frac{\partial \Psi}{\partial \nabla \mathbf{u}}$$

On the other hand, the constitutive law for the plastic strain gives another part of the variational formulation. In the thermodynamical sense, the generalized force related to the plastic strain rate $\dot{\alpha}$ reads :

$$\mathbf{A} = - \rho \frac{\partial \Psi}{\partial \alpha} ,$$

as $- \rho \dfrac{\partial \Psi}{\partial \alpha} \dot{\alpha}$ classically holds for the dissipation.

Then the generalized standart materials assumptions involve a continuous and convex yield surface $\mathcal{F}(\mathbf{A}) = 0$ such that :

$$
(3.3) \qquad
\begin{aligned}
&\text{if } \mathcal{F}(\mathbf{A}) < 0 \text{ then } \dot{\alpha} = 0 \\
&\text{if } \mathcal{F}(\mathbf{A}) = 0 \text{ then } \Rightarrow \text{if } \dot{\mathcal{F}}(\mathbf{A}) < 0 \text{ then } \dot{\alpha} = 0 \\
&\qquad\qquad\qquad\qquad \Rightarrow \text{if } \dot{\mathcal{F}}(\mathbf{A}) = 0 \text{ then } \dot{\alpha} = \mathrm{k}\,\mathbf{n}
\end{aligned}
$$

where $\mathbf{n}$ is the outer normal to the yield surface at the point $\mathbf{A}$. $\alpha$ being the plastic strain, the corresponding k is a positive function called the plastic multiplier.

The latter constitutive assumption can be rewritten in the following variational form :

$$
(3.4) \qquad
\begin{aligned}
&\mathrm{k} \geq 0, \\
&\dot{\mathcal{F}}(\kappa - \mathrm{k}) \leq 0
\end{aligned}
$$

for any positive function $\kappa$ and any point $\mathbf{x}$ at which the yield force is reached.

For the remainder we define two domains $\Omega_e$ and $\Omega_p$ by :

$$
\begin{aligned}
\Omega_p &= \{\, \mathbf{x} \,/\, \mathcal{F}(\mathbf{A}(\mathbf{x})) = 0 \,\}, \\
\Omega_e &= \Omega/\Omega_p .
\end{aligned}
$$

Then, from (3.3) and (3.4), the variational problem formulates as follows :

$$
\text{Find } \{u, k\} \in \mathbf{C} \equiv \left\{ \{\mathbf{v}(\mathbf{x}), \kappa(\mathbf{x})\} \,/\, \mathbf{v}(\mathbf{x}) \in \text{K.A.}, \quad \kappa(\mathbf{x}) \geq 0 \;\forall\, \mathbf{x} \in \Omega_p, \quad \kappa(\mathbf{x}) = 0 \;\forall\, \mathbf{x} \in \Omega_e. \right\}
$$

$$\int_\Omega \left( \frac{\partial^2 \Psi}{\partial u_{i,j}\, \partial u_{k,l}} \dot{u}_{i,j} + \frac{\partial^2 \Psi}{\partial u_{k,l}\, \partial \alpha_I} n_I\, k \right) v_{k,l}\, \rho\, d\Omega = < \dot{f}, v >$$

(3.5)

$$\int_\Omega \left( \frac{\partial^2 \Psi}{\partial u_{i,j}\, \partial \alpha_I} n_I\, \dot{u}_{i,j} + \frac{\partial^2 \Psi}{\partial \alpha_I\, \partial \alpha_J} n_I\, n_J\, k \right) (\kappa - k)\, \rho\, d\Omega \geq 0$$

$$\forall\ (v, \kappa) \in \mathbf{C}$$

K.A. stands for kinematically admissible.

In the following, we will denote this formulation by the classical form of variational inequalities :

(3.6)
$$\mathbf{u} \in \mathbf{C}$$
$$\mathcal{A}\ (\mathbf{u}, \mathbf{v}\text{-}\mathbf{u}) \geq (g \mid \mathbf{v}\text{-}\mathbf{u})$$
$$\forall\ \mathbf{v} \in \mathbf{C}$$

where $\mathbf{u}$ and $\mathbf{v}$ are the pairs velocity-plastic multiplier :

$$\mathbf{u} = \{\dot{u}, k\}\ ;\ \mathbf{v} = \{v, \kappa\}.$$

The main worth noting points are that the bilinear form $\mathcal{A}$ $(\mathbf{u}, \mathbf{v})$ is symmetric, and that $\mathbf{C}$ is a convex cone.

III-2 <u>Bifurcation analysis for the velocity problem</u>

We first use the latter formulation to give some characterization of the loss of uniqueness.

Classically, assume that $\mathbf{v}_1$ and $\mathbf{v}_2$ are two solutions of the problem (3.6). Then, we easily get that their difference is such that :

(3.7) $\mathcal{A}\ (v_1 - v_2\ , v_1 - v_2\ ) \leq 0$

We draw attention to the fact that the difference $v_1 - v_2$ does not generally belong to the cone $\mathbf{C}$ but only to the vector space generated by the cone, say Vect $\mathbf{C}$. This vector space is simply obtained by dropping the positivity condition on the plastic multiplier.

This inequality (3.7) means that a sufficient uniqueness condition follows immediatly :

There is no bifurcation for the problem (3.6) if the quadratic form $\mathcal{A}$ (. , .) remains stritly positive on the vector space Vect $\mathbf{C}$ :

(3.8) $\mathcal{A}\ (v, v) > 0 \qquad \forall\ v \in \text{Vect}\ \mathbf{C}.$

Then, we look for a necessary condition for bifurcation. This problem is still partially open. In the present section we shall get, under some supplementary assumptions, an existence result for a critical load $\lambda_o$ and a buckling mode $\mathbf{U}$ for the problem (3.6).

The first supplementary assumption is the following :

H1 : The quadratic form $\mathcal{A}$ (v, v) is strictly positive for $\lambda = 0$, and remains strictly positive in the whole elastic range

This assumption only means that there is no elastic buckling or elastic instabilities for the problem (3.6). Then, it is quite natural, as we want to deal with plastic instabilities.

For the next assumption, we introduce, as it is usual in buckling problems, the minimum of a Rayleigh quotient :

$$(3.9) \qquad q(\lambda) = \min_{\substack{v \in \text{Vect}\,\mathbf{C} \\ v \neq 0}} \frac{\mathcal{A}(\mathbf{v}, \mathbf{v})}{\|\mathbf{v}\|^2},$$

where $\| \,.\, \|$ denotes a $L^2$ - norm on the vector space Vect $\mathbf{C}$. Notice that this minimum is actually a function of the load. It is known that $q(\lambda)$ is the smallest eigenvalue of the operator associated to the bilinear form $\mathcal{A}(\mathbf{u}, \mathbf{v})$, and that, because of assumption H1, $q(0)$ is positive [21].

Then, the next assumption reads as follows :

H2 : $q(\lambda)$ is a continuous function of the load $\lambda$, at least in the domain where it may change its sign.

Let us notice that, contrary to the first, this second assumption is a little bit restrictive. It would not be satisfied for instance in the case of both a discontinuity of the tangent modulus with respect to the stress or the strain, together with a discontinuous evolution of the boundary between the elastic and the plastic zones. Likewise, this assumption would not be satisfied if (3.6) went to a loss of ellipticity, or a loss of a complementary condition at the boundary [1]. Fortunately we can anticipate that this situation will generally occur for loads significantly higher than the first critical load we are looking for .

Precisely, let us now define the buckling load $\lambda_o$ as the smallest value of the load such that the velocity problem has at least two solutions. Then, the inequality (3.7), associated to the assumption H2, leads to the following result for the buckling load $\lambda_o$ :

$$(3.10) \qquad q(\lambda_o) = 0$$

That means, by the variational characterization of the minimization problem (3.9) , we get, at the buckling load, the existence of a buckling mode $\mathbf{U} = \{V, K\}$ , solution of the following variational problem :

(3.11)

Find $\mathbf{U} \in \text{Vect}\,\mathbf{C}$ such that :

$$\mathcal{A}(\mathbf{U}, \mathbf{v}) = 0 \quad , \quad \forall\, \mathbf{v} \in \text{Vect}\,\mathbf{C}.$$

Remarks :

i) The bifurcation condition given in problem (3.11) has the same form of the second variation criterion in elasticity. Numerically, that means the buckling mode will be got from the singularity of a stiffness matrix.

ii) The stiffness operator is built up by taking the elastic modulus in the elastic zone $\Omega_e$, and the tangent modulus in the plastic zone, that is by neglecting any eventual unloading. Moreover, to obtain problem (3.11) from problem (3.6) we see that we have to put $\dot{\lambda} = 0$ in the initial variational problem. That means we find again Engesser's result [8] for the buckling of non elastic beams, since $\dot{\lambda} = 0$ implies that (3.11) can be interpreted as an existence condition for an "adjacent equilibrium".

iii) The bifurcation condition (3.11) is nothing else than the sufficient uniqueness condition given by Hill [11] . But, by the introduction of supplementary assumptions, and in particular H2, Hill's criterion becomes a necessary and sufficient condition for bifurcation.

iv) This bifurcation result is also in agreement with the non-uniqueness condition given by Q.S. Nguyen [18] [19] [20].

### III-3 Sketch of classification of the generic bifurcations

In the previous section, we have got, by (3.11), a bifurcation result for the velocity problem. Now, we will try to build up actually different initial velocities when (3.11) is true. This should be seen as a kind of converse result, but, moreover, we will see that a classification follows this construction by a somewhat natural way.

Under assumption H1, there exists a fundamental solution. The latter will be denoted by an exponent f, and an exponent o will be used for another solution, different from the fundamental one, at the load $\lambda_o$.

First, we remark that the minimization problem (3.10) implies that, at the value $\lambda_o$ of the load, the inequality (3.7) is actually an equality :

$$\mathcal{A}\,(\dot{\mathbf{u}}^o - \dot{\mathbf{u}}^f\,,\ \dot{\mathbf{u}}^o - \dot{\mathbf{u}}^f\,) = 0. \tag{3.12}$$

For the sake of simplicity, we make another assumption :

H3 : At the smallest eigenvalue $\lambda_o$, the buckling mode **U** is unique.

That way, we get, from (3.11) and H3, the following result for the initial velocity $\dot{\mathbf{u}}^o$ :

$$\dot{\mathbf{u}}^o - \dot{\mathbf{u}}^f = C\,\mathbf{U}. \tag{3.13}$$

In particular, for the plastic multiplier component of the solution, that means :

$$k^o - k^f = C\,K.$$

But let us recall that $k^o$ is precisely a plastic multiplier, that is a positive function which support is in the plastic zone. Then, the bifurcation construction reads :

$$\begin{aligned} &k^o(\mathbf{x}) - k^f(\mathbf{x}) = C\,K(\mathbf{x}), \quad C \in R, \\ &k^f(\mathbf{x}) + C\,K(\mathbf{x}) \geq 0 \qquad\qquad \forall\,\mathbf{x} \in \Omega_p. \end{aligned} \tag{3.14}$$

That means we are led to a discussion on the value of the constant C, since we have to ensure both a proportionality relation given by (3.12) and a positive condition included in (3.3).

<u>Remarks</u> :

i) With the result (3.14), we begin to get some differences between elastic bifurcation theory and elastic-plastic bifurcation theory : even if the operator of the bifurcation result (3.11) is linear because we look precisely at an initial velocity, the corresponding bifurcation problem is non linear because of the inequality.

ii) As a consequence of this difference, the initial velocity on the bifurcated branch is not proportional to the buckling mode, but the difference between the two solutions is.

We begin the analysis of problem (3.14) by an assumption which specifies a very particular case of assumption H1.

H1bis : The fundamental state is homogeneous. That is, as soon as the load level has reached the yield stress, the plastic multiplier is a constant, assumed to be strictly positive, everywhere in the domain.

Notice that this assumption holds for the usual compressed beams with a uniform cross-section, but of course, is typically non generic since it does not persist under any perturbation.

In such a case, the buckling usually occurs with a breaking of symmetry, that is we may assume, in order to get simpler formula, that the buckling mode **U** is such that :

$$\text{Max } K(\mathbf{x}) = - \text{Min } K(\mathbf{x}).$$

Then, by (3.14), we get an interval for the admissible values of the constant C :

$$|C| \leq \frac{k^f}{\text{Max } K(\mathbf{x})} \tag{3.15}$$

This is the Shanley result since that only means the bifurcated solutions we have found are such that the increasing load holds everywhere in the domain, but it is extended to any structure satisfying assumption H1bis. Next, the previous analysis of section III-1 is necessary to eliminate the cases of strict loading everywhere on the bifurcated solution, and to keep only the ends of the interval (3.15).

Now, we drop the homogeneity assumption, and we make the following generic assumption :

H1ter : The fundamental solution is such that, at the critical load, there exist both an elastic and a plastic zone.

Then, there is an elastic-plastic boundary in the domain and, in order to simplify a little bit the discussion, we make another assumption :

H4 : The fundamental solution $k^f(\mathbf{x})$ is continuous across the elastic-plastic interface.

Actually, the latter assumption excludes a class of systems, especially from the standpoint of the constitutive law, but it could be established that the bifurcation results would not be very different if it didn't hold.

The problem (3.14) is not changed by these assumptions. Then, the discussion on the positivity of $k^o$ leads to three possible situations, we will call respectively bifurcation of Case 1, of Case 2 and of Case 3. In the case of only one space variable, these three cases are represented on figure 10 :

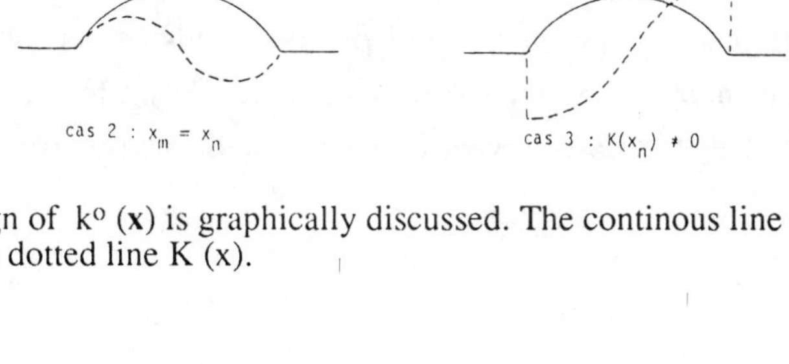


Figure 10. The sign of $k^o$ ($\mathbf{x}$) is graphically discussed. The continous line represents $K^f$ ($\mathbf{x}$), the dotted line K (x).

It is easily seen that the discussion involves both the buckling pattern and the position of the neutral loading point.

If the buckling mode reaches zero at the interface, then, as it can be simply graphically discussed on figure 10 for Cases 1 and 2, there actually exists an interval for the constant C. Of course, as C increases from zero, the first point $\mathbf{x}_n$ at which $k^f(\mathbf{x}) - C\, K(\mathbf{x})$ reaches zero is initially an unknown of the problem. We will get it by this construction. Next the initial velocity results are the following :

a) If $\mathbf{x}_n$ is got strictly inside $\Omega_p$ (Case 1), then the ends of the interval are :

$$C = \pm \frac{k^f(\mathbf{x}_n)}{K(\mathbf{x}_n)}$$

Since neutral loading occurs strictly inside an initially strictly loaded zone, we can guess that, in that case, the bifurcated behavior will be quite close to the homogeneous case, at least in some sense we will analyse for the post-buckling.

b) If $\mathbf{x}_n$ is got on the elastic-plastic boundary (Case 2), then the ends of the admissible values of C are :

$$C = \pm \frac{k^{f'}(\mathbf{x}_n)}{K'(\mathbf{x}_n)},$$

where the ( )' indicates that only the tangents are taken into account. We will see that this case is quite typical of the non homogeneous prebifurcation state, as it will involve, in a very specific way, the motion of the elastic-plastic boundary.

c) If there exists at least one point of the elastic-plastic boundary where the buckling mode K ($\mathbf{x}$) is not equal to zero (Case 3), then, it is immediately seen, at least from figure 10, that (3.14) cannot be satisfied with a non zero constant C. But if C is equal to zero, then, by (3.13), the solution is unique. Only a few mechanical examples seem to exist of that case. Anyway, it can be established that it leads to a tangent bifurcation [25].

Notice that the latter situation is quite particular since we find a case for which the solution is unique, when we have previously computed a critical load and a buckling mode. This kind of tangent bifurcation seems to be in agreement with a result [6] asserting that velocities actually different can exist for loads slightly higher than $\lambda_o$.

Most of the usual examples, simple models such as the continuous Shanley's one due to J.W. Hutchinson [14], or simple structures such as compressed beams, deal with buckling from a homogeneous state. Now, we give a few examples of the possible cases from a non homogeneous state.

Example 1 :

As a simple model, we first recall the one given by N.Triantafillydis [25], and presented on figure 11 :

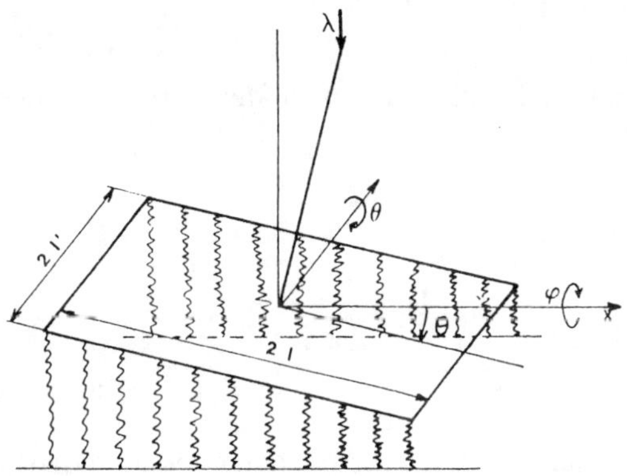


Figure 11. The system of N. Triantafillydis.

A complete analysis of this model has been given in [25]. In this chapter, we only want to point to the fact that it may lead to Case 3 of the previous classification. It is clear on Figure 2 that the solution depends only on three variables : u for the vertical displacement, and the angles θ and φ. On both sides of the structure, the strain velocities are respectively :

$$\dot{\xi}_1 = \dot{u} + x\,\dot{\theta} + l'\dot{\varphi}$$

$$\dot{\xi}_2 = \dot{u} + x\,\dot{\theta} - l'\dot{\varphi}$$

We assume that the external loading is such that there exist a symmetric fundamental branch $\varphi^f(\lambda) \equiv 0$, and, on the x-axis, both a loading and an unloading zones whose boundary is, for small $\theta^f$ :

$$x_n = -\frac{\dot{u}^f}{\dot{\theta}^f}$$

We look at which kind of phenomena we may get if the symmetry $\{u, \theta, \varphi\} \to \{u, \theta, \varphi\}$ is broken. Of course, the corresponding buckling mode is :

$$\{U, \Theta, \Phi\} = \{0, 0, 1\}.$$

Then, by (3.13), the initial strain velocity on the corresponding bifurcated branch would be :

$$\dot{\xi}_1^0 = \dot{\xi}^f + l'C$$

$$\dot{\xi}_2^0 = \dot{\xi}^f - l'C$$

That means any non zero constant C would led to a sudden change from unloading to loading. But, since $\dot{\xi}_1$ and $\dot{\xi}_2$ are bounded, the unknowns u, $\theta$, $\varphi$, and the corresponding strains and stresses are continuous functions of the load and this situation of sudden change is not possible. The constant C is then necessarily equal to zero and, if several branches exist, they are, at the critical load, tangent to the fundamental one.

Example 2 :

The next example anticipates the following sections. But we can assume that the results used to build this example are well known, and the reader may refer to section IV.1 for the preliminary equations.The example is a simply compressed beam with a non uniform cross-section $S(x)$, such that assumption H1ter holds, which is quite possible since the stress of simple compression is $\lambda/S(x)$.

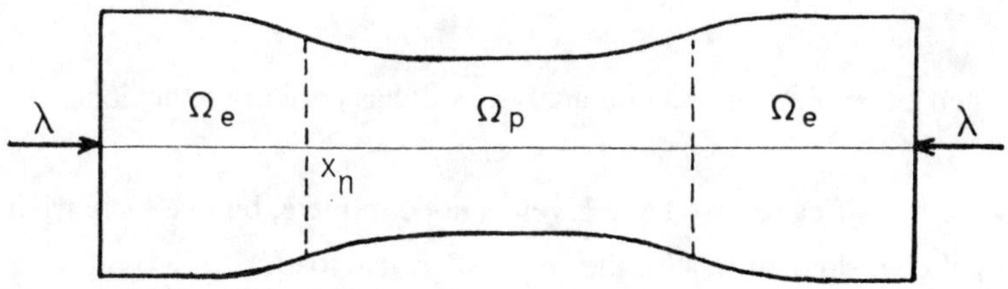


Figure 12. Buckling of a beam of variable cross-section.

The fundamental solution, of simple compression, is :

$$v^f(x) \equiv 0$$
$$k^f(x) = -\frac{1}{S(x)}\frac{E - E_t(x)}{E\,E_t(x)},$$

and $k^f$, continuous because of assumption H4, reaches zero at the elastic-plastic boundary. The buckling mode is solution of :

$$\text{(3.16)}\qquad \begin{aligned} &E\,I(x)\,V''(x) + \lambda_o V(x) = 0\ , \text{ in } \Omega_e \\ &E_t(x)\,I(x)\,V''(x) + \lambda_o V(x) = 0\ , \text{ in } \Omega_p \\ &V(0) = V(l) = 0 \end{aligned}$$

and :

$$K(x, y) = 0, \text{ in } \Omega_e$$
$$K(x, y) = \left(1 - \frac{E_t(x)}{E}\right) y\,V''(x), \text{ in } \Omega_p$$

$I(x)$ is assumed to be continuous. Then (3.16) is a second order ordinary differential equation with continuous coefficients. It means that the solution is two times continuously differentiable and K is continuous over the whole length of the beam. We can consequently get either Case 1 or Case 2 but never Case 3 for such simple structures and loadings.

Comments :

i) The previous bifurcation analysis strongly depends on continuity assumptions (H2 for instance) difficult to remove.

ii) Then a complete bifurcation analysis still has partially to be done.

iii) The classification we have given is not complete, but we have tried to pick up the most significant singularities for the smallest critical load.

## IV - BIFURCATION AND POST-BIFURCATION OF AN ELASTIC PLASTIC COMPRESSED BEAM

### IV-1 Buckling of an elastic-plastic structure

#### IV-1-a Equations of the problem

As it was noticed in the previous section, a two fields variational formulation is quite convenient to study the buckling of an elastic-plastic structure. Such is the pair { velocity, plastic multiplier }. However, for simple structures, the pair { velocity, stress rate } may lead to somewhat simpler computations. The latter is chosen in the following section. The problem at hand is an initially perfectly straight simply supported compressed beam undergoing to bending beyond the yield plastic stress. Anyway, in order to give some precise technical methods without too much loss of generality, we keep assumption H1ter in the form :

H1ter : the beam has a variable cross section such that, at the bifurcation load, it is divided into both an elastic and a plastic part, say $\Omega^o{}_e$ and $\Omega^o{}_p$, which will become $\Omega_e$ and $\Omega_p$ beyond the bifurcation load.

Such a beam is represented on figure 13 with the principal corresponding notations.

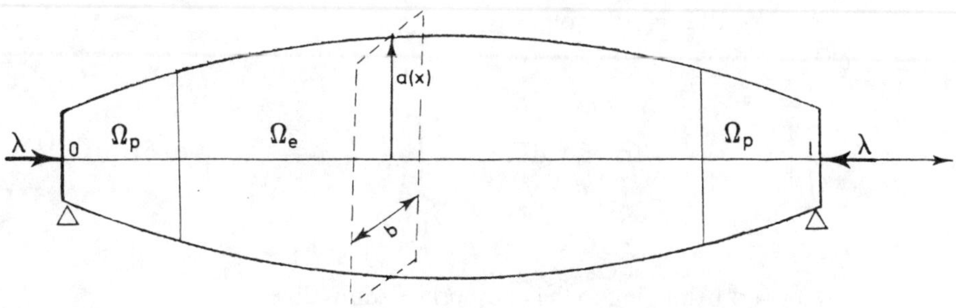


Figure 13. The heterogeneous compressed beam model problem.

An eventual unloading zone will be denoted by $\Omega_u$. The classical case of a beam with a constant cross section uniformly strained at the critical load will naturally come as a particular case.

Let us now precise some points concerning the constitutive law.

H5 : The traction curve is strictly increasing over the whole range of the external applied load.

Of course this assumption is connected to H2. Concerning the moduli, we take the following unusual convention : $\sigma$ and $\gamma$ will be respectively the stress and the strain within the classical beam theory framework ; then we denote by $E_t(\sigma)$ any modulus on the path beginning at $\sigma = 0$ and $\gamma = 0$ under increasing stress, independantly of the stress level. Of course, this definition includes the usual young modulus on the initial loading path. On the other hand, when the corresponding material point has already been under plastic strain, then we shall reserve the notation E for the slope of the elastic traction curve. For the buckling problem, $E^o{}_t$ will denote the value of the modulus at the critical load, and $E^o{}_t\ (\lambda_o,x)$ will recall explicity that it depends on $\lambda_o$, and is a function of x.

The incremental equilibrium equation and the elastic plastic constitutive law give the following variational problem :

$$\int_{\Omega} \sigma . \, \delta\gamma \; d\Omega \; = \; < \dot{f} \, , \delta u >$$

(4.1)

$$\int_{\Omega} \left( \dot{\gamma} \; - \; \frac{1}{E_t(\sigma)} \; \dot{\sigma} \right) \delta\sigma \; d\Omega \; = \; \int_{\Omega_u} \left( \frac{1}{E} \; - \; \frac{1}{E_t(\sigma)} \right) \dot{\sigma} \; \delta\sigma \; d\Omega$$

$\forall$ $\delta u$, $\delta\sigma$ belonging to the appropriate admissible set.

u and v are the axial and transverse displacements, $\gamma$ is the classical non linear measure of the strain :

$$\gamma(x,y) = u'(x) + \frac{[v'(x)]^2}{2} - y \; v''(x),$$

where ( )' denotes a derivative with respect to the longitudinal coordinate x.

In order to expand the equations (4.1), and accounting with the well known results of the beam theory, we restrict $\delta\gamma$ and $\delta\sigma$ to the following test functions :

(4.2)

$$\delta\gamma = \delta u' + v' \; \delta v' + y \; \delta v''$$

$$\delta\dot{\gamma} = \dot{v} \; \delta v'$$

$$\delta\sigma = \frac{\delta \; N(x)}{S(x)} + y \; \frac{\delta \; M(x)}{I(x)}$$

N(x) and M(x) are respectively a resultant and a moment of the stress over the cross section, whose area and moment of inertia are S(x) and I(x).

These variations can be easily inserted in (4.2) and we get :

$$\int_0^1 \{ \dot{N}(\delta u' + v'\,\delta v') - \dot{M}\,\delta v'' + N\,\dot{v}'\,\delta v'\}\,dx = -\,\delta u(1)$$

(4.3)

$$\int_0^1 \left\{ \left( \dot{u}' + v'\dot{v}' - \frac{\dot{N}}{E^o_t S} \right) \delta N - \left( \dot{v}'' + \frac{\dot{M}}{E^o_t I} \right) \delta M \right\} dx$$

$$= \int_{\Omega_u} \left( \frac{1}{E} - \frac{1}{E_t(\sigma)} \right) \dot{\sigma} \left( \frac{\delta N}{S} + y\,\frac{\delta M}{I} \right) d\Omega$$

$$+ \int_{\Omega} \left( \frac{1}{E_t(\sigma)} - \frac{1}{E^o_t} \right) \dot{\sigma} \left( \frac{\delta N}{S} + y\,\frac{\delta M}{I} \right) d\Omega$$

The first aim of this section is to obtain a bifurcation point for this problem. Next, we shall explain in details a specific method for the computation of the post critical behavior. Let us remark for the moment a particularity of equations (4.3) which will appear as specifically important at the end of the post critical analysis. The left hand sides only involve integrals over the longitudinal axis with functions depending only on x. This is a result of the choices (4.2) and of elementary integrations, and it is quite natural within beam theory. But, independantly of the beam theory and of the choices (4.2), the right hand sides involve three dimensional quantities, with integrals over three dimensional domains.

IV-1-b <u>The first order solution.</u>

According to the general results of section III, the first step of the buckling analysis for problem (4.3) is to get the critical load and the buckling mode. We recall that problem (3.11) states that this is obtained by neglecting any eventual unloading and putting $\dot{f} = 0$.

To build up the stiffness operator at the critical load, we have to separate the initial velocity quantities, and the post critical quantities for which we shall establish that they are connected to unloading. To make this distinction, exponents f and 0 always indicates that the quantities are taken respectively on the fundamental solution on the initial bifurcated velocity, while an exponent $*$ will designate a velocity increment for a post critical load.

That way, for any load higher than the critical one, the bifurcated velocity will be written on the form :

$$\begin{aligned} \dot{N} &= \dot{N}^{\circ} + N^{*} \\ \dot{U} &= \dot{U}^{\circ} + U^{*} \qquad (4.4) \\ \dot{M} &= \dot{M}^{\circ} + M^{*} \\ \dot{V} &= \dot{V}^{\circ} + V^{*} \end{aligned}$$

Then, for the initial bifurcated velocity, the formulation (4.3) reduces to the following problem :

$$(4.5) \quad \begin{aligned} &\int_{o}^{1} \left\{ \dot{N}^{\circ}\, \delta u' - \dot{M}^{\circ}\, \delta v'' - \lambda_{o}\, \dot{v}^{\circ}\, \delta v' \right\} dx = -\, \delta u(1) \\ &\int_{o}^{1} \left\{ \left( \dot{u}^{\circ\,\prime} - \frac{\dot{N}^{\circ}}{E_t^{o} S} \right) \delta N - \left( \dot{v}^{\circ\,\prime\prime} + \frac{\dot{M}^{\circ}}{E_t^{o} I} \right) \delta M \right\} dx = 0 \end{aligned}$$

To have a more concise expression, we denote by $\mathbf{L}$ the tangent operator associated to the bilinear form of the left hand side of problem (4.5) :

(4.6) $$< \mathbf{L} u , \delta u > = \int_0^l \{ N \, \delta u' - M \, \delta v'' - \lambda_o \, v' \, \delta v' \} \, dx$$

$$+ \int_0^l \left\{ \left( u' - \frac{N}{E_t^o S} \right) \delta N - \left( v'' + \frac{M}{E_t^o I} \right) \delta M \right\} dx$$

It is elementary to get that, on the vector space of the variation $\{\delta N, \delta u, \delta M, \delta v\}$, the operator $\mathbf{L}$ is singular, and its null-space, that is the buckling mode as defined by (3.11), is generated by a vector $\{N, U, M, V\}$ solution of :

(4.7)
$$N = 0 \qquad \text{(a)}$$
$$U = 0 \qquad \text{(b)}$$
$$M = - E_t^o (x, \lambda_o) \, I(x) \, V''(x) \qquad \text{(c)}$$
$$(E_t^o (x, \lambda_o) \, I(x) \, V''(x))'' + \lambda_o \, V''(x) = 0 \qquad \text{(d)}.$$

Let us point to some remarks concerning these modal quantities, and specially concerning the differential equation (4.7.d).

i) If the cross-section of the beam is uniform, then (4.7) reduces to :

(4.8)
$$N = 0 \qquad \text{(a)}$$
$$U = 0 \qquad \text{(b)}$$
$$U = \quad M = - E_t^o \, I \, V''(x) \qquad \text{(c)}$$
$$E_t^o \, I \, V^{(4)}(x) + \lambda_o \, V''(x) = 0 \qquad \text{(d)}$$

With, for instance, simply supported boundary conditions, this gives the well-known result for the first eigenvalue of the linear problem (4.8.d) :

$$\lambda_o = \frac{\Pi^2 E_t^o I}{l^2},$$

$$V(x) = \sin \frac{\Pi}{l} x.$$

ii) Conversely, if the cross-section is non uniform, satisfying the assumption H1ter, then (4.7.d) is a non linear eigenvalue problem. Of course, since this problem is unidimensional, it is numerically small eventhough non linear. But, for a two dimensional case, such as the buckling of an elastic plastic plate non uniformly loaded at its edges, the problem could become very big. As a matter of fact, we have to make an elastic-plastic computation for the plate under a small in-plane compression loading ; then construct the matrix of the bending problem ; and test the existence of an eigenvalue. As, for an initially small enough external load, this matrix is strictly positive, we make another load increment which gives new shape and position of the elastic-plastic boundary. We then construct a new matrix for the bending problem, and make another test for an eigenvalue ; and so on.

For the one-dimensional case, we give on Figure 5 several examples of the shape of the buckling modes for different aspect ratio and moduli of the beams.

a) $\frac{E}{E_t}$ fixed, l fixed ;
O : $\frac{r\ min}{r\ max} = 0.33$, Δ : $\frac{r\ min}{r\ max} = 0.66$

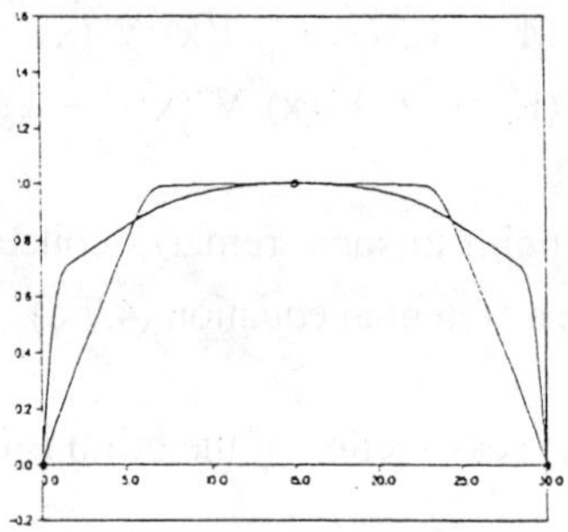

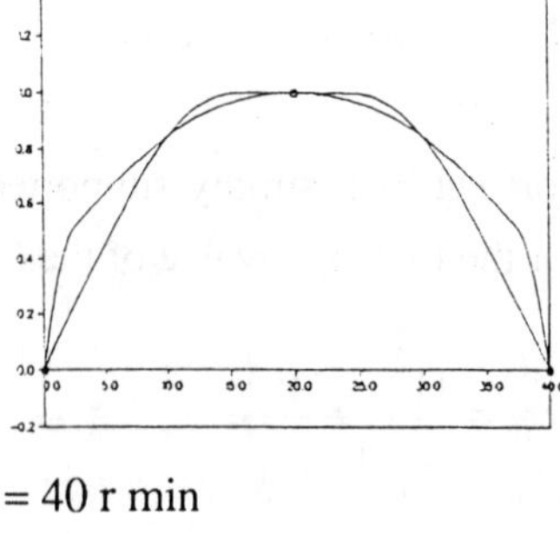

b) $\frac{E}{E_t}$ fixed, $\frac{r\ min}{r\ max}$ fixed ; O : l = 20 r min, Δ : l = 40 r min

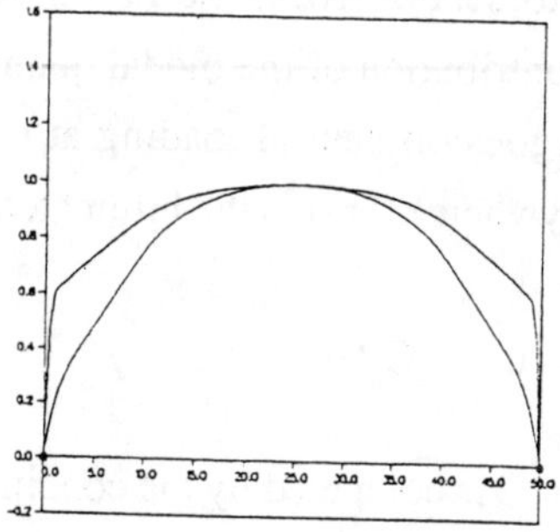

c) $\frac{r\,min}{r\,max}$ fixed, l fixed ; O : $\frac{E}{E_t} = 10$ , Δ : $\frac{E}{E_t} = 100$

Figure 14. Buckling modes for beams with a variable cross-section.

From Figure 14a or 14c remark that the shape of the buckling mode indicates that the buckling mechanism may tend to something like a knee-joint as the parameter taken into account tends to zero, rmin/rmax or $E_t/E$ for instance. On the other hand, for $E_t/E = 0.1$ we have got, for fixed other parameters $\lambda_o = 1559$ N, when, with the same other parameters, we have got $\lambda_o = 839$ N for $E_t/E = 0.01$. We then get that the buckling load is strongly dependant on the value of the tangent modulus, and specially on the modulus immediately beyond the yield stress since the elastic-plastic boundary belongs to the domain. This result is of special interest for engineering applications, more especially as the tangent modulus is usually not well-known in this range.

iii) When looking at the type of bifurcation that may occur from a heterogeneous state, we have got in section III that only Case 1 or Case 2 are possible for a simply supported compressed beam. From the differential equation (4.7.d), it should also be stated that it is possible to discuss explicity the position of the initial neutrally loaded point with respect to the elastic-plastic boundary, and to get that the choice between bifurcation of Case 1 and bifurcation of Case 2 only depends on the axial variation of the cross-section [17].

According to section III.2, the initial velocity is the sum of the fundamental solution and of a contribution of the modal quantities. And this contribution is such that the structure undergoes on neutral loading at one point (from a state non uniform but strictly loaded everywhere). That is the bifurcated solution is of the form :

$$\dot{u}^{\circ} = \dot{u}^{f} + C\,\mathbf{U}$$

The constant C is computed by the condition that the plastic-multiplier component, or the stress-rate component in the one dimensional case, reaches zero at one point denoted in the following $(x_n, y_n)$. For the compressed beam we get, in accordance with (3.1) :

$$\dot{\sigma}^{\circ}(x, y) = \frac{1}{S(x)}\left(1 - \frac{y}{y_N}\left[\frac{r(x_N)}{r(x)}\right]^2 \frac{M(x)}{M(x_N)}\right), \tag{4.9}$$

where r(x) is the radius of gyration of the cross-section ($r^2 = I(x)/S(x)$).

## IV-2 Post-critical analysis

### IV-2-a Qualitative description of the post-buckling

Taking the way we choose when revisiting the Shanley result for the existence of a neutrally loaded point at the bifurcation load, we begin by a short discussion concerning the post-bifurcation solutions. The aim of this part is then not directly a step for the post-critical analysis, but a better understanding of the main qualitative features of the phenomenon. We want to look at which post-critical velocity will be chosen by the structure in the set of admissible velocities, and to get some qualitative characterization of this velocity. But now we do that both with qualitative arguments and with numerical computations.

At first, let us notice that, as we have established that a neutrally loaded point necessarily occurs at the first bifurcation load, we get, by exactly the same analysis, the existence of a non empty unloading zone at any post-bifurcation load. Actually, suppose that beyond the bifurcation load, the initial neutrally loaded point remains on neutral loading, or reverts strict loading again. Then we would have a one-to-one stress-strain relation everywhere in the beam.

That means in particular that the buckling of such a beam could be described by the framework of elastic bifurcation theory and, consequently, we would be allowed to conclude that the bifurcated branch must have begun under constant loading, contrary to the initial velocity analysis.

To show more clearly this statement, we give the results of a numerical computation. We have represented on the Figure 6 the part of a compressed beam of variable cross-section where the elastic-plastic boundary is located. Of course, the load steps and the mesh size must have been very small, and the initial geometry must have included a slight imperfection.

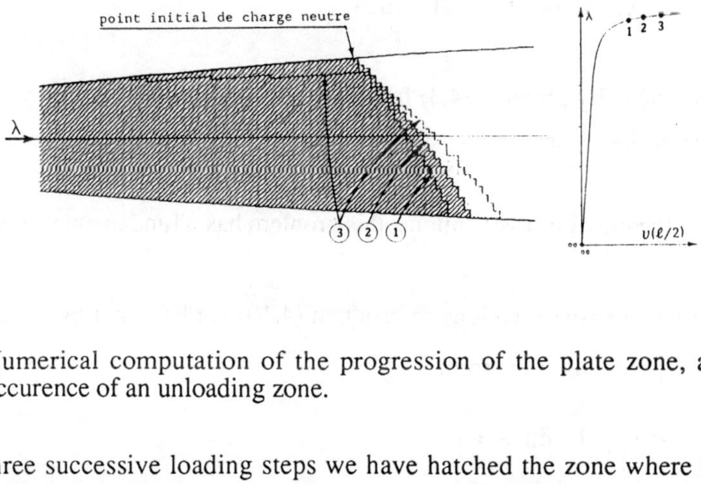


Figure 15. Numerical computation of the progression of the plate zone, and of the occurence of an unloading zone.

For three successive loading steps we have hatched the zone where the plastic strain is not equal to zero, and we have plotted the corresponding values of the load on the load/displacement curve opposite. Comparing steps 1 and 2, we see that the boundary of the plastic loading zone has moved forward everywhere except at one point, that is has rotated around this point. Comparing steps 2 and 3, we see that the boundary has moved back in the previously strictly loaded zone, that is an unloading zone has spread out from the initial neutrally loaded point.

Then, it seems necessary that the post-critical analysis involves a local analysis, in order to get results about the evolution of the boundaries. The numerical computation clearly indicates that, in the generic case of buckling from a heterogeneous state, the post-buckling problem will have two free boundaries, since the prebifurcation elastic-plastic boundary splits into two qualitatively different parts in the post-bifurcation solution.

IV-2-b The bifurcation method.

Because of the moving boundaries, the post-buckling analysis will involve some technical difficulties. Then, in order to distinguish clearly the technical computations and the main steps of the method itself, we begin by a short summary of that method.

The problem (4.3) can be formally written as :

(4.10) $$< A_\lambda \, \dot{\mathbf{u}}, \, \delta\mathbf{u} > \; = \; < \mathbf{f} \, , \delta\mathbf{u} >$$

Such is actually problem (4.3) for which $\mathbf{u}$ is a vector ($\dot{N}$, $\dot{u}$, $\dot{M}$, $\dot{v}$), and $A_\lambda$ is a non linear operator depending on the load $\lambda$, and involving integrals over moving domains.

Under the previous assumptions, this problem has a fundamental solution for any $\lambda \geq 0$.

After some transformations of problem (4.10) we have got two vectors $\mathbf{U}$ and $\dot{\mathbf{u}}^{\circ}$, solution of problems of the form :

(4.11) $$< A_{\lambda o} \, \mathbf{U}, \delta\mathbf{u} > \; = \; 0$$

(4.12) $$< A_{\lambda o} \, \dot{\mathbf{u}}^{\circ}, \delta\mathbf{u} > \; = \; < \mathbf{f}^{\circ} , \delta\mathbf{u} >$$

$A_{\lambda o}$ is obtained from $A_\lambda$ by a linearization, and in particular, according to (3.11), the fundamental inequality character of the constitutive relation has been dropped by neglecting any unloading eventuality. Moreover, this linear operator $A_{\lambda o}$ is self-adjoint. Then, by (4.11), $\mathbf{U}$ is the buckling mode and $\lambda_o$ the bifurcation load. By (4.12), $\dot{\mathbf{u}}^{\circ}$ is the initial velocity for $\lambda = \lambda_o$ on the non fundamental branch. At this step, the first order bifurcation problem is completely solved. Going further is the goal of the post-buckling analysis.

It is now usual to look for post-buckling solutions in the form of power series expansions with respect to a small parameter related to the load increment, such as Koiter has, the first, given in élasticity. Within plasticity, it has also been assumed that the solution $\dot{\mathbf{u}}$ can be expanded into power series.

For the first way suggested [14], it was assumed that power expansions of the following form hold :

(4.13) $$\lambda = \lambda_o + \lambda_1 \xi + \lambda_2 \xi^{1+\beta} + \ldots\ldots\ldots\ldots\ldots,$$

and the same for the solution $\mathbf{u}$. $\xi$ denotes an amplitude of the eigenmode.

A very important particularity, by comparison with elastic-post-buckling theory, is that $\beta$ is a positive rational number strictly lower than 1.

A slightly different choice was to assume integer power series of the form [16] :

(4.14) $$\lambda = \lambda_o + \mu_1 \tau + \mu_2 \tau^2 + \ldots\ldots\ldots\ldots\ldots$$

Of course, since these two expansions (4.13) and (4.14) have to be the same, the kinematic parameter $\tau$ for instance should be something like $\xi^\beta$.

Anyway, the main problem of the initial post-buckling theory within elastic-plasticity is to get the fractional exponent b, and the corresponding coefficient $\lambda_2$. Up to now, this analysis either has received a technically very difficult solution, or has only been partially carried out.

The method we suggest to answer to this question is close to the classical perturbation methods. We choose a positive load increment, and we directly look for a perturbation of the velocity answer.

Precisely, the initial solution $(\dot{\mathbf{u}}^\circ, \lambda_o)$ being given by the first order problem, the post-critical solution is assumed to be :

(4.15) $\quad \dot{\mathbf{u}} = \dot{\mathbf{u}}^{\circ} + \mathbf{u}^{*}$ for $\lambda = \lambda_o + \mu$

That way, we intend to delay, up to the end of the analysis, the discussion on the value of the fractional exponent $\beta$.

Then, inserting (4.15) into (4.10), and accounting with (4.12), we get, for the perturbation $\mathbf{u}^*$, a problem of the form :

(4.16) $\quad < A_{\lambda_o} \mathbf{u}^* , \delta\mathbf{u} > \; = \; < \mathbf{f}^*, \delta\mathbf{u} >$

Moreover, since $A_{\lambda_o}$ is a singular self-adjoint operator which null-space is generated by $\mathbf{U}$, there is a compatibility condition for the equation (4.16) of the following form :

(4.17) $\quad < \mathbf{f}^* , \mathbf{U} > \; = \; 0 \;,$

where the braket denotes the appropriate scalar product.

Of course, the difficult step of this analysis is to obtain equation (4.16) from equation (4.10). As suggested by the numerical computation of chapter IV-2-a, the motion of the boundaries will require special attention. For instance the unloading zone grows inside the initially strictly loaded plastic zone, but the latter goes on moving forward into the elastic zone. That means this perturbation method will need a kind of perturbation of the geometry of the domains. As a matter of fact, only that point is technically a little bit difficult and involves mathematical justifications.

Anyway, accounting with estimates of the motion of the boundaries, the right hand side of equation (4.16) appears as a function of the load increment $\mu$ and of a measure, say $\tau$, of the variations of the domains, that is, precisely here, of the extension of the unloading zone. Then, the condition (4.17) naturally appears as a relation between the load increment and of the size of the unloading zone. This relation will be established, anticipating the following results, in the form :

(4.18) $\quad \mu \; = \; \text{Coef} \;\; \tau^r \; + \; \text{h.o.t.}$

where h.o.t. stands for terms which tend to zero faster than $\tau^r$.

To get the exponent r, and the coefficient, we shall have to estimate the equations of the boundaries, the order of the unknowns in the moving domains, and so on. But it is very important to notice that this relation is got without any assumption concerning the kind of power expansion we have to choose, and the kind of kinematic parameter allowing such an expansion. The length of the unloading zone naturally appears as the good candidate for that.

Then, since we shall get, by the way, that the solution should be written as :

$$\dot{u} = \dot{u}^{\circ} + \dot{u}^{1}\,\tau + \;.....,$$

which we could write :

$$\dot{u} = \dot{u}^{\circ} + \hat{\dot{u}}^{1}\,\mu^{1/r} + \text{h.o.t.}$$

Then, after an integration with respect to $\mu$, that leads to an expansion with the specific fractional exponent we are looking for :

$$u = u^{\circ} + \dot{u}^{\circ}\,\mu + \hat{\dot{u}}^{1}\,\mu^{1+1/r} + \text{h.o.t.}$$

Our first aim was to delay, as far as possible, the difficult discussion on the fractional exponents. The result is that this discussion is completely dropped, and only remains one simple mathematical tool concerning the properties of equations with a singular self-adjoint operator.

### IV-2-c Local analysis, and description of the boundaries

As suggested in the previous chapters, we see that the free boundary character of the post.buckling problem is a very important difference between elastic and elastic-plastic post buckling theories. The goal of the present local analysis is an estimate of the equations, and of the motions of the boundaries.

In order to give results with a sufficient generality, we essentially study a typical case of the post buckling from a heterogeneous state. According to section III-2, we choose Case 2 of the classification. As we made for the first order problem, we will point to the difference between heterogeneous and homogeneous case when this difference will be significant. Moreover this difference will sometimes be very close to the difference between Case 2 and Case 1.

The corresponding free boundary problems are qualitatively described by the figure 16 on which we have represented, first a beam with an adequate aspect ratio such than Case 2 occurs, second a beam with a uniform cross section. We have drawn only the part of the beam where the initial neutrally loaded point is located.

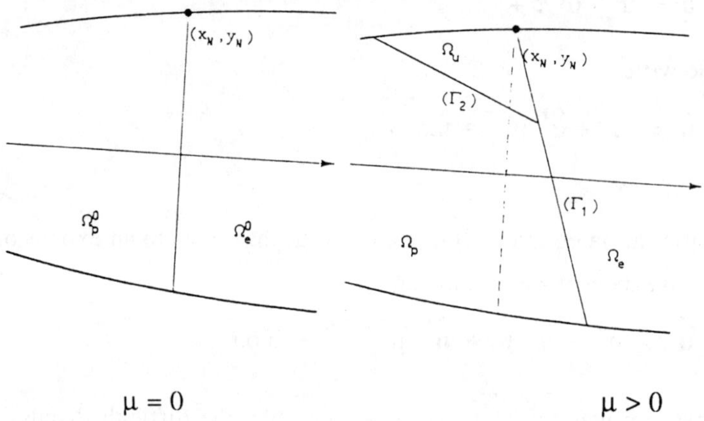


a - From a heterogeneous state, the post-buckling problem has two free boundaries.

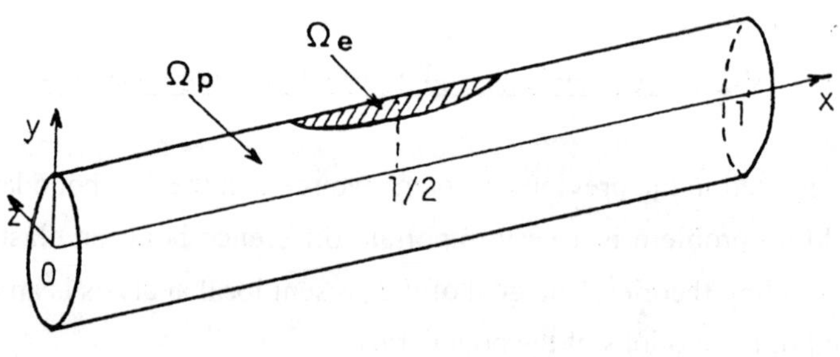


b - From a homogeneous state, the post-buckling problem has only one free boundary.

Figure 16. Local evolution of the boundaries.

The qualitative shapes of the boundaries, and especially of the domain $\Omega_u$, anticipate the following results. Of course these boundaries are unknowns of the problem.

However, it will be useful to keep in mind a few simple results :

i) The stress $\sigma$ is continuous accross the boundary $(\Gamma_1)$ between elastic and plastic loading zones.

ii) The stress rate $\dot{\sigma}$ is continuous across the boundary $(\Gamma_2)$ between plastic loading and elastic unloading zones.

iii) The initial stress velocity $\dot{\sigma}^o$ has derivatives up to the order we need.

Only point ii needs a little care. It is straightforward that $\dot{\sigma}$ is continuous on each side of the boundary $(\Gamma_2)$. Because the strain rate $\dot{\gamma}$ is continuous in y, and because unloading and strict loading respectively mean $\dot{\gamma} < 0$ and $\dot{\gamma} > 0$, the strain rate $\dot{\gamma}$ is equal to zero on $(\Gamma_2)$. Then the stress rate $\dot{\sigma}$ is also zero on $(\Gamma_2)$ since it is $E\dot{\gamma}$ and $E_t\dot{\gamma}$ on both sides. So $\dot{\sigma}$ has no discontinuity through $(\Gamma_2)$, which establishes the result.

In the remainder of this paper, we will refer to these simple results as "regularity results".

As a consequence of this regularity, the post-critical evolution will be described by two equations :

* The equation of the boundary $(\Gamma_1)$ is :

(4.19) $$\sigma\,(x, y) = -\sigma_y$$

* The equation of the boundary $(\Gamma_2)$ is :

(4.20) $$\dot{\sigma}(x, y) = 0$$

We then try to get asymptotic approximations for the equations of the curves $(\Gamma_1)$ and $(\Gamma_2)$.

Although this is unessential for the analysis, we make two supplementary assumptions concerning the geometry of the beam :

H6 : As shown on figure 4, the beam is assumed to have a rectangular cross-section which thicknesses are respectively 2a and b.

The generalization to any cross-section will given at the end, and will not be difficult.

H6bis : The cross-section varies slowly along the longitudinal axis.

The latter assumption is made in order to simplify the discussion on the relative contributions of the derivatives of the cross-section with respect to x. This derivative at the point $(x_N, y_N)$ is obviously of the order of the variation $2(a_{max} - a_{min})/l$, denoted by $\eta$.

<u>The first free boundary is $(\Gamma_1)$</u>

The stress $\sigma(x, y, \mu)$ can be deduced from the stress rate :

$$\sigma(x, y, \mu) = \sigma^0(x, y) + \mu \dot{\sigma}^0(x, y) + \int_0^{\mu} \sigma^*(x, y, s)\, ds.$$

As $\dot{\sigma}$ and therefore $\sigma^*$ are continuous, and as $\dot{\sigma}^{\circ}$ is a smooth function, we can write the following Taylor expansions of the stress close to the neutrally loaded point $(x_N, y_N)$ :

$$\sigma(x, y, \mu) = \sigma_y + (x - x_N) \left.\frac{\partial \dot{\sigma}^{\circ}}{\partial x}\right)_N + \mu (x - x_N) \left.\frac{\partial \dot{\sigma}^{\circ}}{\partial x}\right)_N$$

$$+ \mu (y - y_N) \left.\frac{\partial \dot{\sigma}^{\circ}}{\partial y}\right)_N + \mu\, \sigma_N^* + \text{h.o.t},$$

where the subscript N indicates that the value of the function is taken at $(x_N, y_N)$ .

Then, from (4.19) we get the equation of $(\Gamma_1)$ :

$$(4.21)\quad x - x_N = -\frac{1}{S(x_N)\, y_N} \frac{\mu}{\left.\dfrac{\partial \sigma^0}{\partial x}\right)_N + \mu \left.\dfrac{\partial \dot{\sigma}^0}{\partial x}\right)_N} (y - y_N) + 0\,(\mu\sigma^* + \mu^2).$$

We will speak of immediate post-buckling for such a structure if the following estimate holds :

$$(4.22)\qquad \frac{\mu}{\lambda_0} < \eta .$$

Then, for the immediate post-buckling, the equation (4.21) indicates that the slope of the boundary $(\Gamma_1)$ is of the order of $\mu^{-1}$. We will limit ourselves to this case for the explicit computations, in order to get some simplications about unessential technical points.

<u>The second free boundary is $(\Gamma_2)$</u>

The method we use to study this boundary, and to describe both the geometry of the unloading zone and the value of the solution in this unloading zone, essentially deals with the introduction of small parameter, by mean of a small parameter $\tau$, which is the longitudinal length of the unloading zone.

Precisely, let us change the scale on the x-coordinate of any point of $\Omega_u$ in the following way :

$$x = x_N - \tau X , \qquad X \in [0, 1].$$

Then, it does not seem too restrictive to write the equation of the boundary $(\Gamma_2)$ as :

$$y = y_N - g\,(X, \tau).$$

Recall that, for the homogeneous case, the post-buckling involves only the latter free boundary ($\Gamma_2$). Moreover, this function g (X, $\tau$) will not be the same for the homogeneous and for the heterogeneous cases.

Actually, for the homogeneous case, the first order problem is such that the buckling mode is locally quadratic in a neighbourghood of the neutrally loaded point. Consequently, it is natural to look for a function g (X, $\tau$) of the form :

$$g(X, \tau) = \tau^2 \left[f_2(X) + \tau f_3(X) + \ldots\right]$$

This choice equally holds for post-buckling from a heterogeneous state, in the Case 1 of the classification.

But, from Case 2 of the classification, the point $(x_N, y_N)$ is such that the mode is locally linear. Then, with respect to the extend $\tau$ of $\Omega_u$, there is no reason to drop a priori the linear term, and g (X, $\tau$) will be written as :

$$g(X, \tau) = \tau \left[f_1(X) + \tau f_2(X) + \ldots\right].$$

Thus we see that another stretched coordinate Y can be defined on the y-axis in such a way that any point (x, y) belonging to $\Omega_u$ is mapped to (X, Y) belonging to [0, 1] X [0, 1] by :

$$x = x_N - \tau X$$

(4.23) $$y = y_N - \tau Y \left[f_1(X) + \tau f_2(X)\right] + 0(\tau^3)$$

$$0 \leq X \leq 1 \ , \ 0 \leq Y \leq 1 \ ,$$

and the corresponding expressions for the homogeneous case.

These expressions are close to the one used by Hutchinson [14]. The main difference is that we do not want to deal with a singular perturbation expansion, as the unloading zone is not a boundary layer, but only to map the small variable domain $\Omega_u$ to a fixed domain.

The shape of the unloading zone will then be described by the functions $f_i$ (X) introduced in (4.23) but, of course, this description depends on a parameter $\tau$ which remains unknown. That means the next step of the post-buckling analysis consists in connecting this parameter $\tau$ with the load increment $\mu$.

As a consequence of the previous regularity results, the following Taylor expansion holds :

$$\dot{\sigma}(x, y) = (x - x_N) \left.\frac{\partial \dot{\sigma}^0}{\partial x}\right)_N + \frac{1}{2} (x - x_N)^2 \left.\frac{\partial^2 \dot{\sigma}^0}{\partial x^2}\right)_N$$

$$+ (y - y_N) \left.\frac{\partial \dot{\sigma}^0}{\partial y}\right)_N + \sigma_N^* + \text{h.o.t,}$$

and can be inserted in the equation of the boundary ($\Gamma_2$). Using the stretched coordinates, this yields :

$$0 = -\tau X \left.\frac{\partial \dot{\sigma}^0}{\partial x}\right)_N + \frac{1}{2}\tau^2 X^2 \left.\frac{\partial^2 \dot{\sigma}^0}{\partial x^2}\right)_N \tag{4.24}$$

$$- \tau f_1(X) \left.\frac{\partial \dot{\sigma}^0}{\partial y}\right)_N - \tau^2 f_2(X) \left.\frac{\partial^2 \dot{\sigma}^0}{\partial y^2}\right)_N + \sigma_N^* + \text{h.o.t.}$$

The variables X and Y have been chosen, by (4.23), in such a way that (X = 1, Y = 1) is the intersection between the boundary ($\Gamma_2$) and the external surface of the beam. For the beam with a rectangular cross-section, this holds at a point such that :

$$y_o - \tau f_1(1) = y_o + \tau \left.\frac{1}{b}\frac{dS}{dx}\right)_N + \text{h.o.t.}$$

That means $f_1$ (1) is naturally of the order of $\eta$. Putting X = 1 in the equation (4.24) indicates that the stress perturbation is of the order of $\tau$ at the initially neutrally loaded point :

$$\sigma^*(x_N, y_N) = \tau\,[(\sigma_1^*(x_N, y_N) + 0(\tau + \eta)],$$

(4.25) with

$$\sigma_1^*(x_N, y_N) = \left.\frac{\partial \dot{\sigma}^0}{\partial x}\right)_N .$$

Then, by (4.24) and (4.25), we get an analytical expression of the function $f_1(X)$ for the first order of the expansion of the boundary $(\Gamma_2)$ :

$$f_1(X) = (1 - X)\left[\left.\frac{\partial \dot{\sigma}^0}{\partial x}\right)_N \Big/ \left.\frac{\partial \dot{\sigma}^0}{\partial y}\right)_N\right] + 0(\eta). \tag{4.26}$$

Since the variational formulation (4.3) includes integrals over $\Omega_u$, we also need an estimate of the velocity $\dot{\sigma}$ in $\Omega_u$.

By the regularity results, we will be allowed to expand the velocity $\dot{\sigma}$ near the initial neutrally loaded point. But we have to take care of the change of moduli across the boundary $(\Gamma_2)$. As indicated on the figure 17, this difficulty can be easily overcome by extending the stress velocity from the plastic loading zone to the unloading zone. This extension is actually possible as $\dot{\sigma}^0$ and $\sigma^*$ have an affine dependence on y.

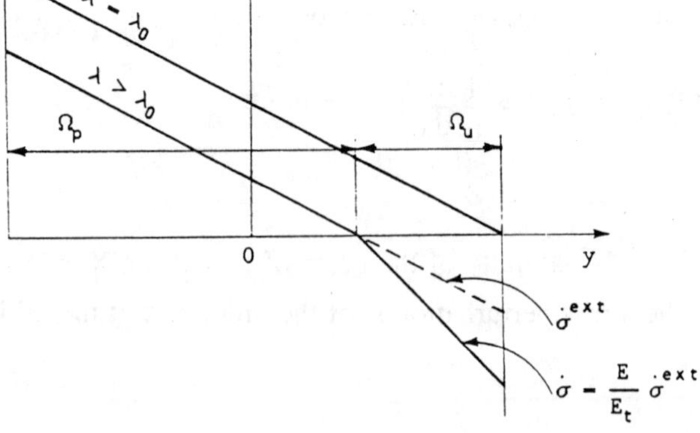


Figure 17. Variation of the stress-rate in a cross-section meeting the unloading zone.

Then, we have to expand not the stress rate itself, but the extended stress rate near $(x_N, y_N)$. Of course, this procedure modifies the result by a factor $E_t/E$. Thus the extended stress velocity in $\Omega_u$, say $\dot{\sigma}^{ext}$, can be expanded in the form :

$$\dot{\sigma}^{ext}(X, Y) = -\tau X \left.\frac{\partial \dot{\sigma}^0}{\partial x}\right)_N - \tau Y f_1(X) \left.\frac{\partial \dot{\sigma}^0}{\partial y}\right)_N$$

$$+ \tau \overset{*}{\sigma_1}(x_N, y_N) + 0(\tau^2 + \tau\eta),$$

where the perturbation $\overset{*}{\sigma_1}(x_N, y_N)$ and the function $f_1(X)$ are respectively given by (4.25) and (4.26).

Consequently, we get in $\Omega_u$ :

(4.27) $$\dot{\sigma}(X, Y) = \frac{E_t}{E}(1 - X)(1 - Y) \left.\frac{\partial \dot{\sigma}^0}{\partial x}\right)_N \tau + 0(\tau^2 + \tau\eta).$$

For the homogeneous case, we would have got, by the same way :

(4.28) $$f_1(X) = 0$$
$$f_2(X) = -\frac{1}{2}(1 - X^2)\left[\left.\frac{\partial^2 \dot{\sigma}^0}{\partial x^2}\right)_N \Big/ \left.\frac{\partial \dot{\sigma}^0}{\partial y}\right)_N\right],$$

and :

(4.29) $$\dot{\sigma}(X, Y) = \frac{1}{2}\frac{E_t}{E}(1 - X^2)(1 - Y) \left.\frac{\partial^2 \dot{\sigma}^0}{\partial x^2}\right)_N \tau^2 + O(\tau^3).$$

To sum up, the local evolution of the plastic and elastic zones is described by figure 18 :

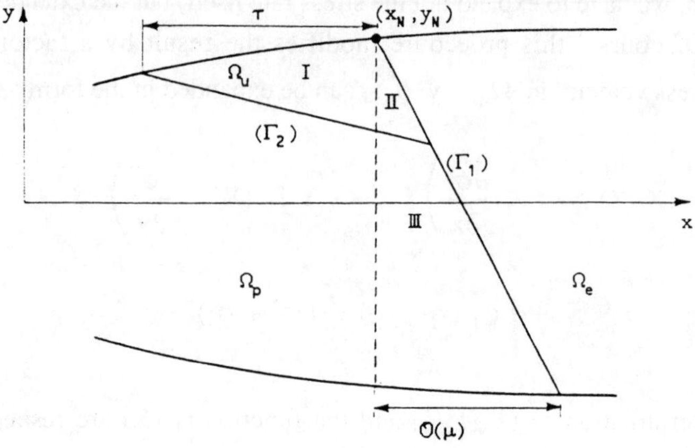


Figure 18. Summary of the different zones and boundaries for a post-buckling state.

By (4.21), (4.23) and (4.26), a qualitative description of the boundaries indicates that ;

* ($\Gamma_1$) is about a straight line, having a slope of the order of $\mu^{-1}$, and which rotates around the initial neutral loading point ;

* ($\Gamma_2$) also is about a straight line, which moves back into the plastic zone $\Omega_p$ from ($x_N$, $y_N$), and which remains parallel to itself during this motion.

Then, as indicated on the figure 18, the post-critical evolution involves three zones, denoted by I, II, and III, which surfaces are respectively of the order of $\tau^2$, of $\mu\tau^2$, and of $\mu$. Hence the zone II is small by comparison with I or III.

IV-2-a Global bifurcation analysis

From the local analysis, we have got estimates for the equations of the boundaries, that is for the size of the moving domains, and for the unknowns increments. Inserting (4.4) into (4.3) we have, at first, obtained a problem for the mode and the initial velocity. For the perturbation $\mathbf{u}^*$, the remaining problem reads :

$$\int_0^1 \{ \dot{N}^* \delta u' - \dot{M}^* \delta v'' - \lambda_o \dot{v}^{*} {}' \delta v'' \} dx$$

$$= \int_0^1 \{ \mu \dot{v}^{o} {}' \delta v' - (\dot{N}^o + N^*) v' \delta v' \} dx$$

$$\int_0^1 \left\{ \left( \dot{u}^{*}{}' - \frac{\dot{N}^*}{E_t^0 S} \right) \delta N - \left( \dot{v}^{*}{}'' + \frac{\dot{M}^*}{E_t^0 I} \right) \delta M \right\} dx$$

(4.30)
$$= \int_{\Omega_u} \left( \frac{1}{E} - \frac{1}{E_t(\sigma)} \right) \dot{\sigma} \left( \frac{\delta N}{S} + y \frac{\delta M}{I} \right) d\Omega$$

$$+ \int_{\Omega} \left( \frac{1}{E_t(\sigma)} - \frac{1}{E_t^0} \right) \dot{\sigma} \left( \frac{\delta N}{S} + y \frac{\delta M}{I} \right) d\Omega$$

$$- \int_0^1 v' (\dot{v}^{o}{}' + v^{*}{}') \delta N \, dx.$$

Then, we apply the method summurized in chapter IV-2-b. Using the notation (4.6), (4.30) can be formally written as :

(4.31) $$< \mathcal{L} \, u^* , \delta u > = < \mathcal{F} (v, \sigma, \tau, \mu) , \delta u >.$$

Moreover, as defined by (4.6), the operator $\mathcal{L}$ is obviously self-adjoint, and its null-space is known by (4.7). That means, the expression $\mathcal{F}$ in the right hand side of (4.31) has to satisfy one solvability condition of the form :

$$< \mathcal{F} , U > = 0.$$

As we know almost explicitly the right-hand side of (4.30) this condition can be written down :

$$2\int_0^1 \mu\, \dot{v}^{0\prime}\, V'\, dx + \int_{\Omega}\left(\frac{1}{E_t(\sigma)} - \frac{1}{E^0_t}\right)\dot{\sigma}\, y\, \frac{M}{I}\, d\Omega$$

(4.32)

$$+ \int_{\Omega_u}\left(\frac{1}{E} - \frac{1}{E_t(\sigma)}\right)\dot{\sigma}\, y\, \frac{M}{I}\, d\Omega + \text{h.o.t} = 0$$

In order to give explicit values, we make another constitutive assumption :

H7 : The stress-strain relation is bilinear, that is involves a constant modulus E before the yield stress, and a constant tangent modulus $E_t$ beyond.

Let us call attention to the fact that, contrary to several previous simplifying assumptions, this one may have important consequences on the bifurcation results. We will have to look at the problems we get when it does not hold.

Under this assumption, the compatibility condition becomes :

$$2\int_0^1 \mu\, \dot{v}^{0\prime}\, V'\, dx + \int_{\Omega_p \cap \Omega_e^0}\left(\frac{1}{E_t} - \frac{1}{E}\right)\dot{\sigma}\, y\, \frac{M}{I}\, d\Omega$$

(4.33)

$$+ \int_{\Omega_u}\left(\frac{1}{E} - \frac{1}{E_t}\right)\dot{\sigma}\, y\, \frac{M}{I}\, d\Omega + \text{h.o.t.} = 0$$

So we have to compute three integrals. With all the previous local results, there is nothing difficult for that, but before, it is interesting to give some qualitative remarks :

* the first integral which appears in (4.33) is obviously of the order of the load increment $\mu$ ;

* the second integral is also of the order of $\mu$ because of the order of the measure of the domain $\Omega_p \cap \Omega_e^0$;

* the third is of the order of $\tau^3$ because, by (4.27), $\dot{\sigma}$ is of the order of $\tau$ in $\Omega_u$, and the measure of $\Omega_u$ is of the order of $\tau^2$.

It means that condition (4.33) takes the form of a relation between $\mu$ and some power of $\tau$. This, therefore, solves the free boundary problem, as the growth of the unloading zone becomes explicitly related to the load increment.

It is interesting to pay attention to the particular meaning of this qualitative result, especially in view of a comparison with the homogeneous case.

Results for the generic case number 2

We compute explicitly the condition (4.33).

Since the value of the intitial velocity is given by the first order problem, the first integral in (4.33) reads :

$$I = 2\mu \int_0^1 \dot{v}^{o\,\prime} \; V' \; dx$$

$$= 2\mu \, \frac{1}{M(x_N)} \, \frac{[r(x_N)]^2}{y_N} \int_0^1 [V'(x)]^2 \, dx \, ,$$

where V, and the new integral over (0, 1), can be computed numerically from (4.7).

By the previous local results, and of course by the assumed smoothness of the variation of the cross-section, the following expressions hold in $\Omega_p \cap \Omega_e^0$ :

$$\dot{\sigma}(x, y) = - \frac{1}{S(x_N)} \left(1 - \frac{y}{y_N}\right) + O(\mu + \tau) \, ,$$

$$\frac{M(x)}{I(x)} = \frac{M(x_N)}{I(x_N)} \, (1 + O(\mu)).$$

Consequently, the second integral in (4.33) is :

$$J = \int_{\Omega_p \cap \Omega_e^0} \left(\frac{1}{E_t} - \frac{1}{E}\right) \dot{\sigma}\, y \frac{M}{I}\, d\Omega$$

$$= -\frac{2b}{S(x_N)} \frac{M(x_N)}{I(x_N)} \frac{E - E_t}{E\, E_t} \iint_{\Omega_p \cap \Omega_e^0} \left\{ y \left(1 - \frac{y}{y_N}(1 + 0(\mu + \tau))\right)\right\} dx\, dy ,$$

where the computation of the new integral over $\Omega_p \cap \Omega_e^0$ is elementary thanks to equation (4.21).

The result is :

$$J = -\frac{3}{8} b \left(\frac{y_N}{S(x_N)}\right)^2 \frac{M(x_N)}{I(x_N)} \left.\frac{1}{\frac{\partial \sigma}{\partial x}}\right)_N \frac{E - E_t}{E\, E_t} \mu + O(\mu\tau + \mu^2)$$

The last integral involves (4.23), (4.26), (4.27), and the regularity results :

$$K = \int_{\Omega_u} \left(\frac{1}{E} - \frac{1}{E_t}\right) \dot{\sigma}\, y \frac{M}{I}\, d\Omega$$

$$= 2b \frac{E - E_t}{E\, E_t} \int_{x_N - \tau}^{x_N} \int_{y_N - \tau f_1(X) + O(\eta)}^{y_N + O(\eta)} y\, \dot{\sigma} \frac{M}{I}\, dx\, dy$$

$$= 2b \frac{E - E_t}{E \; E_t} \int_0^1 \int_0^1 y_N \left(\frac{\partial \dot{\sigma}^0}{\partial x}\right)_N \frac{M(x_N)}{I(x_N)} f_1(X)(1-X)(1-Y)\tau^3 dX\, dY$$

That is, omitting elementary integrations :

$$K = -\frac{b}{3} \frac{E - E_t}{E \; E_t} \frac{1}{S(x_N)\, I(x_N)} \frac{\left[y_N \; M'(x_N)\right]^2}{M(x_N)} \tau^3 + O(\tau^4).$$

The condition (4.33) is then explicited in the form :

(4.34) $\mu = A\, \tau^3 + O(\tau^4)$ ,

where the coefficient A is a numerical constant given by the previous computations of integrals I, J, and K.

Taking account of this result, the perturbation problem (4.31) is formally rewritten as :

$$< \mathcal{L} \; \mathbf{u}^* , \delta \, \mathbf{u} > \; = \; < O(\tau^3) , \delta \, \mathbf{u} > .$$

But we know, from the local analysis, that the perturbation $\mathbf{u}^*$ is of the order of $\tau$ at the neutrally loaded point. Assuming this order of $\mathbf{u}^*$ holds everywhere in $\Omega$, since the growth of $(\Gamma_2)$ is the only cause of perturbation, we get that the first order perturbation $\mathbf{u}_1^*$ is proportional to the buckling mode $\mathbf{U}$, and the corresponding proportionality constant, say B, is deduced from (4.25).

(4.35) $\mathbf{u}^* = B \; \mathbf{U} \; \tau + O(\tau^2).$

In particular, the transverse velocity $\dot{v}$ is of the form :

$$\dot{v}(x, \tau) = \dot{v}^o(x) + v_1^*(x)\, \tau + O(\tau^2).$$

We can put (4.34) and (4.35) in this result. Then the integration with respect to $\mu$ gives :

$$v(x, \mu) = C\,V(x)\;\mu + B\,V(x)\;\frac{1}{A^{1/3}}\,\mu^{4/3} + O(\mu^{5/3}).$$

This expression can be easily inverted. Setting for instance $\xi = v(l/2)$, we get a load-displacement curve of the classical following form :

(4.36) $$\lambda = \lambda_o + \mu = \lambda_o + \lambda_1\,\xi + \lambda_2\,\xi^{4/3} + \text{h.o.t.}$$

Of course the exponent 4/3 refers to a rectangular cross-section. The generalization to any cross-section will be made in the following.

The computation of the coefficients $\lambda_1$ and $\lambda_2$ needs a numerical solution of the eigenvalue problem (4.7.d), which is the only part needing a little care. Then, the computation of integrals I, J and K is elementary. The main worth seeing point is that these coefficients depend completely on the material and on the shape of the beam. That means we have not an explicit formula, but a numerical value for any given beam. For instance, let us choose the beam of figure 19 :

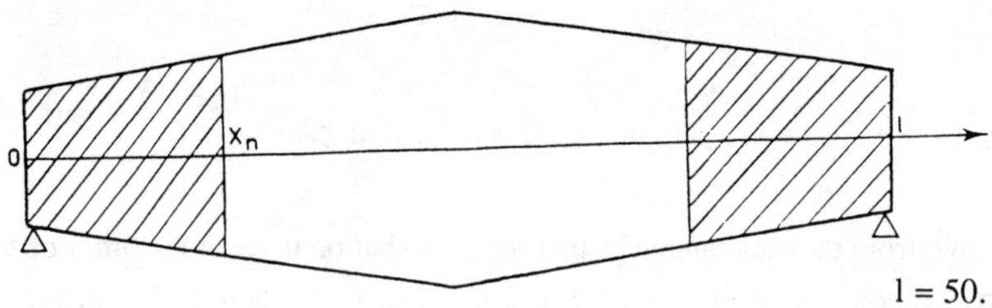


$l = 50.$
$a_{min} = 1$
$a_{max} = 3$
$E = 1.9\,D{+}05$
$E_T = 3.8\,D{+}04$
$\sigma_y = 3.0\,D{+}02$

Figure 19. Example of beam for a numerical computation.

Then, (4.7), (4.34) and (4.35) lead to :

$x_N = 8.5$

$\lambda_o = 1007$

$$\frac{\lambda}{\lambda_o} = 1 + 0.39\,\xi - 14.1\,\xi^{4/3} + \text{h.o.t.}$$

Results for the bifurcation from a homogeneous state

In the case of a uniform cross-section, the comptability condition (4.32) reduces to the following, containing only two terms :

$$(4.37) \qquad 2\mu \int_0^1 \dot{V}^{0'} V' \, dx - \int_{\Omega_u} \left( \frac{1}{E} - \frac{1}{E_t} \right) \dot{\sigma} \; y \, E_t \, V'' \, d\Omega = 0$$

Of course, the tangent modulus is constant in that case. Although non generic, the bifurcation from a homogeneous state is interesting since, because of (4.8.d) instead of (4.7.d), we will get explicit formula.

With the solution of (4.8), the results (4.28) for the geometry of the domain $\Omega_u$ and (4.29) for the velocity $\dot{\sigma}$ in this domain, (4.37) becomes, after elementary integrations :

$$\mu = \frac{2}{15} \left( \frac{\pi^2}{l^2} \; a \right)^3 \frac{b}{l} \; (E - E_t) \; \tau^5 + O\,(\tau^6).$$

Then we can get an expansion for the load $\lambda$ with respect to a displacement parameter exactly by the same way we use to obtain (4.36) from (4.34), but with explicit formula for $\lambda_1$ and $\lambda_2$.

Let us previously use this simpler case to point to the effect of the shape of the cross-section. On figure 20, we have hatched the unloading zone on the longitudinal axis and in a cross-section intersecting the unloading zone for three types of cross-section.

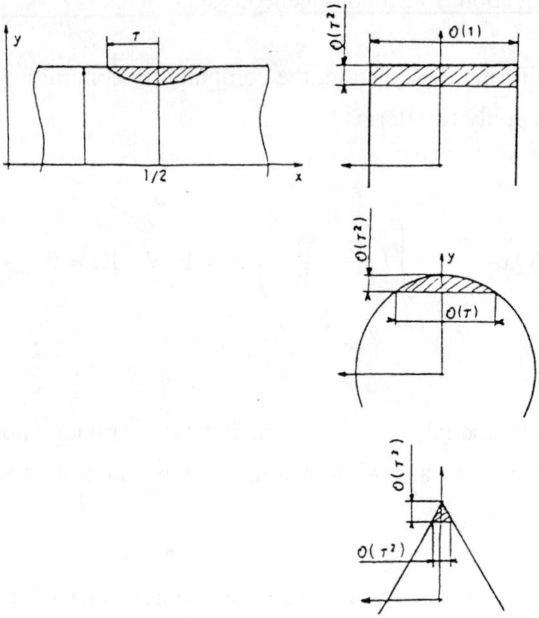


Figure 20. Different shapes of the unloading zone.

Then, it is straightforward that the shape of the cross-section has a simple geometrical effect by the size of the domain $\Omega_u$ in (4.33) or (4.37). For a uniform cross-section the results are plotted in table1.

| | $\lambda_1$ | $\nu$ | $\lambda_2$ |
|---|---|---|---|
| 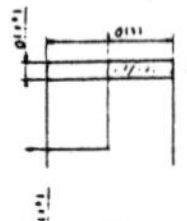 | 3 | $\frac{2}{5}$ | $-3,165\left(\pi\frac{E_t}{E - E_t}\right)^{\frac{2}{5}}$ |
| 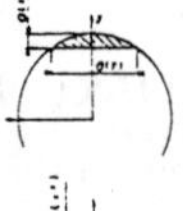 | 4 | $\frac{1}{3}$ | $-4,327\left(\pi\frac{E_t}{E - E_t}\right)^{\frac{1}{3}}$ |
| 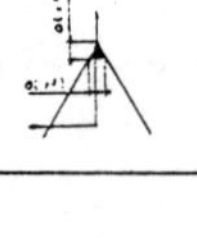 | 2,83 | $\frac{2}{7}$ | $-2,621\left(\pi\frac{E_t}{E - E_t}\right)^{\frac{2}{7}}$ |

Table 1 : Results for 3 types of cross-section, for expansions of the form : $\lambda = \lambda_o + \lambda_1 \xi + \lambda_2 \xi^{1+\nu}$ + h.o.t.

Remarks :

i) These results for the homogeneous case are in agreement with [14] and [27].

ii) As the post-buckling analysis is usually made in order to get estimates of the maximal support load, it is important to assert that a maximal load obtained by dropping the higher order terms in the expansions of $\lambda$ is not an asymptotic result. Nevertheless, it is qualitatively intersting to look at the shape of the corresponding load/displacement curves, such as the one given on figure 21.

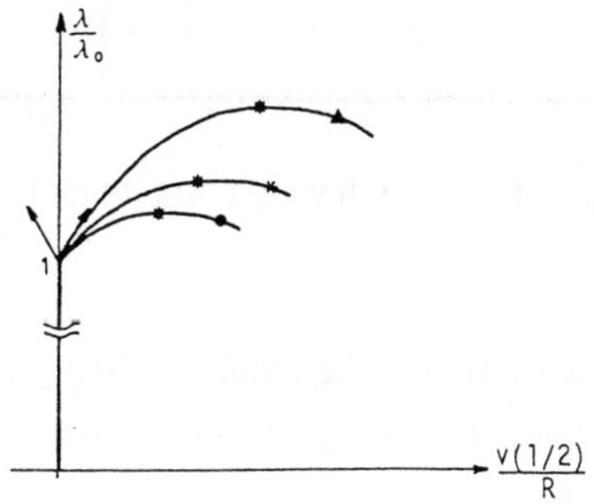


: maximal load
Σ : $E_t/E = 0.5$
* : $E_t/E = 0.3$
Δ : $E_t/E = 0.2$

$\xi > 0$ : • : $\lambda m/\lambda 0 = 1.108$, * : $\lambda m/\lambda 0 = 1.253$, Δ : $\lambda m/\lambda 0 = 1.434$

$\xi < 0$ : • : $\lambda m/\lambda 0 = 1.090$, * : $\lambda m/\lambda 0 = 1.211$, Δ : $\lambda m/\lambda 0 = 1.362$

Figure 21. The post-buckling of a beam with a triangular cross-section by dropping the higher order terms.

iii) The same discussion on the effect of the shape of the cross-section can be made for bifurcations from a heterogeneous state. By the same way, the condition (4.33) leads to the result that the fractional exponents only depend on the local geometry of the cross-section in the neighbourhood of the initial neutrally loaded point. This is summarized by table 2 for the value of the fractional exponent.

| | Homogeneous case or Generic case 1 | Generic case 2 |
|---|---|---|
| Neutral loading appears on a straight line | 7/5 | 4/3 |
| Neutral loading appears at one point of a curved line | 4/3 | 9/7 |
| Neutral loading appears at a vertex | 9/7 | 5/4 |

Table 2 : Different values of the exponent $1 + \nu$ of the first non linear term.

iv) We have got, and used many times, the result asserting that elastic-plastic buckling occurs under increasing load. By condition (4.32) or (4.37), we have an interpretation of the corresponding mechanism. As a matter of fact, we see, quantitatively, that this buckling under increasing load is related to the fact that the load increment is balanced by a local increase of rigidity in an unloading zone. Moreover, we find for the heterogeneous case, that the unloading zone has to balance both the load increment, and the loss of rigidity in the zone spanned by the extension, into the initially elastic zone, of the plastic zone.

IV-2-e Comments upon some difficulties and assumptions

This analysis involves several assumptions. Some of them have been made only in order to specify the kind of problem we want to deal with. Then no problem arises about those. But some other are actually restrictive for the bifurcation results. That means they suggest some mathematical problems remain open to get a more complete necessary and sufficient condition for bifurcation, or a more complete classification of the generic bifurcation within elastic-plasticity.

Moreover, it is still necessary to clear some mathematical aspects related for instance to the regularity of the free boundaries in the class of problems we deal with. This is important essentially in order to specify first the kind of problems for which the previous local analysis might not be allowed, and second to begin the analysis of imperfection sensitivity.

In this course, we have not approached several effects of the constitutive law such as, for multiaxial problems, the occurence of vertex on the yield surface. The bifurcation analysis of an elastic-plastic cruciform column remains, for that, a typical example [14]. Here, we limit ourselves to conclude only with some qualitative indications about the choices of the constitutive assumptions for a uniaxial traction curve.

For the homogeneous case, the prebuckling stress is uniform. So, as soon as this stress is higher than the yield stress, there is no fundamental difference between the case of a constant plastic modulus, and the laws of Figure 13b and 13c. Actually, it has been established ([27] and [16]) that the post-buckling analysis involves only the local curvature of the stress-strain law, and that means the precise shape of that law implies only a new term, of the order of $\mu$, in the solvability equation. Between the law of figure 22a and the one of figure 22b or 22c, we then get only a small change in the numerical coefficients of the final load-displacement curve.

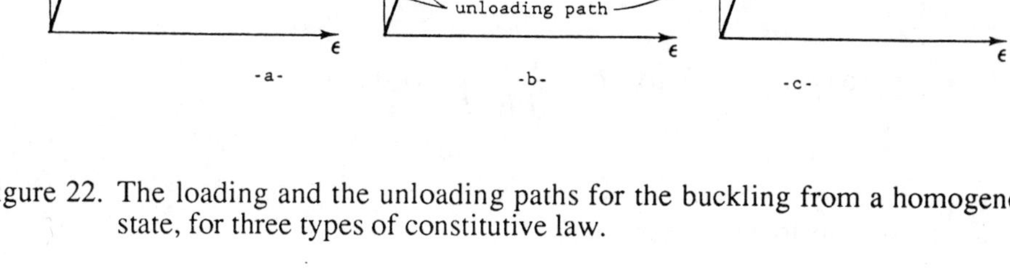


Figure 22. The loading and the unloading paths for the buckling from a homogeneous state, for three types of constitutive law.

But for the problem at hand, the prebuckling stress is not uniform and, at the buckling load, this stress $\sigma^f(x)$ is equal to the initial yield stress $\sigma^y$ somewhere in the domain, precisely on the elastic-plastic boundary. Then, the question of the presence, or not, of a corner on the stress-strain law at the initial yield level will change significantly the post-buckling analysis. That means the case plotted on figure 23c would lead to very different post-buckling behavior than what we would get with the laws of figure 23a or 23b.

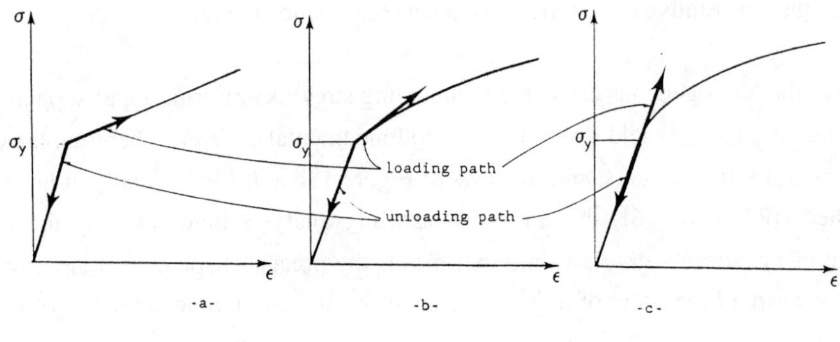


Figure 23. The loading and the unloading paths for the buckling from a heterogeneous partially plastic state, for three types of constitutive law.

Actually, the law plotted on figure 23b is quite similar to the one of figure 23a, for which we gave explicit results in the preceeding sections.

But, looking at the stress-strain curve of figure 23c for any stress slightly higher than $\sigma^o$, we see that we can write the tangent modulus in the following way :

$$E_t(\sigma) = E_t(\sigma^o) + (\sigma - \sigma^o) \left.\frac{\partial E_t}{\partial \sigma}\right)_{\sigma^o} \text{s+ h.o.t.}$$

Then, using the following estimate :

$$\sigma - \sigma^o = \int_{\lambda_o + \mu}^{\lambda_o} (\dot{\sigma} - \dot{\sigma}^o)\, d\lambda = O(\mu\tau) \quad \text{in } \Omega_u ,$$

we get a comptability condition (4.33) of the form :

$$O(\mu\tau^4) + O(\mu) = 0 \ ,$$

contrary to (4.34).

Obviously, as $\tau^4$ is small, this condition can be satisfied only with $\mu = 0$. That means, in this case, we should only get a bifurcation under constant loading. But let us recall again that such a solution, corresponding to a formally elastic buckling, is not plastically admissible. Then we can guess that the bifurcated branch should not persist with such a constitutive law.

But it is not straightforward to establish the non existence of a branch of solutions. Let us only mention for the moment that this situation can be compared with the results established in section II for simple systems, according to which, within plasticity, several critical points occur generically without branching [22].

## REFERENCES

1. BENALLAL, A., BILLARDON, R. and GEYMONAT, G. : *C.R. Acad. Sci.*, **308**, II, 1989, 893-897.
2. BUDIANSKY, B. : Theory of buckling and postbuckling behaviour of elastic structures, Advances in *Appl. Mech.*, **14**, 1974, 1-65.
3. CAMOTIM, D.R. : To appear in *Dynamics Stability of Systems*, 1992.
4. CHOW, S.N., HALE, J.K., MALLET-PARET, J. : Applications of generic bifurcation, I *Arch. Rat. Mech. Anal.*, **59**, 1975, 159-188, II *id.* **62**, 1976, 209-235.
5. CIMETIERE, A. : Modèle incrémental pour la forte flexion des plaques élastoplastiques. *C.R. Acad. Sci. Paris*, **298**, série II, 1984, 99-102.
6. CIMETIERE, A. : Flambage élastoplastique des plaques : *Thèse d'Etat*, Université de Poitiers, 1987.

7. DUVAUT, G. et LIONS, J.-L. : Equations et inéquations en mécanique et en physique, Dunod, Paris, 1972.
8. ENGESSER, F. : Ueber die Knickfestigkeit Gerader Stäbe, *Z. Architektur Ing.*, vol. 35, 1889, p. 455.
9. GOLUBITSKY, M., SCHAEFFER, D.G. : A theory for imperfect bifurcation via singularity theory, **32**, *Comm. Pure Appl. Match*, 1979, 21-98.
10. GOLUBITSKY, M., SCHAEFFER, D.G. : Singularities and Groups in Bifurcation Theory, vol. 1, Springer-Verlag, New York, 1985.
11. HILL, R. : A general theory of uniqueness and stability in elastic/plastic solids. *J. Mech. Phys. Solids*, **6**, 1958, 236-249.
12. HUTCHINSON, J.W. : Post-bifurcation behavior in the plastic range, *J. Mech. Phys. Solids*, **21**, 1973, 163-190.
13. HUTCHINSON, J.W. : Imperfection sensitivity in the plastic range, *J. Mech. Phys. Solids*, **21**, 1973, 191-204.
14. HUTCHINSON, J.W. : Plastic buckling, *Adv. Appl. Mech.*, **14**, 1974, 67-144.
15. KOITER, W.T. : On the Stability of Elastic Equilibrium, *Thesis*, Delft, 1945, English translation NASA Techn. Trans. F-10, 833, 1967.
16. LEGER, A. and POTIER-FERRY, M. : Sur le flambage plastique, *J. Mécanique Théor. Appl.*, **7**, 1988, 819-857.
17. LEGER, A. and POTIER-FERRY, M. : Elastic-plastic post-buckling from a heterogeneous state, to appear in *J. Mech. Phys. Solids*, 1992.
18. NGUYEN, Q.S. : Stabilité et bifurcation en rupture et en plasticité, *C.R. Acad. Sci. Paris*, **292**, série II, 1981, 817-821.
19. NGUYEN, Q.S. : Bifurcation et stabilité des systèmes irréversibles obéissant au principe de dissipation maximale, *J. Méc. Théo. Appl.*, **3**, 1984, 41-61.
20. NGUYEN, Q.S. : Bifurcation and Postbifurcation Analysis in Plasticity and Brittle Fracture, *J. Mech. Phys. Solids*, **33**, 1987, 303-324.
21. POTIER-FERRY, M. : On the Mathematical Foundations of Elastic Stability Theory, *Arch. Rat. Mech. Anal.*, **78**, 1982, 301-320.
22. POTIER-FERRY, M. : Towards a Catastroph Theory for the Mechanics of Plasticity and Fracture, *Int. J. Engng. Sci.*, **23**, 1985, 821-837.
23. POTIER-FERRY, M. : Buckling and Postbuckling, Lecture Notes in *Physics*, 288, Springer-Verlag, Heidelberg, 1987, 1-82.

24. SHANLEY, F.R. : Inelastic Column Theory, *J. Aeronaut. Sci.*, **14**, 1947, 261-267.
25. TRIANTAFYLLIDIS N. : On the Bifurcation and Postbifurcation Analysis of Elastic-Plastic Solids under General Prebifurcation Conditions, *J. Mech. Phys. Solids*, **31**, 1983, 499-510.
26. THOMPSON, J.M.T. and HUNT, G.W. : A General Theory of Elastic Stability, Wiley, London, 1973.
27. TVERGAARD, V. and NEEDLEMAN, A. : On the buckling of Elastic-Plastic Columns with Asymmetric Cross-Sections, *Int. J. Mech. Sci.*, **17**, 1975, 419-424.
28. VON KARMAN, Th. : Untersuchungen über knickfestigkeit, mitteilungen über forschungsarbeiten, *VDI (Ver. Deut. Ing.) Forschungsh.* **81**, 1910.

# ON THE BIFURCATION AND POSTBIFURCATION THEORY FOR A GENERAL CLASS OF ELASTIC-PLASTIC SOLIDS

N. Triantafyllidis
The University of Michigan, Ann Arbor, Mich., USA

## ABSTRACT

The present work is concerned with the bifurcation and postbifurcation analysis of a class of rate independent plasticity models obeying Hill's maximum dissipation principle. A variational inequality approach, which differs from the classical formulation of the plastic bifurcation problem, is employed. The rate n bifurcation problem is formulated and sufficient conditions for uniqueness of the corresponding boundary value problem are given. A connection is made with Hill's nonbifurcation criterion. In addition the issue of the postbifurcation behavior of the solid is addressed in this more general context showing the possiblity of angular as well as smooth bifurcations of rate $n > 1$.

Finally an example, capable of exhibiting both an angular as well as a smooth bifurcation is analyzed using the general formulation derived in this work. The presentation is concluded with some comments and comparisons of the present methodology with the classical approach.

## 1. INTRODUCTION

The plastic buckling of structures is a very interesting topic in mechanics, both for its practical importance as well as for its mathematical challenge. Following some early work of CONSIDERE (1), VON KARMAN (2) and SHANLEY (3), HILL (4,5) was the first to put the bifurcation instability problem for a rate independent elastoplastic continuum on a sound mathematical basis.

More specifically Hill has addressed the issue of loss of uniquness in the velocities (i.e. displacement increments) for the incremental boudary value problem of an elastoplastic solid and gave sufficient conditions for the exclusion of such a bifurcation (termed rate one or angular bifurcation in the present work). He then went on to generalise the validity of his criterion by postulating the absence of a bifurcated branch at a load where a bifurcation in velocity was excluded. Subsequently, and based on Hill's formulation of the problem, HUTCHINSON (6,7) has studied in considerable generality the postbifurcation and imperfection sensitivity issues for the aforementioned (angular) bifurcation problem. His approach is restricted to problems with total loading throughout their principal solution. In addition, fractional asymptotic expansions (in terms of the time like parameter) of the various field quantities are required, in contrast to the (now classical) postbifurcation analysis for elastic systems introduced by KOITER (8) (on this subject see also BUDIANSKY's (9) comprehensive review article and references quoted therein).

The present work is an outgrowth of some initially independent efforts by the authors to circumvent some of the problems of the plastic bifurcation and postbifurcation theory mentioned above. A mathematically simpler approach to the plastic bifurcation problem (which is also applicable to a wide range of rate independent dissipative phenomena, besides plastic stability) has been proposed by NGUYEN (10) and is based on the generalized standard material formalism for rate independent solids, a concept initially introduced by NGUYEN (11) (see also HALPHEN and NGUYEN (12)). The issue of higher order (smooth) bifurcations was initially addressed by TRIANTAFYLLIDIS (13) using a somewhat restrictive approach based on a class of CRISTOFFERSEN and HUTCHINSON (14) type constitutive models (i.e. models with a regularized dependence of the incremental moduli on the stress or strain rates). The work presented here can be viewed as the logical continuation of the aformentioned efforts. The outline of the present work, which follows closely NGUYEN and TRIANTAFYLLIDIS (15) is as follows:

The incremental boundary value problem of rate n (with respect to the time like parameter) is formulated and sufficient conditions for its uniquness are explored (the present results are a generalization of some preliminary ones obtained by NGUYEN and STOLZ (16)). In addition a connection is made between an exclusion principle for bifurcation of arbitrary rate and Hill's exclusion of bifurcation principle, which is shown to preclude bifurcation of any order (and not merely the first rate as initially proved by Hill). The presentation continues with the formulation of the postbifurcation expansion problem in a unified approach (and in terms of an integral power series with respect to the time like parameter) that is valid for bifurcations at any rate. The general theory is followed by a thoroughly analyzed example in which both angular (rate one) and smooth (higher than rate one) bifurcations occur. Finally some concluding remarks are made about the advantages and limitations of the proposed approach.

## 2. GENERAL FORMULATION OF THE BIFURCATION AND POSTBIFURCATION PROBLEM

The generalized standard material formalism introduced by NGUYEN (11) ( see also HALPHEN and NGUYEN) (12) ) will be the basis for the stability analysis presented here. Accordingly, if $\phi(\underset{\sim}{\varepsilon}, \underset{\sim}{\alpha})$ is the free energy density function of the solid at a material point with strain $\underset{\sim}{\varepsilon}$ and internal variables $\underset{\sim}{\alpha}$, the associated stress $\underset{\sim}{s}$ and internal forces $\underset{\sim}{A}$ are given by:

$$\underset{\sim}{s} = \frac{\partial \phi}{\partial \underset{\sim}{\varepsilon}} \quad , \qquad \underset{\sim}{A} = - \frac{\partial \phi}{\partial \underset{\sim}{\alpha}} \tag{2.1}$$

In addition, the internal forces $\underset{\sim}{A}$ are assumed to lie inside or on the surface of a convex set C defined by $f(\underset{\sim}{A}) \leq 0$. The evolution law for the internal variables is then found from the maximum dissipation principle

$$(\underset{\sim}{A} - \underset{\sim}{A}^*) \bullet \dot{\underset{\sim}{\alpha}} \geq 0^{\dagger} \quad \forall \underset{\sim}{A}^* \in C \quad , \; C \equiv \{ \underset{\sim}{A}^* \mid f(\underset{\sim}{A}^*) \leq 0 \} \tag{2.2}$$

where a quantity surmounted by a dot $(\cdot)$ denotes differentiation with respect to the monotonically increasing time-like parameter $\tau$.

The majority of the commonly employed small strain, rate independent plasticity laws can be cast in the above formalism. In the applications considered here, the small strain - moderate displacement and rotation hypothesis will be made, i.e. it will be assumed that $\underset{\sim}{\varepsilon} = \underset{\sim}{\varepsilon}(\underset{\sim}{u}, \underset{\sim}{u}\underset{\sim}{\nabla})$ is a nonlinear operator of the displacement field $\underset{\sim}{u}$ while $\underset{\sim}{s}$ is the second Piola - Kirchhoff stress and $\phi$ is the material's free energy per unit reference volume.

From (2.2) and assuming that $f(A)$ is adequately smooth, one obtains:

$$\dot{\underset{\sim}{\alpha}} = \mu \frac{\partial f}{\partial \underset{\sim}{A}} \text{ with } \begin{cases} \mu = 0 \text{ if } f(\underset{\sim}{A}) < 0 \\ \\ \mu \geq 0 \text{ if } f(\underset{\sim}{A}) = 0 \end{cases} \tag{2.3}$$

† Note: For notational convenience the inner product between tensors of arbitrary order will be denoted by (•)

The constitutive equations (2.1), (2.2) (or (2.1) and (2.3)) will be subsequently employed in the formulation of the boundary value problem for the rate independent elastoplastic materials considered in this work.

In addition to (2.1) - (2.3) the following useful relations are recorded: by subsequently taking $\underset{\sim}{A}^*$ to be $\underset{\sim}{A}(\tau) \pm \dot{\underset{\sim}{A}}(\tau)\,\Delta\tau$ and assuming a smooth dependence of $f$ on $\underset{\sim}{A}$, one deduces from (2.2):

$$\dot{\underset{\sim}{A}} \bullet \dot{\underset{\sim}{\alpha}} = 0 \;\; \Rightarrow \mu \, \frac{\partial f}{\partial \underset{\sim}{A}} \bullet \dot{\underset{\sim}{A}} = \mu \, \dot{f} = 0 \tag{2.4}$$

The n rates of (2.4) as well as (2.1) - (2.3) will be required for the determination of the solution to the boundary value problem of rate n.

This section is only meant as a brief reminder for the governing equations of the generalized standard material model. For a more detailed account of the theory accompanied by appropriate illustrative examples, the interested reader is referred to NGUYEN (17).

## 2.1 UNIQUENESS FOR THE BOUNDARY VALUE PROBLEM OF RATE N

Consider a generalized standard solid with free energy $\phi\,(\underset{\sim}{\varepsilon}, \underset{\sim}{\alpha})$ per unit reference volume occupying a volume $\Omega$ with boundary $\partial\,\Omega$. Moreover assume that on a part of the boundary, say $\partial\Omega_u$, the displacements $\underset{\sim}{u}$ are prescribed while on the remaining part of the boundary say $\partial\Omega_t$ surface tractions $\lambda \underset{\sim}{T}^0$ are applied in proportion to a scalar parameter $\lambda$. Neglecting body forces for simplicity and assuming U to be the space of all admissible displacement functions, the solid's potential energy is given by:

$$\mathcal{E} = \int_{\Omega} \varphi\,(\underset{\sim}{\varepsilon}, \underset{\sim}{\alpha})\, d\,V - \lambda \int_{\partial\Omega} \underset{\sim}{T}^0 \bullet \underset{\sim}{u}\, d\,S \tag{2.5}$$

where for simplicity it has been assumed that $\underset{\sim}{u} = 0$ on $\partial\,\Omega_u$.

The equilibrium equation, i.e.the boundary value problem of rate 0, is given by extremizing $\mathcal{E}$ with respect to $\underset{\sim}{u}$

$$\mathcal{E}_{,u}(\delta \underset{\sim}{u}) = \int_{\Omega} \left[ \frac{\partial \phi}{\partial \underset{\sim}{\varepsilon}} \bullet \delta \underset{\sim}{\varepsilon} \right] dV - \lambda \int_{\partial\Omega} \underset{\sim}{T}^0 \bullet \delta \underset{\sim}{u}\, dS = 0 \tag{2.6}$$

where $\delta \underset{\sim}{\varepsilon} = \underset{\sim}{\varepsilon}_{,u}(\delta u)$ is the first variation of the (nonlinear) strain - displacement operator $\underset{\sim}{\varepsilon}(\underset{\sim}{u}, \underset{\sim}{u}\underset{\sim}{\nabla})$ with respect to $\underset{\sim}{u}$. Assuming at this stage that the displacement $\underset{\sim}{u}$ (and hence the strain $\underset{\sim}{\varepsilon}$) and the internal variable $\underset{\sim}{\alpha}$ fields are known at an instant $\tau$, one is interested in determining the rates of the aforementioned fields $\dot{\underset{\sim}{u}}$ and $\dot{\underset{\sim}{\alpha}}$, or equivalently from (2.3) $\dot{\underset{\sim}{u}}$ and $\mu$. It will be shown that $(\dot{\underset{\sim}{u}}, \mu)$ is the solution to the following variational inequality:

$$\int_{\Omega} \left[ \left( \frac{\partial \phi}{\partial \underset{\sim}{\varepsilon}} \bullet \delta \underset{\sim}{\varepsilon} \right)^{\cdot} + (\beta - \mu)(-\dot{f}) \right] dV - \dot{\lambda} \int_{\partial\Omega} \underset{\sim}{T}^0 \bullet \delta \underset{\sim}{u}\, dS \geq 0 \quad \forall (\delta \underset{\sim}{u}, \beta) \in U \times K_1$$

$$K_1 \equiv \{ \beta(\underset{\sim}{x}) \mid \beta \geq 0 \text{ if } \underset{\sim}{x} \in I_1^0 \,,\ \beta = 0 \text{ if } \underset{\sim}{x} \in I_1^- \} \tag{2.7}$$

$$I_1^+ \equiv \varnothing \,,\quad I_1^0 \equiv \{ \underset{\sim}{x} \in \Omega \mid f(\underset{\sim}{A}(\underset{\sim}{x})) = 0 \} \,,\quad I_1^- \equiv \{ \underset{\sim}{x} \in \Omega \mid f(\underset{\sim}{A}(\underset{\sim}{x})) < 0 \}$$

Note that by definition the sets $I_1^+$, $I_1^0$, $I_1^-$ are disjoint and that their union gives $\Omega$, i. e. $I_1^+ \cup I_1^0 \cup I_1^- = \Omega$ (see Fig. 1).

At this point it is tacitly assumed that no moving discontinuities of the quantity $(\partial\phi / \partial\underset{\sim}{\varepsilon}) \bullet \delta\underset{\sim}{\varepsilon}$ exist. Consequently, the first rate of the equilibrium equation (2.6), when written out explicitly, does not contain any surface terms. This assumption is satisfied in most applications of interest for it corresponds to problems with continuous stress and strain fields, or more generally to problems with stationary stress and strain discontinuities. This assumption is no longer valid for the rates of the aforementioned quantities, as will be discussed later.

By taking $\delta \underset{\sim}{u} \neq 0$ and $\beta = \mu$, one recovers the first rate of the equilibrium equation (2.6). Conversely, for $\delta \underset{\sim}{u} = 0$, $\beta \neq \mu$ and noting that $\beta, \mu \in K_1$ one has:

$$\int_{\Omega} (\beta - \mu)(-\dot{f})\, dV \geq 0 \Rightarrow \int_{I_1^+ \cup I_1^0} (\beta - \mu)(-\dot{f})\, dV \geq 0 \Rightarrow \text{For } \underset{\sim}{x} \in I_1^0 \begin{cases} \text{If } \mu = 0 \Rightarrow \dot{f} \leq 0 \\ \\ \text{If } \mu > 0 \Rightarrow \dot{f} = 0 \end{cases} \tag{2.8}$$

exactly as predicted from (2.3) and (2.4).

Having formulated, the rate 1 boundary value problem, attention is focused on the uniqueness of its solution $(\dot{\underset{\sim}{u}}, \mu)$. To this end, consider the following quadratic in $(\hat{\underset{\sim}{u}}, \beta)$ functional $\mathcal{F}[\hat{\underset{\sim}{u}}, \beta]$ defined by:

$$\mathcal{F}[\hat{\underset{\sim}{u}}, \beta] \equiv \int_\Omega \Big[ \underset{\sim}{\varepsilon}_{,u}(\hat{\underset{\sim}{u}}) \bullet \frac{\partial^2\phi}{\partial\underset{\sim}{\varepsilon}\partial\underset{\sim}{\varepsilon}} \bullet \underset{\sim}{\varepsilon}_{,u}(\hat{\underset{\sim}{u}}) + \underset{\sim}{\varepsilon}_{,u}(\hat{\underset{\sim}{u}}) \bullet \frac{\partial^2\phi}{\partial\underset{\sim}{\varepsilon}\partial\underset{\sim}{\alpha}} \bullet \frac{\partial f}{\partial\underset{\sim}{A}}\beta +$$

$$\beta\frac{\partial f}{\partial\underset{\sim}{A}} \bullet \frac{\partial^2\phi}{\partial\underset{\sim}{\alpha}\partial\underset{\sim}{\varepsilon}} \bullet \underset{\sim}{\varepsilon}_{,u}(\hat{\underset{\sim}{u}}) + \beta\frac{\partial f}{\partial\underset{\sim}{A}} \bullet \frac{\partial^2\phi}{\partial\underset{\sim}{\alpha}\partial\underset{\sim}{\alpha}} \bullet \frac{\partial f}{\partial\underset{\sim}{A}}\beta + \frac{\partial\phi}{\partial\underset{\sim}{\varepsilon}} \bullet \underset{\sim}{\varepsilon}_{,uu}(\hat{\underset{\sim}{u}}, \hat{\underset{\sim}{u}}) \Big] dV \qquad (2.9)$$

If $\mathcal{F}[\hat{\underset{\sim}{u}}, \beta]$ is positive definite $\forall\, (\hat{\underset{\sim}{u}}, \beta) \in U \times \mathcal{V}K_1$ (with $\mathcal{V}K$ denoting the linear hull of the convex set K), it will be shown that a rate 1 bifurcation is excluded. For if not, assume $(\dot{\underset{\sim}{u}}_1, \mu_1)$ and $(\dot{\underset{\sim}{u}}_2, \mu_2)$ to be two different solutions of the variational inequality in (2.7). Employing (2.7) and (2.8) in (2.9) one obtains the contradiction:

$$\mathcal{F}[\dot{\underset{\sim}{u}}_1 - \dot{\underset{\sim}{u}}_2, \mu_1 - \mu_2] = \int_{I_1^0} (\mu_1 - \mu_2)(-\dot{f}_1 + \dot{f}_2)\, dV \le 0 \qquad (2.10)$$

The positive definitness of $\mathcal{F}$ on $U \times \mathcal{V}K_1$ is a sufficient condition for the uniqueness of the rate 1 problem and hence too restrictive. One expects a unique solution to the rate 1 problem in (2.7) under less stringent conditions.

Once a unique solution to (2.7) is found, i.e. given $(\underset{\sim}{u}, \underset{\sim}{\alpha})$, $(\dot{\underset{\sim}{u}}, \dot{\underset{\sim}{\alpha}})$, the rate 2 problem can be formulated. It will be shown that $(\ddot{\underset{\sim}{u}}, \dot{\mu})$ is the solution to the variational inequality:

$$\int_\Omega \Big[ \Big( \frac{\partial\phi}{\partial\underset{\sim}{\varepsilon}} \bullet \delta\underset{\sim}{\varepsilon} \Big)^{\cdot\cdot} + (\beta - \dot{\mu})(-\ddot{f}) \Big] dV - \ddot{\lambda} \int_{\partial\Omega} \underset{\sim}{T}^0 \bullet \delta\underset{\sim}{u}\, dS \ge 0 \quad \forall\, (\delta\underset{\sim}{u}, \beta) \in U \times K_2$$

$$K_2 \equiv \{ \beta(\underset{\sim}{x}) \mid \beta \in R \text{ if } \underset{\sim}{x} \in I_1^+ \cup I_2^+,\ \beta \ge 0 \text{ if } \underset{\sim}{x} \in I_2^0,\ \beta = 0 \text{ if } \underset{\sim}{x} \in I_2^- \cup I_1^- \}$$

$$I_2^+ \ \{ \underset{\sim}{x} \in I_1^0 \mid \dot{\mu}(\underset{\sim}{x}) > 0\ ,\ \dot{f}(\underset{\sim}{A}(\underset{\sim}{x})) = 0 \} \qquad (2.11)$$

$$I_2^0 \equiv \{ \underset{\sim}{x} \in I_1^0 \mid \dot{\mu}(\underset{\sim}{x}) = 0\ ,\ \dot{f}(\underset{\sim}{A}(\underset{\sim}{x})) = 0 \}$$

$$I_2^- \equiv \{ \underset{\sim}{x} \in I_1^0 \mid \dot{\mu}(\underset{\sim}{x}) = 0\ ,\ \dot{f}(\underset{\sim}{A}(\underset{\sim}{x})) < 0 \}$$

Note also that the sets $I_2^+, I_2^0, I_2^-$ are disjoint and that their union gives $I_1^0$; $I_2^+ \cup I_2^0 \cup I_2^- = I_1^0$ (Fig.1).

Unlike $(\partial\phi/\partial\underset{\sim}{\varepsilon}) \bullet \delta\underset{\sim}{\varepsilon}$, its rate one derivative with respect to $\tau$ can have moving discontinuities. Hence the term $((\partial\phi/\partial\underset{\sim}{\varepsilon})^{\cdot\cdot} \bullet \delta\underset{\sim}{\varepsilon})$ in (2.11) has to be interpreted in the generalized function sense in order to account for the time dependent discontinuities in this quantity. Had a more explicit notation been adopted, some additional surface integrals would have also been included in (2.11). Knowing that the discontinuity in the rate of a field quantity, say $[[\dot{g}]]$, propagating with velocity $v_n$, along the outward normal $\underset{\sim}{n}$ to the discontinuity surface in question satisfies $( [[g\underset{\sim}{\nabla}]] \bullet \underset{\sim}{n} ) v_n + [[\dot{g}]] = 0$, one deduces that the surface discontinuity terms in (2.11) involve field quantities up to rate one. This remark is important when considering bifurcation of the rate two problem for only the volume term given in (2.9) appears in the final calculation.

Similarly to the rate one case, by taking $\delta \underset{\sim}{u} \neq 0$, $\beta = \dot{\mu}$ one recovers the second rate of the equilibrium equation (2.6). Moreover, for $\delta \underset{\sim}{u} = 0$, $\beta \neq \dot{\mu}$ and since $\beta, \mu \in K_2$ (2.11) yields:

$$\int_{\Omega} (\beta - \dot{\mu})(-\ddot{f})\, dV \geq 0 \Rightarrow \int_{I_1^+ \cup I_2^+ \cup I_2^0} (\beta - \dot{\mu})(-\ddot{f})\, dV \geq 0 \Rightarrow \begin{cases} \text{For } \underset{\sim}{x} \in I_2^+ \cup I_1^+ & \Rightarrow \ddot{f} = 0 \\ \text{For } \underset{\sim}{x} \in I_2^0 & \begin{cases} \text{if } \dot{\mu} > 0 \Rightarrow \ddot{f} = 0 \\ \text{if } \dot{\mu} = 0 \Rightarrow \ddot{f} \leq 0 \end{cases} \\ \text{For } \underset{\sim}{x} \in I_2^- \cup I_1^- & \Rightarrow \dot{\mu} = 0 \end{cases} \tag{2.12}$$

which can be easily verified by taking rates of (2.2) - (2.4). Notice also that for the rate 2 problem $\dot{\mu}\, \ddot{f} = 0 \quad \forall\, x \in \Omega$ (compare with (2.4)).

The natural question arising next pertains to the uniqueness of the solution $(\ddot{\underset{\sim}{u}}, \dot{\mu})$ to the rate 2 boundary value problem. It will be shown that a sufficient condition for the uniqueness of a solution to (2.11) is the positive definitness $\mathcal{F}[\hat{\underset{\sim}{u}}, \beta]$ on $U \times V_{K_2}$. Considering again $\mathcal{F}\,[\ddot{\underset{\sim}{u}}_1, -\ddot{\underset{\sim}{u}}_2, \dot{\mu}_1, -\dot{\mu}_2]$ for the two different solutions $(\ddot{\underset{\sim}{u}}_1, \dot{\mu}_1)$ and $(\ddot{\underset{\sim}{u}}_2, \dot{\mu}_2)$ to the rate 2 problem (2.11) (with the tacit assumption that up to the time of interest the solutions to the rate 0 and 1 problems are unique) and making use of (2.9), (2.11) and (2.12) one has:

$$\mathcal{F}[\ddot{\underset{\sim}{u}}_1 - \ddot{\underset{\sim}{u}}_2, \dot{\mu}_1 - \dot{\mu}_2] = \int_{I_2^0} (\dot{\mu}_1 - \dot{\mu}_2)(-\ddot{f}_1 + \ddot{f}_2) \leq 0 \tag{2.13}$$

which obviously contradicts the assumption about the positive definitness of $\mathcal{F}$.

The rate 3 problem, which will be examined next, suggests the algorithm for the generalization to the problem of rate n. Once more, it is assumed that a unique solution has been found to the problems of rates 0, 1 and 2 up to the time of interest. The wanted solution ($\dddot{\underset{\sim}{u}}$, $\ddot{\mu}$) is determined from the following variational inequality:

$$\int_{\Omega}\Big[\Big(\frac{\partial\phi}{\partial\underset{\sim}{\varepsilon}}\bullet\delta\underset{\sim}{\varepsilon}\Big)^{\dddot{}}+(\beta-\ddot{\mu})(-\dddot{f})\Big]dV-\dddot{\lambda}\int_{\partial\Omega}\underset{\sim}{T}^0\bullet\delta\underset{\sim}{u}\geq 0 \qquad \forall(\delta\underset{\sim}{u},\beta)\in U\times K_3$$

$$K_3\equiv\{\beta(\underset{\sim}{x})\mid\beta\in R \text{ if } \underset{\sim}{x}\in I_1^+\cup I_2^+\cup I_3^+,\ \beta\geq 0 \text{ if } \underset{\sim}{x}\in I_3^0,\ \beta=0 \text{ if } \underset{\sim}{x}\in I_3^-\cup I_2^-\cup I_1^-\}$$

$$I_3^+\equiv\{\underset{\sim}{x}\in I_2^0\mid\dot{\mu}(\underset{\sim}{x})>0,\ \ddot{f}(\underset{\sim}{A}(\underset{\sim}{x}))=0\} \tag{2.14}$$

$$I_3^0\equiv\{\underset{\sim}{x}\in I_2^0\mid\dot{\mu}(\underset{\sim}{x})=0,\ \ddot{f}(\underset{\sim}{A}(\underset{\sim}{x}))=0\}$$

$$I_3^-\equiv\{\underset{\sim}{x}\in I_2^0\mid\dot{\mu}(\underset{\sim}{x})=0,\ \ddot{f}(\underset{\sim}{A}(\underset{\sim}{x}))<0\}$$

As in (2.7) and (2.11) by construction $I_3^+$, $I_3^0$, $I_3^-$ are disjoint with $I_3^+\cup I_3^0\cup I_3^-=I_2^0$ (see Fig. 1).

Similarly to the rate 2 problem, the rate 3 derivative of $(\partial\phi/\partial\underset{\sim}{\varepsilon})\bullet\delta\underset{\sim}{\varepsilon}$, with respect to $\tau$ that appears in (2.14) has to be interpreted in the generalized function sense in view of the possible moving discontinuities. In writing (2.14) explicitly, only the volume integral contains rate 3 terms while the surface discontinuity integrals contain terms up to rate two. Hence in the bifurcation calculation ofthe rate 3 problem, only the volume term in (2.9) appears in the final expression.

Once more, the third rate of the equilibrium equation (2.6) is recovered from (2.14) by taking $\delta\underset{\sim}{u}\neq 0$, $\beta=\ddot{\mu}$, while for $\delta\underset{\sim}{u}=0$, $\beta\neq\ddot{\mu}$ and given that $\beta,\ddot{\mu}\in K_3$ one obtains:

$$\int_{\Omega}(\beta-\ddot{\mu})(-\dddot{f})\,dV\geq 0\Rightarrow\int_{I_1^+\cup I_2^+\cup I_3^+\cup I_3^0}(\beta-\ddot{\mu})(-\dddot{f})\geq 0\Rightarrow\begin{cases}\text{For } \underset{\sim}{x}\in I_3^+\cup I_2^+\cup I_1^+\Rightarrow\dddot{f}=0\\ \text{For } \underset{\sim}{x}\in I_3^0\begin{cases}\text{if } \ddot{\mu}>0\Rightarrow\dddot{f}=0\\ \text{if } \ddot{\mu}=0\Rightarrow\dddot{f}\leq 0\end{cases}\\ \text{For } \underset{\sim}{x}\in I_3^-\cup I_2^-\cup I_1^-\Rightarrow\ddot{\mu}=0\end{cases} \tag{2.15}$$

which also follows from (2.2), (2.3) and (2.4). As for the previous two cases, one has from (2.15) $\ddot{\mu}\,\dddot{f}=0 \quad \forall\,\underset{\sim}{x}\in\Omega$.

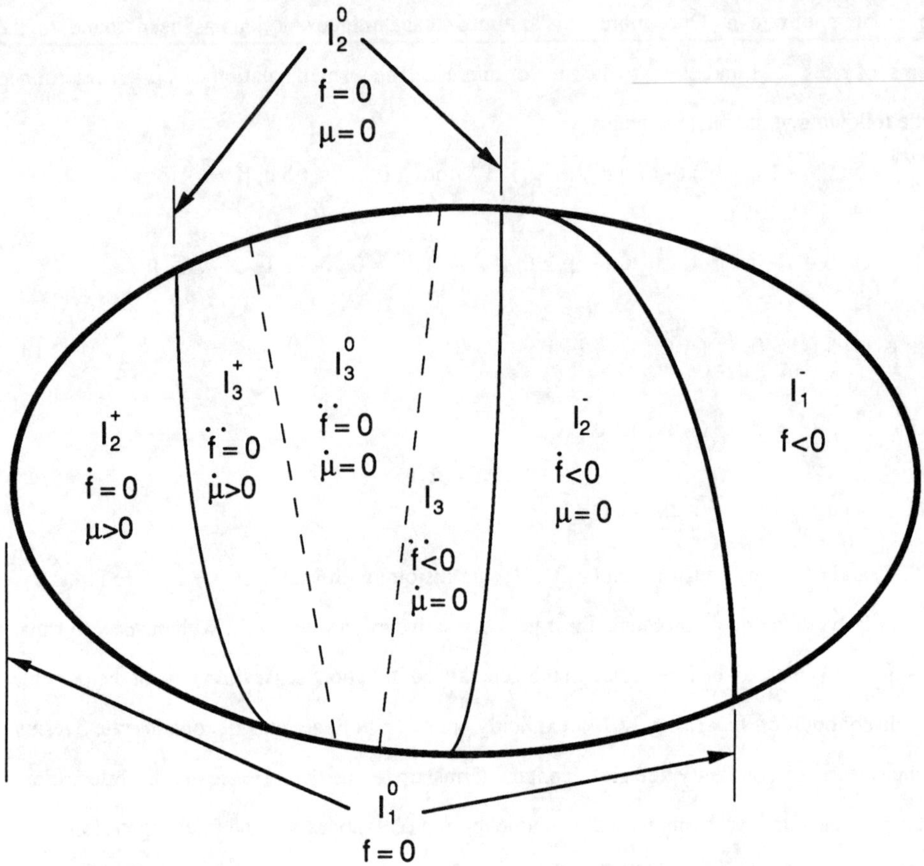


Fig.1. Subdivision of domain Ω in the various sets defined in Section 2.1

As expected, a sufficient condition for the uniqueness of the solution ($\dddot{\tilde{u}}$, $\ddot{\mu}$) to (2.14) is the positive definitness of the functional $\mathfrak{F}$ in (2.9) over the set $U \times V\, K_3$. Indeed, by a similar construction, if ($\dddot{\tilde{u}}_1, \ddot{\mu}_1$) and ($\dddot{\tilde{u}}_2, \ddot{\mu}_2$) are two different solutions to (2.14), from (2.9), (2.14) and (2.15) follows the contradiction:

$$\mathfrak{F}\,[\,\ddot{\tilde{u}}_1 - \ddot{\tilde{u}}_2,\ \ddot{\mu}_1 - \ddot{\mu}_2\,] = \int_{I_3^0} (\ddot{\mu}_1 - \ddot{\mu}_2)\,(-\dddot{f}_1 + \dddot{f}_2) \leq 0 \tag{2.16}$$

The generalization of the above algorithm for the rate n boundary value problem can be accomplished as follows: assuming a unique response ($\dot{\tilde{u}}, \mu$), ($\ddot{\tilde{u}}, \dot{\mu}$),..., ($\overset{(n-1)}{\tilde{u}}$ , $\overset{(n-2)}{\mu}$) to all the boundary value problems of order 1, 2,... (n-1), the variational inequality corresponding to the rate n problem is given by:

$$\int_{\Omega}\Big[\Big(\frac{\partial\phi}{\partial\underset{\sim}{\varepsilon}}\bullet\delta\underset{\sim}{\varepsilon}\Big)^{(n)}+(\beta-\overset{(n-1)}{\mu})(-\overset{(n)}{f})\Big]dV-\overset{(n)}{\lambda}\int_{\partial\Omega}\underset{\sim}{T}^{0}\bullet\delta\underset{\sim}{u}\geq 0\quad\forall(\delta\underset{\sim}{u},\beta)\in U\times K_n$$

$$K_n\equiv\{\ \beta(\underset{\sim}{x})\mid\beta\in R\ \text{if}\ \underset{\sim}{x}\in\bigcup_{k=1}^{n}I_k^{+},\ \beta\geq 0\ \text{if}\ \underset{\sim}{x}\in I_n^{0},\ \beta=0\ \text{if}\ \underset{\sim}{x}\in\bigcup_{k=1}^{n}I_k^{-}\}$$

$$I_n^{+}\equiv\{\underset{\sim}{x}\in I_{n-1}^{0}\mid\overset{(n-2)}{\mu}(\underset{\sim}{x})>0\ ,\ \overset{(n-1)}{f}(\underset{\sim}{A}(\underset{\sim}{x}))=0\}\tag{2.17}$$

$$I_n^{0}=\{\underset{\sim}{x}\in I_{n-1}^{0}\mid\overset{(n-2)}{\mu}(\underset{\sim}{x})=0\ ,\ \overset{(n-1)}{f}(\underset{\sim}{A}(\underset{\sim}{x}))=0\}$$

$$I_n^{-}=\{\underset{\sim}{x}\in I_{n-1}^{0}\mid\overset{(n-2)}{\mu}(\underset{\sim}{x})=0\ ,\ \overset{(n-1)}{f}(\underset{\sim}{A}(\underset{\sim}{x}))<0\}$$

Note that $I^{+}_{n}$, $I^{0}_{n}$, $I^{-}_{n}$ are disjoint while $I^{+}_{n}\cup I^{0}_{n}\cup I^{-}_{n}=I^{0}_{n-1}$

As in all the previous cases for $\delta\underset{\sim}{u}=0$ , $\beta=\overset{(n-1)}{\mu}$ the variational inequality (2.17) yields the rate n of equilibrium equation (2.6) while for $\delta\underset{\sim}{u}=0$ , $\beta\neq\overset{(n-1)}{\mu}$ and since $\beta$, $\overset{(n-1)}{\mu}\in K_n$

$$\int_{\Omega}(\beta-\overset{(n-1)}{\mu})(-\overset{(n)}{f})\geq 0\Rightarrow\int_{(\bigcup_{k=1}^{n}I_k^{+})\cup I_n^{0}}(\beta-\overset{(n-1)}{\mu})(-\overset{(n)}{f})\geq 0\quad\begin{cases}\text{For}\ \underset{\sim}{x}\in\bigcup_{k=1}^{n}I_k^{+}\Rightarrow\overset{(n)}{f}=0\\[2ex]\text{For}\ \underset{\sim}{x}\in I_n^{0}\begin{cases}\text{if}\ \overset{(n-1)}{\mu}>0\Rightarrow\overset{(n)}{f}=0\\[2ex]\text{if}\ \overset{(n-1)}{\mu}=0\Rightarrow\overset{(n)}{f}\leq 0\end{cases}\\[2ex]\text{For}\ \underset{\sim}{x}\in\bigcup_{k=1}^{n}I_k^{-}\Rightarrow\overset{(n-1)}{\mu}=0\end{cases}\tag{2.18}$$

which similarly follows from (2.2), (2.3) and (2.4) after taking appropriate rates. Also note that $\overset{(n-1)}{\mu}\ \overset{(n)}{f}=0\quad\forall\,\underset{\sim}{x}\in\Omega$.

A sufficient condition for the uniqueness of the solution $(\overset{(n)}{\underset{\sim}{u}},\ \overset{(n-1)}{\mu})$ to (2.17) is again the positive definitness of the functional $\mathfrak{F}$ defined in (2.9) over the set $U\times\mathbf{V}K_n$. For if not, assuming the existence of two different solutions $(\overset{(n)}{\underset{\sim}{u}_1},\ \overset{(n-1)}{\mu_1})$, $(\overset{(n)}{\underset{\sim}{u}_2},\ \overset{(n-1)}{\mu_2})$ and making use of (2.9), (2.17) and (2.18) one is lead to the contradiction:

$$\mathfrak{F}[\overset{(n)}{\underset{\sim}{u}}_1-\overset{(n)}{\underset{\sim}{u}}_2,\ \overset{(n-1)}{\mu_1}-\overset{(n-1)}{\mu_2}]=\int_{I_n^{0}}(\overset{(n-1)}{\mu_1}-\overset{(n-1)}{\mu_2})(-\overset{(n)}{f_1}+\overset{(n)}{f_2})\leq 0\tag{2.19}$$

Note that in the derivation of (2.19), use was made of the property that only the volume part in the incremental equilibrium equation of rate n contains terms of order n. All surface discontinuity terms are of rate n-1 or lower and hence do not appear in (2.19). This property is a straightforward generalization of similar remarks made after (2.11) and (2.14) about the problems of rate 2 and 3.

The development of the variational inequality method for the rate problem of order n is a straightforward generalization of the work by NGUYEN and STOLZ (16) for the problems of rates 1 and 2. Note that in contrast to the aforementioned work, in the present development no explicit expression for the $n^{th}$ rate of the virtual work is given. (see equs (2.7), (2.11), (2.14), (2.17) ) in view of the resulting cumbersome expressions. It should also be mentioned at this point that special care has to be taken when considering the explicit expressions for the rates of the equilibrium equation (2.6) in view of the surfaces of discontinuity in $\overset{(n)}{\underset{\sim}{\varepsilon}}$ and $\overset{(n)}{\underset{\sim}{\alpha}}$; this difficulty will become apparent in the next subsection as well as in the example. The important point to retain is that the surface discontinuity terms involve rates of order n-1 or lower, a key property required in the derivation of (2.19).

It is not difficult to see that $\mathcal{V}\ K_1 \supset \mathcal{V}\ K_2 \supset ... \supset \mathcal{V}\ K_n$... Hence positive definitness of $\mathcal{F}\ [\hat{\underset{\sim}{u}}, \beta]\ \forall\ (\hat{\underset{\sim}{u}}, \beta) \in U \times \mathcal{V}\ K_1$ and $\forall\ \lambda \in [0, \lambda_c]$ (where $\mathcal{F}$ is evaluated on the principal solution $\overset{0}{\underset{\sim}{u}}(\lambda)$, $\overset{0}{\underset{\sim}{\alpha}}(\lambda)$ ) implies the impossibility of bifurcation in <u>any</u> order in the load parameter interval $[0, \lambda_c)$. If one makes the conjecture that any possible bifurcation is amenable to a polynomial type one, i.e. that there exists a parametrization at a neighborhood of the critical load $\lambda_c$ such that (see NGUYEN (10)):

$$\lambda(\tau) = \lambda_c + \frac{\tau^p}{p!}\lambda_p + 0(\tau^{p+1}) \qquad 0 < \tau << 1\ ,\ \underset{\sim}{u}_q \neq 0$$

$$\underset{\sim}{u}(\tau) = \overset{0}{\underset{\sim}{u}}(\tau) + \frac{\tau^q}{q!}\underset{\sim}{u}_q + 0(\tau^{q+1})\ ,\ \underset{\sim}{\alpha}(\tau) = \overset{0}{\underset{\sim}{\alpha}}(\tau) + \frac{\tau^q}{q!}\underset{\sim}{\alpha}_q + 0(\tau^{q+1}) \tag{2.20}$$

where $p, q \in \mathcal{N} \setminus \{0\}$, then the positive definitness of $\mathcal{F}$ on $U \times \mathcal{V}\ K_1$ for a solution ($\overset{0}{\underset{\sim}{u}}(\lambda)$, $\overset{0}{\underset{\sim}{\alpha}}(\lambda)$ ) on an interval $[0, \lambda_c)$ excludes bifurcation at that interval.

At this point a connection is made between Hill's sufficiency criterion for nonbifurcation and the one presently proposed. To this end, according to HILL (5), the following functional $\mathcal{H}[\hat{\underset{\sim}{u}}]$ (quadratic in $\hat{\underset{\sim}{u}}$) is defined:

$$\mathcal{H}[\hat{\tilde{u}}] \equiv \int_\Omega \Big\{ \tilde{\varepsilon}_{,u}(\hat{\tilde{u}}) \bullet \Big[ \frac{\partial^2\phi}{\partial\tilde{\varepsilon}\partial\tilde{\varepsilon}} - \chi_{I_1^0}\Big( \frac{\partial^2\phi}{\partial\tilde{\varepsilon}\partial\tilde{\alpha}} \bullet \frac{\partial f}{\partial\tilde{A}} \Big)\Big( \frac{\partial f}{\partial\tilde{A}} \bullet \frac{\partial^2\phi}{\partial\tilde{\alpha}\partial\tilde{\alpha}} \bullet \frac{\partial f}{\partial\tilde{A}} \Big)^{-1}\Big( \frac{\partial f}{\partial\tilde{A}} \bullet \frac{\partial^2\phi}{\partial\tilde{\alpha}\partial\tilde{\varepsilon}} \Big)\Big] \bullet \tilde{\varepsilon}_{,u}(\hat{\tilde{u}})$$

$$+ \frac{\partial\phi}{\partial\tilde{\varepsilon}} \bullet \tilde{\varepsilon}_{,uu}(\hat{\tilde{u}}, \hat{\tilde{u}}) \Big\} dV, \quad \hat{\tilde{u}} \in U \tag{2.21}$$

with $\chi_{I_1^0}$ the characteristic function of $I_1^0$ (i.e. $\chi(\tilde{x}) = 1$ if $\tilde{x} \in I_1^0$ and $\chi(\tilde{x}) = 0$ if $\tilde{x} \notin I_1^-$ ).

HILL (5) has shown that positive definitness of $\mathcal{H}$ in an interval $[0, \lambda_c)$ of the load parameter $\lambda$ excludes the bifurcation in the rate one problem there and without addressing the question of any higher order bifurcation, he postulated that bifurcation in the interval $[0, \lambda_c)$ (where $\mathcal{H}$ is positive definite) is excluded. The result presented here provides the justification for Hill's intuition. It will be shown that positive definitness of $\mathcal{H}$ in $[0, \lambda_c)$ implies (assuming certain additional properties of $\mathcal{F}$ which are met in the applications of interest ) the positive definitness of $\mathcal{F}$ on the same interval and hence bifurcation in any rate is excluded.

Indeed, had this not been the case, i.e. if $\mathcal{F}$ is not positive definite at some point of $[0, \lambda_c)$ and given that $\mathcal{F}$ is positive definite at $\lambda = 0$ (otherwise the problem is ill posed since a bifurcation is possible at a vanishing externally applied load) from the continuity of $\mathcal{F}$ with respect to $\lambda$ there exists $\lambda^* \in [0, \lambda_c)$ such that $\lambda^*$ is the point of the first loss of positive definitness of $\mathcal{F}$ as $\lambda$ increases away from zero. Hence at $\lambda = \lambda^*$, one can find $(\dot{\tilde{u}}^*, \mu^*) \in U \times \mathcal{V}K_1$ such that:

$$\int_\Omega \Big\{ \tilde{\varepsilon}_{,u}(\hat{\tilde{u}}) \bullet \Big[ \frac{\partial^2\phi}{\partial\tilde{\varepsilon}\partial\tilde{\varepsilon}} \bullet \tilde{\varepsilon}_{,u}(\dot{\tilde{u}}^*) + \frac{\partial^2\phi}{\partial\tilde{\varepsilon}\partial\tilde{\alpha}} \bullet \frac{\partial f}{\partial\tilde{A}} \mu^* \Big] + \beta \Big[ \frac{\partial f}{\partial\tilde{A}} \bullet \frac{\partial^2\phi}{\partial\tilde{\alpha}\partial\tilde{\varepsilon}} \bullet \tilde{\varepsilon}_{,u}(\dot{\tilde{u}}^*) + \frac{\partial f}{\partial\tilde{A}} \bullet \frac{\partial^2\phi}{\partial\tilde{\alpha}\partial\tilde{\alpha}} \bullet \frac{\partial f}{\partial\tilde{A}} \mu^* \Big]$$

$$+ \frac{\partial\phi}{\partial\tilde{\varepsilon}} \bullet \tilde{\varepsilon}_{,uu}(\hat{\tilde{u}}, \dot{\tilde{u}}^*) \Big\} dV = 0, \quad \forall (\hat{\tilde{u}}, \beta) \in U \times \mathcal{V}K_1 \tag{2.22}$$

For $\hat{\tilde{u}} = 0$ and $\beta \neq 0$, from (2.22) follows that:

$$\mu^* = -\chi_{I_1^0}\Big( \frac{\partial f}{\partial\tilde{A}} \bullet \frac{\partial^2\phi}{\partial\tilde{\alpha}\partial\tilde{\varepsilon}} \bullet \tilde{\varepsilon}_{,u}(\dot{\tilde{u}}^*) \Big)\Big( \frac{\partial f}{\partial\tilde{A}} \bullet \frac{\partial^2\phi}{\partial\tilde{\alpha}\partial\tilde{\alpha}} \bullet \frac{\partial f}{\partial\tilde{A}} \Big)^{-1} \tag{2.23}$$

since $\beta(\tilde{x}) = 0 \quad \forall \tilde{x} \in I_1^-$, for $\beta \in \mathcal{V}K_1$. Upon substitution of (2.23) into (2.22) and after taking

$\hat{\underset{\sim}{u}} = \dot{\underset{\sim}{u}}^*$ , one easily obtains the contradiction that at $\lambda^* \in [0, \lambda_c)$ , $\mathcal{H} [\dot{\underset{\sim}{u}}^*] = 0$.

A discussion on why Hill's uniqueness criterion excludes not only rate 1 bifurcations but higher order ones as well, was given in a different setting by TRIANTAFYLLIDIS (13). Some subsequent work by PETRYK and THERMANN (18) has also addressed the smooth bifurcation problem but only up to the second rate and by employing a completely different approach than the one proposed here.

## 2.2 POSTBIFURCATION ANALYSIS. SMOOTH VS. ANGULAR BIFURCATION

Having established a sufficient condition for non-bifurcation, attention is next focussed on its failure and more specifically on the postbifurcation asymptotic expansions for generalized standard solids. Following NGUYEN (10), a polynomial expansion in terms of integral powers of the time like parameter $\tau$ is postulated near the critical load $\lambda_c$ (i.e. the load at which the positive definitness of $\mathcal{F}$ is lost for the first time as $\lambda$ moves away from zero), as indicated by (2.20). A bifurcation is termed "angular" if $p = q$* and "smooth" if $p < q$ the terminology being motivated by the geometry of the corresponding $\lambda$ - $\| u \|$ diagram where the bifurcated branch emerges at an angle ($\lim_{\lambda \to \lambda_c} \| \underset{\sim}{u} - \overset{0}{\underset{\sim}{u}} \| / (\lambda - \lambda_c) \neq 0$ ) or tangently ( $\lim_{\lambda \to \lambda_c} \| \underset{\sim}{u} - \overset{0}{\underset{\sim}{u}} \| / (\lambda - \lambda_c) = 0$ ) from the principal one $\overset{0}{\underset{\sim}{u}} (\lambda)$.

Unlike the postbifurcation analysis of elastic solids where (in general) adequate regularity of the postbifurcation solution is available and permits a straightforward and general scheme for the calculation of all the higher order terms as found by KOITER (8) , no such regularity is present even in the simplest meaningful problems in plasticity in view of the presence of propagating discontinuities in $\overset{(n)}{\underset{\sim}{\alpha}}$. Hence only the first terms (i.e. $\lambda_p$, $\underset{\sim}{u}_q$, $\underset{\sim}{\alpha}_q$) in the postbifurcation expansions in (2.20) will be seeked via the general method proposed here.

---

* Note: The case $p > q$ can be reduced to an angular bifurcation by taking $p' = q$ with $\lambda_{p'} = 0$. In (2.20) $\underset{\sim}{u}_q$ always $\neq 0$, but $\lambda_p$ may vanish

It is not difficult to deduce from (2.20) that $\overset{0}{I}_1 = \overset{0}{I}_2 = \dots \overset{0}{I}_p$. Indeed, by assuming (without loss of generality) $(d\overset{0}{\underset{\sim}{u}}/d\lambda)_c \neq 0$, one has $\dot{\underset{\sim}{u}} = \ddot{\underset{\sim}{u}} = \dots \overset{(p-1)}{\underset{\sim}{u}} = 0$ and hence $\mu = \dot{\mu} = \dots \overset{(p-2)}{\mu} = 0$, $\dot{f} = \ddot{f} = \dots \overset{(p-1)}{f} = 0$ on $\overset{0}{I}_p = \overset{0}{I}_1$. At bifurcation from (2.19) one obtains:

$$0 = \mathcal{F}[\overset{(p)}{\underset{\sim}{u}}_1 - \overset{(p)}{\underset{\sim}{u}}_2, \overset{(p-1)}{\mu_1} - \overset{(p-1)}{\mu_2}] = \int_{\overset{0}{I}_p} (\overset{(p-1)}{\mu_1} - \overset{(p-2)}{\mu_2})(-\overset{(p)}{f_1} + \overset{(p)}{f_2})\, dV \Rightarrow \overset{(p-1)}{\mu_1}\overset{(p)}{f_2} = \overset{(p-1)}{\mu_2}\overset{(p)}{f_1} = 0 \quad \forall \underset{\sim}{x} \in \overset{0}{I}_p = \overset{0}{I}_1 \tag{2.24}$$

since $\overset{(p-1)}{\mu} \geq 0$ and $\overset{(p)}{f} \leq 0$ ( from (2.16) $\overset{+}{I}_{p+1} \cup \overset{0}{I}_{p+1} \cup \overset{-}{I}_{p+1} = \overset{0}{I}_p$ ). At this stage, and without impairing the problem's generality it is additionally assumed that the principal solution satisfies $\overset{0}{I}_p = \overset{+}{I}_{p+1}$*. This assumption, which is equivalent to the statement that (in the principal solution) all points of the plastic zone - except perhaps the boundary points - are plastically loading i.e. $\overset{(p-1)}{\mu} > 0$ almost everywhere on $\overset{0}{I}_p = \overset{0}{I}_1$, is met by the vast majority of applications. Hence from (2.24) follows:

$$\overset{(p)}{f_i} = 0 \text{ on } \overset{0}{I}_p \Rightarrow \overset{(p-1)}{\mu_i} = -\frac{\partial f}{\partial \underset{\sim}{A}} \bullet \frac{\partial^2 \phi}{\partial \underset{\sim}{\alpha} \partial \underset{\sim}{\varepsilon}} \bullet \underset{\sim}{\varepsilon}_{,u}(\overset{(p)}{\underset{\sim}{u}}_i) \Big/ \left( \frac{\partial f}{\partial \underset{\sim}{A}} \bullet \frac{\partial^2 \phi}{\partial \underset{\sim}{\alpha} \partial \underset{\sim}{\alpha}} \bullet \frac{\partial f}{\partial \underset{\sim}{A}} \right) \quad (i = 1, 2) \tag{2.25}$$

Supposing for simplicity that at bifurcation, the eigenmode ( $\bar{\underset{\sim}{u}}, \bar{\mu}$ ) of the functional $\mathcal{F}$ in (2.9) is unique, and employing (2.20) and (2.25) (note $\underset{\sim}{u}_q = \bar{\underset{\sim}{u}}$), one has for the bifurcated solution, assuming at first an angular bifurcation, i.e. p = q

$$-\left\{ \lambda_p \left[ \frac{\partial f}{\partial \underset{\sim}{A}} \bullet \frac{\partial^2 \phi}{\partial \underset{\sim}{\alpha} \partial \underset{\sim}{\varepsilon}} \bullet \underset{\sim}{\varepsilon}_{,u} \left( \frac{d\overset{0}{\underset{\sim}{u}}}{d\lambda} \right)_c \right] + \left[ \frac{\partial f}{\partial \underset{\sim}{A}} \bullet \frac{\partial^2 \phi}{\partial \underset{\sim}{\alpha} \partial \underset{\sim}{\varepsilon}} \bullet \underset{\sim}{\varepsilon}_{,u} (\bar{\underset{\sim}{u}}) \right] \right\} \Big/ \left( \frac{\partial f}{\partial \underset{\sim}{A}} \bullet \frac{\partial^2 \phi}{\partial \underset{\sim}{\alpha} \partial \underset{\sim}{\alpha}} \bullet \frac{\partial f}{\partial \underset{\sim}{A}} \right) \geq 0 \quad \forall \underset{\sim}{x} \in \overset{0}{I}_p \tag{2.26}$$

since (for the bifurcated solution) $\overset{(p)}{\mu_2} \geq 0$ on $\overset{0}{I}_p$.

* Note: The equality of sets in this case is in the almost everywhere sense i.e. the two sets differ by a set of Lebesque measure zero where i = 1 is the principal and i = 2 is the bifurcated solution respectively.

Two cases are distinguished:

$$\text{i) case: } \min_{\underset{\sim}{x} \in I_p^0} \Big[ -\frac{\partial f}{\partial \underset{\sim}{A}} \bullet \frac{\partial^2 \phi}{\partial \underset{\sim}{\alpha} \partial \underset{\sim}{\varepsilon}} \bullet \underset{\sim}{\varepsilon}_{,u} \left( \frac{d\overset{0}{\underset{\sim}{u}}}{d\lambda} \right)_c \Big/ \frac{\partial f}{\partial \underset{\sim}{A}} \bullet \frac{\partial^2 \phi}{\partial \underset{\sim}{\alpha} \partial \underset{\sim}{\alpha}} \bullet \frac{\partial f}{\partial \underset{\sim}{A}} \Big] > 0 \tag{2.27}$$

This eventuality occurs in structures with no unloading in their principal solution and which satisfy the loading criterion everywhere where the yield condition is satisfied. Then from (2.26)

$$\lambda_p \geq -\frac{\partial f}{\partial \underset{\sim}{A}} \bullet \frac{\partial^2 \phi}{\partial \underset{\sim}{\alpha} \partial \underset{\sim}{\varepsilon}} \bullet \underset{\sim}{\varepsilon}_{,u} (\bar{\underset{\sim}{u}}) \Big/ \frac{\partial f}{\partial \underset{\sim}{A}} \bullet \frac{\partial^2 \phi}{\partial \underset{\sim}{\alpha} \partial \underset{\sim}{\varepsilon}} \bullet \underset{\sim}{\varepsilon}_{,u} \left( \frac{d\overset{0}{\underset{\sim}{u}}}{d\lambda} \right)_c \tag{2.28}$$

for $\underset{\sim}{x}$ in the aforementioned set $\overset{0}{I}_p$.

Given that the bifurcated solution exhibits no unloading in $\overset{0}{I}_p$, such a solution can be constructed if one introduces a priori the value:

$$\tilde{\lambda}_p = \max_{\underset{\sim}{x} \in I_p^0} \Big\{ -\frac{\partial f}{\partial \underset{\sim}{A}} \bullet \frac{\partial^2 \phi}{\partial \underset{\sim}{\alpha} \partial \underset{\sim}{\varepsilon}} \bullet \underset{\sim}{\varepsilon}_{,u} (\bar{\underset{\sim}{u}}) \Big/ \frac{\partial f}{\partial \underset{\sim}{A}} \bullet \frac{\partial^2 \phi}{\partial \underset{\sim}{\alpha} \partial \underset{\sim}{\varepsilon}} \bullet \underset{\sim}{\varepsilon}_{,u} \left( \frac{d\overset{0}{\underset{\sim}{u}}}{d\lambda} \right)_c \Big\} \tag{2.29}$$

where the maximum exists and is finite in view of (2.27).

Since at bifurcation the sign of the eigenmode can be either positive or negative, one might expect two different values for $\tilde{\lambda}_p$ associated respectively with $\pm \bar{\underset{\sim}{u}}$, one of which is necessarily positive. In most applications, due to the symmetries of the structure, both these maxima coincide and are positive. Hence of the two possible (corresponding to $\pm (\bar{\underset{\sim}{u}}, \bar{\mu})$ respectively) branches at $\lambda = \lambda_c$, at least one will take place under an increasing applied load, i.e. the corresponding $\tilde{\lambda}_p > 0$ as first discussed by SHANLEY (3) for the column case and subsequently generalized by HILL (5).

The bifurcation analysis proceeds in the following way: If $\lambda_p > \tilde{\lambda}_p$, then the value of $\lambda_p$ is found from the higher order terms expansion of the equilibrium equation (2.6). Of course, one has to check the validity of the priori assumption $\lambda_p > \tilde{\lambda}_p$. For additional details in the special case of $p = 1$, the interested reader is referred to HUTCHINSON (6) (who follows Hill's approach instead of the generalized standard formalism employed here). The other possible alternative is $\lambda_p = \tilde{\lambda}_p$, which implies the existence of a point $\underset{\sim}{x}^* \in \overset{0}{I}_p$ for which unloading starts in the bifurcated branch of the solution, this being the case for the majority of the applications treated to date.

It should also be remarked here that in most of the applications presented in the literature thus far which satisfy (2.27), their principal solutions have the additional property $\overset{0}{I}_1 = \overset{+}{I}_p = \overset{0}{I}_{p+1} = \Omega$

(i.e. the principal solution satisfies everywhere in the solid the total loading condition). This property however is not necessary for (2.27) to be valid and one can easily find such applications (for example the buckling of a tapered column).

Hence, if at the critical load the princial solution satisfies (2.27), an angular bifurcation is possible (with $p = q$) and $\lambda_p$, the first term in the asymptotic expansion for the load parameter $\lambda$ has to satisfy (2.28). The value of the integer p, as well as the information on whether (2.28) is satisfied as a strict inequality or as an equality depend on considerations from higher order terms.

The second case of interest is obviously:

$$\text{ii case)}\quad \min_{\underset{\sim}{x}\in \overset{0}{I}_p}\left[-\frac{\partial f}{\partial \underset{\sim}{A}}\bullet\frac{\partial^2\phi}{\partial\underset{\sim}{\alpha}\,\partial\underset{\sim}{\varepsilon}}\bullet\underset{\sim}{\varepsilon}_{,u}\left(\frac{d\overset{0}{\underset{\sim}{u}}}{d\lambda}\right)_c \Big/ \frac{\partial f}{\partial\underset{\sim}{A}}\bullet\frac{\partial^2\phi}{\partial\underset{\sim}{\alpha}\,\partial\underset{\sim}{\alpha}}\bullet\frac{\partial f}{\partial\underset{\sim}{A}}\right]=0 \qquad (2.30)$$

i.e. there is neutral loading in the principal solution at certain points $\underset{\sim}{x}_0 \in \overset{0}{I}_p$. In the majority of applications, these points belong to a portion $\Gamma$ of the boundary $\partial\overset{0}{I}_p$.

If one continues to adopt the angular bifurcation assumption of $p = q$ as in the previous case, the search for a bifurcated branch with the same loading zone $\overset{0}{I}_p$ leads again to inequality (2.28) which is available $\forall\ \underset{\sim}{x} \in \overset{0}{I}_p$. If $\underset{\sim}{x}_0 \in \Gamma$, then (2.30) implies that:

$$\frac{\partial f}{\partial \underset{\sim}{A}}\bullet\frac{\partial^2\phi}{\partial\underset{\sim}{\alpha}\,\partial\underset{\sim}{\varepsilon}}\bullet\underset{\sim}{\varepsilon}_{,u}\left(\frac{d\overset{0}{\underset{\sim}{u}}}{d\lambda}\right)_c \Bigg|_{\underset{\sim}{x}_0} = 0 \qquad (2.31)$$

in view of the denominator's strict positivity. The inequality (2.28) leads, when $\underset{\sim}{x} \to \underset{\sim}{x}_0 \in \Gamma$, necessarily to a contradiction if:

$$\lim_{\substack{\underset{\sim}{x}\to\underset{\sim}{x}_0\\ \underset{\sim}{x}\in\Gamma}} \frac{\partial f}{\partial \underset{\sim}{A}}\bullet\frac{\partial^2\phi}{\partial\underset{\sim}{\alpha}\,\partial\underset{\sim}{\varepsilon}}\bullet\underset{\sim}{\varepsilon}_{,u}(\bar{\underset{\sim}{u}})\Big|_{\underset{\sim}{x}} \neq 0 \qquad (2.32)$$

since the right hand is side of (2.28) is unbounded in $\overset{0}{I}_p$.

In this condition, one concludes that no such $\lambda_p$ can be found, rendering the angular bifurcation impossible and hence a tangent bifurcation with $p < q$ has to be investigated.

Note that for this case (i.e. if (2.30) is satisfied) an angular bifurcation is still possible if the eigenmode $\bar{\underset{\sim}{u}}$ satisfies:

$$\frac{\partial f}{\partial \underset{\sim}{A}} \bullet \frac{\partial^2 \phi}{\partial \underset{\sim}{\alpha} \partial \underset{\sim}{\varepsilon}} \bullet \underset{\sim}{\varepsilon}_{,u} (\bar{\underset{\sim}{u}}) \Big|_{\underset{\sim}{x}_0} = 0 \quad \forall \underset{\sim}{x}_0 \in \Gamma \tag{2.33}$$

and is such that

$$\lim_{\underset{\sim}{x} \to \underset{\sim}{x}_0} \left[ -\frac{\partial f}{\partial \underset{\sim}{A}} \bullet \frac{\partial^2 \phi}{\partial \underset{\sim}{\alpha} \partial \underset{\sim}{\varepsilon}} \bullet \underset{\sim}{\varepsilon}_{,u} (\bar{\underset{\sim}{u}}) \Big/ \frac{\partial f}{\partial \underset{\sim}{A}} \bullet \frac{\partial^2 \phi}{\partial \underset{\sim}{\alpha} \partial \underset{\sim}{\varepsilon}} \bullet \underset{\sim}{\varepsilon}_{,u} \left( \frac{d\overset{0}{\underset{\sim}{u}}}{d\lambda} \right)_c \right] < +\infty \, , \quad \forall \underset{\sim}{x}_0 \in \Gamma \tag{2.34}$$

In the interest of simplicity, it will be assumed for the following general discussion that $p = 1$, $q = 2$. Thus the general methodology for the determination of the post-bifurcation expansion, in the case of a tangent bifurcation, as well as the related difficulties will be explained in a simpler context without significant loss of generality.

In view of the assumption $p = 1$, $q = 2$ the solution to the rate 1 problem (see (2.8)) is unique, but the solution to the rate 2 problem (see (2.11)) is not unique at $\lambda = \lambda_c$. The explicit form of the rate 2 equilibrium equation i.e. the $0\,(\tau^2)$ term in the expansion of the equilibrium equ. (2.6) is:

$$\begin{aligned}
&\int_\Omega \Big\{ \underset{\sim}{\varepsilon}_{,u} (\delta \underset{\sim}{u}) \bullet \Big[ \frac{\partial^2 \phi}{\partial \underset{\sim}{\varepsilon} \partial \underset{\sim}{\varepsilon}} \bullet \ddot{\underset{\sim}{\varepsilon}} + \frac{\partial^2 \phi}{\partial \underset{\sim}{\varepsilon} \partial \underset{\sim}{\alpha}} \bullet \frac{\partial f}{\partial \underset{\sim}{A}} \dot{\mu} + \Big( \frac{\partial^3 \phi}{\partial \underset{\sim}{\varepsilon} \partial \underset{\sim}{\varepsilon} \partial \underset{\sim}{\varepsilon}} \bullet \dot{\underset{\sim}{\varepsilon}} + \frac{\partial^3 \phi}{\partial \underset{\sim}{\varepsilon} \partial \underset{\sim}{\varepsilon} \partial \underset{\sim}{\alpha}} \bullet \frac{\partial f}{\partial \underset{\sim}{A}} \mu \Big) \bullet \dot{\underset{\sim}{\varepsilon}} \\
&+ \Big( \frac{\partial^3 \phi}{\partial \underset{\sim}{\varepsilon} \partial \underset{\sim}{\alpha} \partial \underset{\sim}{\varepsilon}} \bullet \dot{\underset{\sim}{\varepsilon}} + \frac{\partial^3 \phi}{\partial \underset{\sim}{\varepsilon} \partial \underset{\sim}{\alpha} \partial \underset{\sim}{\alpha}} \bullet \frac{\partial f}{\partial \underset{\sim}{A}} \mu \Big) \bullet \frac{\partial f}{\partial \underset{\sim}{A}} \mu - \frac{\partial^2 \phi}{\partial \underset{\sim}{\varepsilon} \partial \underset{\sim}{\alpha}} \bullet \frac{\partial^2 f}{\partial \underset{\sim}{A} \partial \underset{\sim}{A}} \bullet \Big( \frac{\partial^2 \phi}{\partial \underset{\sim}{\alpha} \partial \underset{\sim}{\varepsilon}} \bullet \dot{\underset{\sim}{\varepsilon}} + \frac{\partial^2 \phi}{\partial \underset{\sim}{\alpha} \partial \underset{\sim}{\alpha}} \bullet \frac{\partial f}{\partial \underset{\sim}{A}} \mu \Big) \mu \Big] \\
&+ 2\, \underset{\sim}{\varepsilon}_{,uu} (\dot{\underset{\sim}{u}}, \delta \underset{\sim}{u}) \bullet \Big[ \frac{\partial^2 \phi}{\partial \underset{\sim}{\varepsilon} \partial \underset{\sim}{\varepsilon}} \bullet \dot{\underset{\sim}{\varepsilon}} + \frac{\partial^2 \phi}{\partial \underset{\sim}{\varepsilon} \partial \underset{\sim}{\alpha}} \bullet \frac{\partial f}{\partial \underset{\sim}{A}} \mu \Big] + \frac{\partial \phi}{\partial \underset{\sim}{\varepsilon}} \bullet \Big[ \underset{\sim}{\varepsilon}_{,uuu} (\dot{\underset{\sim}{u}}, \dot{\underset{\sim}{u}}, \delta \underset{\sim}{u}) + \\
&+ \underset{\sim}{\varepsilon}_{,uu} (\ddot{\underset{\sim}{u}}, \delta \underset{\sim}{u}) \Big] \Big\} \, dV + \int_{\partial I_2^+} \underset{\sim}{\varepsilon}_{,u} (\delta \underset{\sim}{u}) \bullet \frac{\partial^2 \phi}{\partial \underset{\sim}{\varepsilon} \partial \underset{\sim}{\alpha}} \bullet \frac{\partial f}{\partial \underset{\sim}{A}} [[\, \mu \,]] \, v_n \, dS
\end{aligned} \tag{2.35}$$

where

$$\mu = -\chi_{I_2^+} \Big( \frac{\partial f}{\partial \underset{\sim}{A}} \bullet \frac{\partial^2 \phi}{\partial \underset{\sim}{\alpha} \partial \underset{\sim}{\varepsilon}} \bullet \dot{\underset{\sim}{\varepsilon}} \Big/ \frac{\partial f}{\partial \underset{\sim}{A}} \bullet \frac{\partial^2 \phi}{\partial \underset{\sim}{\alpha} \partial \underset{\sim}{\alpha}} \bullet \frac{\partial f}{\partial \underset{\sim}{A}} \Big) \, , \quad \dot{\underset{\sim}{\varepsilon}} = \underset{\sim}{\varepsilon}_{,u} (\dot{\underset{\sim}{u}})$$

with analogous expressions for $\dot{\mu}$ and $\ddot{\underset{\sim}{\varepsilon}}$. In addition $[[\, \mu \,]]$ denotes the jump in $\mu$ and $v_n$ denotes the velocity of the boundary $\partial I_2^+$ of $I_2^+$ in the direction of its outward normal $\underset{\sim}{n}$. Note that in the derivation of the rate two equilibrium equation, account is taken of the fact that in general $\mu\,(\underset{\sim}{x})$ is a discontinuous function on the surface $\partial I_2^+$ (recall from (2.11) that $\mu > 0$ on $I_2^+$ and $\mu = 0$ on $\Omega \setminus I_2^+$). The strain rate $\dot{\underset{\sim}{\varepsilon}}$ on the other hand is assumed to be continuous, for a hypothesis to the contrary means the occurrence of strain localization in the solid possibility which is excluded in this work.

In the interest of computational simplicity and without impairing significantly the problem's generality, it is further assumed that the strain rates $\dot{\underset{\sim}{\varepsilon}}, \ddot{\underset{\sim}{\varepsilon}}$, ... are continuous functions of $\underset{\sim}{x}$. In contrast the rates of the internal variable $\dot{\underset{\sim}{\alpha}}, \ddot{\underset{\sim}{\alpha}}$,... have moving discontinuities which are being accounted for.

Applying successively (2.35) to the principal and the bifurcated solutions respectively one obtains as expected $\left( \text{recalling that } \lambda = \lambda_c + \lambda_1 \tau + \ldots, \ \underset{\sim}{u} = \underset{\sim}{u}_0(\lambda) + \underset{\sim}{u}_2 \frac{\tau^2}{2} + \underset{\sim}{u}_3 \frac{\tau^3}{6} + \ldots \right)$

$$\int_\Omega \Big\{ \varepsilon_{,u}(\delta \underset{\sim}{u}) \bullet \Big[ \frac{\partial^2 \phi}{\partial \underset{\sim}{\varepsilon} \partial \underset{\sim}{\varepsilon}} - \chi_{I_2^+} \Big( \frac{\partial^2 \phi}{\partial \underset{\sim}{\varepsilon} \partial \underset{\sim}{\alpha}} \bullet \frac{\partial f}{\partial \underset{\sim}{A}} \Big) \Big( \frac{\partial f}{\partial \underset{\sim}{A}} \bullet \frac{\partial^2 \phi}{\partial \underset{\sim}{\alpha} \partial \underset{\sim}{\alpha}} \bullet \frac{\partial f}{\partial \underset{\sim}{A}} \Big)^{-1} \Big( \frac{\partial f}{\partial \underset{\sim}{A}} \bullet \frac{\partial^2 \phi}{\partial \underset{\sim}{\alpha} \partial \underset{\sim}{\varepsilon}} \Big) \Big] \bullet \varepsilon_{,u}(\underset{\sim}{u}_2)$$

$$+ \frac{\partial \phi}{\partial \underset{\sim}{\varepsilon}} \bullet \varepsilon_{,uu}(\underset{\sim}{u}_2, \delta \underset{\sim}{u}) \Big\} \, dV = 0 \qquad (2.36)$$

with the functional evaluated on the principal branch at $\lambda = \lambda_c$.

The determination of $\lambda_1$ will require the $0(\tau^3)$ term in the expansion of the equilibrium equation (2.6). By taking rates in (2.35) and applying the result successively to the principal and bifurcated branch of the solution one obtains after some lengthy but straightforward calculations (in which (2.36) is also taken into account):

$$3\lambda_1 \int_\Omega \frac{d}{d\lambda} \Big\{ \varepsilon_{,u}(\underset{\sim}{u}_2) \bullet \Big[ \frac{\partial^2 \phi}{\partial \underset{\sim}{\varepsilon} \partial \underset{\sim}{\varepsilon}} - \chi_{I_2^+} \Big( \frac{\partial^2 \phi}{\partial \underset{\sim}{\varepsilon} \partial \underset{\sim}{\alpha}} \bullet \frac{\partial f}{\partial \underset{\sim}{A}} \Big) \Big( \frac{\partial f}{\partial \underset{\sim}{A}} \bullet \frac{\partial^2 \phi}{\partial \underset{\sim}{\alpha} \partial \underset{\sim}{\alpha}} \bullet \frac{\partial f}{\partial \underset{\sim}{A}} \Big)^{-1} \Big( \frac{\partial f}{\partial \underset{\sim}{A}} \bullet \frac{\partial^2 \phi}{\partial \underset{\sim}{\alpha} \partial \underset{\sim}{\varepsilon}} \Big) \Big] \bullet \varepsilon_{,u}(\underset{\sim}{u}_2)$$

$$+ \frac{\partial \phi}{\partial \underset{\sim}{\varepsilon}} \bullet \varepsilon_{,uu}(\underset{\sim}{u}_2, \underset{\sim}{u}_2) \Big\} \, dV + \Delta \Big[ \int_{\partial I_2^+} \varepsilon_{,u}(\underset{\sim}{u}_2) \bullet \frac{\partial^2 \phi}{\partial \underset{\sim}{\varepsilon} \partial \underset{\sim}{\alpha}} [[\mu]] \, v_n \, dS \Big]^{\cdot} + \qquad (2.37)$$

$$\Delta \Big[ \int_{\partial I_2^+} \varepsilon_{,u}(\underset{\sim}{u}_2) \bullet \frac{\partial^2 \phi}{\partial \underset{\sim}{\varepsilon} \partial \underset{\sim}{\alpha}} [[\dot{\mu}]] \, v_n \, dS \Big] = 0$$

where $\Delta f \equiv \lim_{\tau \to 0} (f_2(\tau) - f_1(\tau))$ where subscripts 1 and 2 denote quantities evaluated respectively on the principal and bifurcated branch. Obviously the surface integrals also depend on $\lambda_1$ but in a highly complicated fashion which requires the details of motion near $\lambda = \lambda_c$ of the principal and bifurcated surfaces $\partial I_2^+$ which separate the plastically loading zones from the corresponding elastic ones.

An application of the above analysis to the bifurcation of an elastic-plastic structure will be presented in the next section, where the computational complications due to the propagating discontinuites become apparent.

It should also be mentioned here, that for the special case of an angular bifurcation (in which also $I_p^0 = \Omega$) a comprehensive post bifurcation analysis based on the propagation of the unloading zone was first presented by HUTCHINSON (6) using Hill's plasticity formulation and requiring fractional expansions. More recently some of the problems treated by HUTCHINSON (7), were repeated by NGUYEN (10) using the standard generalized material formulation and an integral series expansion of the type proposed on (2.20). All the aforementioned work concerned angular bifurcations with $I_p^0 = \Omega$, a fact which simplifies the analysis considerably.

While a first attempt to address the problem of smooth bifurcation was made by TRIANTAFYLLIDIS (13) using a different line of attack (i.e. considering a set of rate invariant constitutive laws proposed by CHRISTOFFERSEN and HUTCHINSON (14) ) the present methodology provides a unified approach for the study of elastic-plastic bifurcation problems without any of the restrictive assumptions adopted in the literature thus far.

## 3. ILLUSTRATIVE EXAMPLE

An example is considered which is capable of exhibiting both an angular as well as a smooth bifurcation. This example, which is in the spirit of the finite degree of freedom models used by SHANLEY (3) ( see also HUTCHINSON (7), NGUYEN (10) ) has been employed by TRIANTAFYLLIDIS (13) to show the possibility of a smooth bifurcation in an elastic-plastic solid.

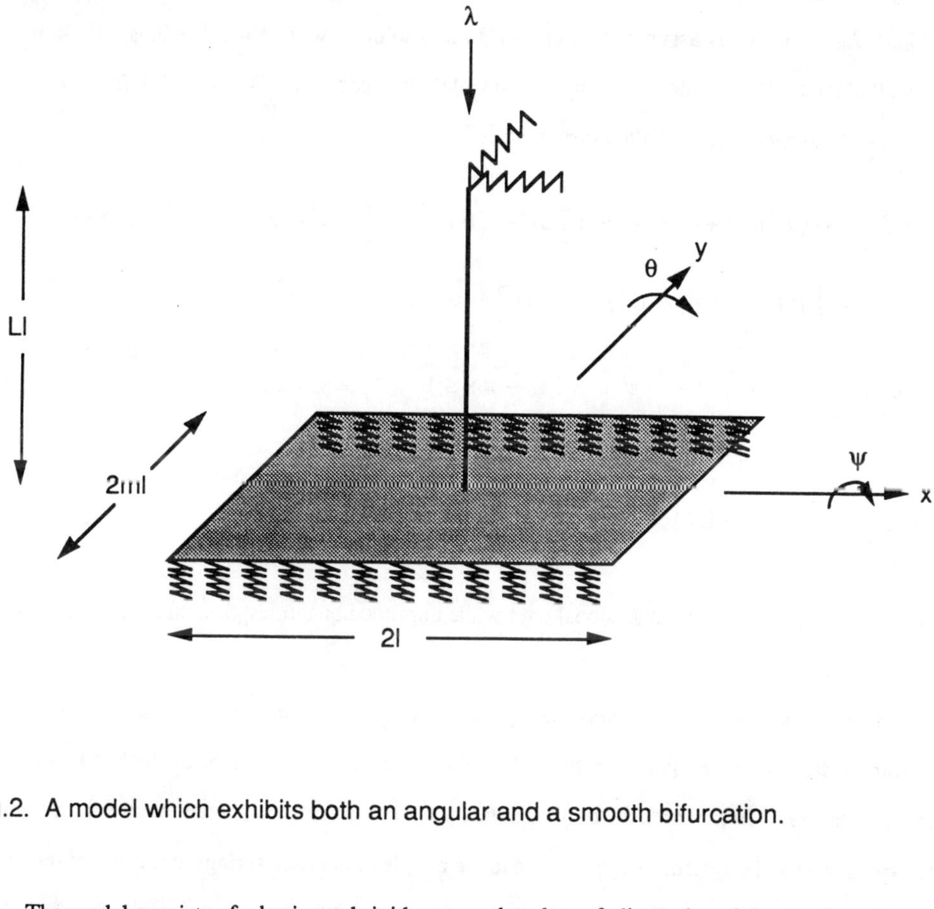


Fig.2. A model which exhibits both an angular and a smooth bifurcation.

The model consists of a horizontal rigid rectangular plate of dimensions $2\,l \times 2\,ml$ resting on two lines of continuously distributed elastic - plastic springs as shown in Fig. 2. A straight rod of length $Ll$ is rigidly attached perpendicularly to the plate's center on one end while the other end is attached to two coupled nonlinear springs (i.e. such that the displacement in one of them induces a force in the other) one in the $x'$ and the other in the $y'$ direction as shown in Fig. 2.

The externally applied vertical load is $\lambda E l^2$ where $\lambda$ is the problem's load parameter and E is the elastic region modulus for the bilinear hardening elastic-plastic springs. The structure has three degrees of freedom, namely the two rotations $\theta$ and $\psi$ about the $+y'$ and $-x'$ axes respectively and the vertical displacement $ul$ (note that $\lambda, u, \theta, \psi$ are dimensionless ). Under the action of the vertical load $\lambda$, the structure which initially follows a path $u \neq 0, \theta = 0, \psi = 0$ bifurcates first at $\lambda = \lambda_\theta$ to a solution with $u \neq 0, \theta \neq 0, \psi = 0$ and upon a further increase of $\lambda$ it bifurcates again at $\lambda = \lambda_\psi > \lambda_\theta$ ( it is tacitly assumed that $m^2 > 1/3$ ) to a solution with $u \neq 0, \theta \neq 0, \psi \neq 0$. It will be shown that the first bifurcation is an angular one while the second one is a smooth bifurcation.

The total energy stored in the system is:

$$\mathcal{E} = \frac{1}{2} \int_{-l}^{l} [\, \phi(x', ml) + \phi(x', -ml)\,]\, dx' - \lambda E l^3 [\, u + \frac{L}{2}(\theta^2 + \psi^2)\,] - E l^3 K L \theta \frac{\psi^2}{2}$$

$$\phi(x', y') = \frac{1}{2} E \{ [\, \varepsilon'(x', y') - \alpha'(x', y')\,]^2 + h [\, \alpha'(x', y')\,]^2 \}$$

$$\frac{\varepsilon'}{l} \equiv \varepsilon(x', y') = u + \frac{x}{l}\theta + \frac{y'}{l}\psi \;, \quad -1 \leq \frac{x'}{l} \equiv x \leq 1 \;, \quad \frac{y'}{l} \equiv y = \pm m \tag{3.1}$$

$$\frac{\alpha'}{l} \equiv \alpha \;, \qquad A' = -\frac{\partial \phi}{\partial \alpha'} = E l \,[\,(1+h)\alpha - \varepsilon\,] = l E A$$

$$A' \in C' \equiv [-kEl \;,\; kEl] \Leftrightarrow A \in C \equiv [-k, k]$$

where primed quantities are dimensional ones while unprimed quantities designate their dimensionless counterparts.

The internal variable $\alpha'$ in this case represents the plastic strain in the kinematically hardening bilinear model, while the associated force $A'$ is the corresponding backstress which is bounded by the yield stress $k' = El\,k$, with k the dimensionless yield stress. It is also noted that the third term in the expression for the system's energy is due to the coupled nonlinear springs at the end of the vertical rod.

The constitutive and equilibriums equs. (2.1), (2.3) and (2.6) employed in (3.1) yield:

$$\lambda = 4u - \int_{-1}^{1} \left[ \alpha(x, m) + \alpha(x, -m) \right] dx$$

$$\lambda L\theta = -KL\frac{\psi^2}{2} + \frac{4}{3}\theta - \int_{-1}^{1} \left[ \alpha(x, m) + \alpha(x, -m) \right] x\, dx \tag{3.2}$$

$$\lambda L\psi = -KL\psi\theta + 4m^2\psi - \int_{-1}^{1} \left[ \alpha(x, m) - \alpha(x, -m) \right] m\, dx$$

$$\dot{\alpha} = 0 \ \text{ if } |(1+h)\alpha - \varepsilon| < k \ , \ \ \dot{\alpha} \geq 0 \ \text{ if } |(1+h)\alpha - \varepsilon'| = k$$

In view of the relative analytic simplicity of the problem an exact solution for the primary ($\theta$) bifurcation (branch $u \neq 0, \theta \neq 0, \psi = 0$) is available and a complete asymptotic solution for the secondary ($\psi$) bifurcation (branch $u \neq 0, \theta \neq 0, \psi \neq 0$) can be constructed. The specialization of the general results presented in Section 2.2 to the present example will also be discussed.

### 3.1 PRIMARY ($\theta$) BIFURCATION - (ANGULAR CASE, p=q=2)

The primary ($\theta$) bifurcation of the plate is identical to the rigid T bifurcation example presented repeatedly in the literature (see for example HUTCHINSON (7), NGUYEN (10)). However, only the asymptotic behavior near the critical load $\lambda_\theta$ was of interest and the full solution for the primary ($\theta$) branch, which is necessary for calculations in the secondary ($\psi$) bifurcation has not been presented in the literature thus far. For this reason, as well as for the completeness of the presentation the primary ($\theta$) bifurcation of the studied example will be given here.

Starting with the rate form of (3.2) and considering the tilted equilibrium position with $u \neq 0$, $\theta \neq 0$, $\psi = 0$ for which the length of the unloading zone is $\tau$, i.e. $\varepsilon = 0$ at $x_n = \tau - 1$ where $x_n$ is the coordinate of the neutral loading zone, one deduces (assuming for simplicity $\theta$ to be the time like parameter for this particular derivation i.e. $(\dot{\ }) \equiv d(\ )/d\theta$ .

$$\dot{\lambda} = -\frac{1}{1+h}(\dot{u}^2 + 1) + 2\,\frac{1+2h}{1+h}\,\dot{u}$$

$$L\,(\lambda\theta)^{\cdot} = \frac{1}{3}\,\frac{1}{1+h}\,\dot{u}^3 - \frac{1}{1+h}\,\dot{u} + \frac{2}{3}\,\frac{1+2h}{1+h} \tag{3.3}$$

$$\text{at } \theta = 0,\ \ u = u_\theta = \frac{1}{3L} - \frac{k}{h},\ \ \lambda = \lambda_\theta = \frac{4}{3L}\,\frac{h}{1+h}$$

assuming of course $Lk > h/3(1+h)$ in order for the ($\theta$) bifurcation to occur in the plastic range. The above system can be integrated by considering $\theta$ as the dependent variable and $\tau' \equiv 1 - \dot{u}$ as the independent one to give:

$$\theta(\tau) = \frac{1}{2L}\left[4h(1-\tau)-\tau^2\right]^{-1/2} \int_0^\tau \tau'(2-\tau')\left[4h(1-\tau')-\tau'^2\right]^{-1/2} d\tau'$$

$$u(\tau) = \frac{1}{3L} - \frac{k}{h} + \int_0^\tau (1-\tau')\frac{d\theta}{d\tau'}\, d\tau'$$

$$\lambda(\tau) = \frac{4}{3L}\frac{h}{1+h} + \int_0^\tau \left[\frac{4h(1-\tau')-\tau'^2}{1+h}\right]\frac{d\theta}{d\tau'}\, d\tau' \tag{3.4}$$

$$\alpha(x,m) = \alpha(x,-m) = \begin{cases} \left[u(1+x)+x\,\theta(1+x)-k\right]/(1+h) & \text{for } \tau-1 \geq x \text{ (unloading zone)} \\ \left[u(\tau)+x\,\theta(\tau)-k\right]/(1+h) & \text{for } \tau-1 \leq x \text{ (loading zone)} \end{cases}$$

The above equations constitute the full solution for the primary ($\theta$) bifurcation of the model which occurs at the critical load $\lambda_\theta = 4h/3L(1+h)$.

Given that the size of the unloading zone $\tau$ increases with increasing $\lambda$ (at least near $\lambda = \lambda_\theta$) and that $\tau = 0$ at $\lambda = \lambda_\theta$, one can consider $\tau$ as a new time like parameter. Expanding (3.4) about $\tau = 0$ yields:

$$\begin{aligned} \theta &= 0 + \frac{1}{8Lh}\tau^2 + 0(\tau^3) \\ u &= u_\theta + \frac{1}{8Lh}\tau^2 + 0(\tau^3) \\ \lambda &= \lambda_\theta + \frac{1}{2L(1+h)}\tau^2 + 0(\tau^3) \end{aligned} \tag{3.5}$$

This reconfirms that the primary ($\theta$) bifurcation of this example (which coincides with the bifurcation of the literature's popular rigid T model) is an angular one with $p = q = 2$ (see 2.20). A different derivation of (3.5) following the general asymptotic method in terms of integral powers of $\tau$ proposed in Section 2.2 has already been given in NGUYEN (10) and hence the higher order terms in (3.5) will not be of concern here. Of course, according to the general analysis presented in Section 2, a sufficient condition for angular bifurcation (provided that a bifurcated solution exists) is given by (2.27). Indeed, using (3.1) and (3.2) in (2.27) one obtains (note also that $I_1^0 = I_1^+ = \Omega$ in the principal solution):

$$\min_{\underset{\sim}{x}\in\Omega}\Big[-\frac{\partial f}{\partial \underset{\sim}{A}}\bullet\frac{\partial^2\phi}{\partial\underset{\sim}{\alpha}\partial\underset{\sim}{\varepsilon}}\bullet\underset{\sim}{\varepsilon}_{,u}\Big(\frac{\overset{0}{du}}{d\lambda}\Big)_c\Big/\frac{\partial f}{\partial \underset{\sim}{A}}\bullet\frac{\partial^2\phi}{\partial\underset{\sim}{\alpha}\partial\underset{\sim}{\alpha}}\bullet\frac{\partial f}{\partial \underset{\sim}{A}}\Big]=\min_{-1\le x\le 1}\Big[\frac{(\overset{0}{du}/d\lambda)_c}{1+h}\Big]=\frac{1}{4h}>0 \qquad (3.6)$$

as expected from (3.5). One can also easily verify in this case that $\lambda_2=\tilde{\lambda}_2$ in view of the unloading starting at x = -1 at the onset of the bifurcation.

Indeed, recalling that the eigenmode $\bar{\underset{\sim}{u}} = \underset{\sim}{u}_2 = (u_2, \theta_2, \psi_2) = (0, (8Lh)^{-1}, 0)$, from (2.29):

$$\tilde{\lambda}_2=\max_{\underset{\sim}{x}\in\Omega}\Big\{-\frac{\partial f}{\partial \underset{\sim}{A}}\bullet\frac{\partial^2\phi}{\partial\underset{\sim}{\alpha}\partial\underset{\sim}{\varepsilon}}\bullet\underset{\sim}{\varepsilon}_{,u}(\bar{\underset{\sim}{u}})\Big/\frac{\partial f}{\partial \underset{\sim}{A}}\bullet\frac{\partial^2\phi}{\partial\underset{\sim}{\alpha}\partial\underset{\sim}{\varepsilon}}\bullet\underset{\sim}{\varepsilon}_{,u}\Big(\frac{\overset{0}{d\underset{\sim}{u}}}{d\lambda}\Big)_c\Big\}=\max_{-1\le x\le 1}\Big[\frac{-x/8Lh}{(1+h)/4h}\Big]=\frac{1}{2L(1+h)}=\lambda_2 \qquad (3.7)$$

with the maximum occuring at x = -1, where unloading starts at bifurcation, exactly as expected by the general theory since $\lambda_2=\tilde{\lambda}_2$. In the derivation of (3.7) one of the two modes was considered (the one with $\theta_2 > 0$). Had the other eigenmode being used, the value of $\tilde{\lambda}_2$ would have remained the same (in view of the structure's symmetry), but unloading would have started at x = +1.

## 3.2 SECONDARY ($\psi$) BIFURCATION - (SMOOTH CASE, p = 1, q = 2)

Next attention is focussed on the secondary ($\psi$) bifurcation of the model. The results have already been presented by TRIANTAFYLLIDIS (13) using a different approach but the derivations will be repeated here using the standard generalized material formulation. The results will also be compared with those deduced from the general analysis in Section 2.2.

Without loss of generality (in view of the structure's symmetry) it will be assumed that $\psi \ge 0$, in which case from (3.1) follows:

$$y=-m\begin{cases}\dot{\alpha}=0 & \text{for } -1\le x\le x_-\\[2ex] \dot{\alpha}=\dfrac{\dot{u}+x\,\dot{\theta}-m\,\dot{\psi}}{1+h}>0 & \text{for } x_-<x\le 1\end{cases}$$

$$y=m\begin{cases}\dot{\alpha}=0 & \text{for } -1\le x\le x_+^r\\[2ex] \dot{\alpha}=\dfrac{\dot{u}+x\,\dot{\theta}+m\,\dot{\psi}}{1+h}>0 & \text{for } x_+^r<x\le 1\end{cases} \qquad (3.8)$$

with: $x_-=(-\dot{u}+m\dot{\psi})/\dot{\theta}$ , $x_+=(-\dot{u}-m\dot{\psi})/\dot{\theta}$

and $x_+^r(t)$ the root of the following equation:

$$[\, u(t) - u_0(1 + x_+^r)\,] + x_+^r\,[\, \theta(t) - \theta_0(1 + x_+^r)\,] + m\psi(t) = 0$$

where $x_-(t)$ and $x_+^r(t)$ are the coordinates of the boundaries separating the elastic from the plastic zone at $y = -m$ and $y = +m$ respectively, while $x_-$ and $x_+$ are the corresponding coordinates of the points where the strain rate $\dot{\varepsilon} = 0$. The principal solution for the secondary ($\psi$) bifurcation is obviously given by (3.4) and to avoid confusion is characterized by a zero subscript in (3.8) and subsequently. In addition, the time like parameter for this case is denoted by t and hence $(\dot{\ }) \equiv d(\ )/dt$ in (3.8).

Combining (3.8) with the rate form of the equilibrium equations (3.2), one obtains the following expressions governing the secondary ($\psi$) bifurcation of the model:

$$\dot{\lambda}\,\dot{\theta} = 4\dot{u}\dot{\theta} + \frac{1}{1+h}\left[\frac{\dot{\theta}^2}{2}(x_+^r - x_+)^2 - (\dot{u} + \dot{\theta})^2 - (m\dot{\psi})^2\right]$$

$$\left[(\lambda L\theta)^{\cdot} + KL\psi\dot{\psi}\right]\dot{\theta}^2 = \frac{2}{3}\frac{1+2h}{1+h}\dot{\theta}^3 + \frac{1}{1+h}\left[\frac{\dot{u}^3}{3} + \dot{u}\,((m\dot{\psi})^2 - \dot{\theta}^2_+) + \frac{\dot{\theta}^3}{6}(x_+^r - x_+)^2(2x_+^r + x_+)\right]$$

$$\left[(\lambda + K\theta)L\psi\right]\dot{\theta} = 4m^2\dot{\psi}\dot{\theta} + \frac{m}{1+h}\left[\frac{\dot{\theta}^2}{2}(x_+^r - x_+)^2 - 2m\dot{\psi}(\dot{u} + \dot{\theta})\right] \tag{3.9}$$

For convenience, the length of the elastic unloading zone $\tau$ in the principal solution (see (3.4) will be adopted as the time like parameter (in reality $\lambda$ is controlled but in the vicinity of the critical load $\lambda = \lambda_\psi$, $d\lambda/d\tau > 0$, and hence $\tau$ can be chosen to play the role of the time like parameter).

Using the general results obtained in Section 2, it is not difficult to see that an angular bifurcation in ($\psi$) is impossible in this case. Indeed, noticing from (3.4) that the ($\psi$) bifurcation eigen mode is $\bar{\underset{\sim}{u}} = (0, 0, 1)$ and that from the principal solution in (3.4) $\Gamma = \{(-1 + \tau, m), (-1 + \tau, -m)\}$ (the set is made up of the two neutral loading points separating the loading from the unloading zones), one can easily see that both (2.31) and (2.32) are satisfied and hence no angular bifurcation in ($\psi$) is possible. Therefore, a tangent one with, p = 1, q = 2 will be adopted, with the following asymptotic expansions in $\tau$ ( $0 < t << 1$ ):

$$\tau(t) = \tau_c + \tau_1 t + \tau_2 \frac{t^2}{2} + \ldots$$

$$u(t) = u_0(\tau(t)) + u_2 \frac{t^2}{2} + u_3 \frac{t^3}{6} + \ldots$$

$$\theta(t) = \theta_0(\tau(t)) + \theta_2 \frac{t^2}{2} + \theta_3 \frac{t^3}{6} + \ldots \tag{3.10}$$

$$\psi(t) = 0 + \psi_2 \frac{t^2}{2} + \psi_3 \frac{t^3}{6} + \ldots$$

$$\lambda(t) = \lambda_0(\tau(t))$$

where $(u_2, \theta_2, \psi_2) = (0, 0, 1) = \underset{\sim}{u}_2$, the structure's normalized eigenmode, as it easily follows from (3.7). Hence (3.8) is the specialization of (2.20) to the finite degree of freedom example treated here with $p = 1$ and $q = 2$.

Employing (3.4), (3.8), and (3.10) into the rate 1 equilibrium equation (3.9) one obtains from the $0(t)$ term:

$$[\lambda_0(\tau_c) + K\,\theta_0(\tau_c)]\,L = 2m^2 \left(\frac{2h + \tau_c}{1 + h}\right) \tag{3.11}$$

which is the equation providing the critical load $\lambda_\psi = \lambda_0(\tau_c)$ (or equivalently $\tau_c$) for the secondary ($\psi$) bifurcation of the model.

For the higher order terms in the asymptotic expansion of the equilibrium equations, the following expressions for the boundary coordinates of the plastic zones at $y = \pm m$ will be needed (see (3.8), (3.10)):

$$x_+^r(t) = [\tau(t) - 1] + x_1^r t + x_2^r \frac{t^2}{2} + \ldots \; ; \; x_1^r = -\left(m \Big/ \frac{d\theta_0}{d\tau}\right)^{1/2} \tag{3.12}$$

$$x_+(t) = [\tau(t) - 1] + x_1^+ t + x_2^+ \frac{t^2}{2} + \ldots \; ; \; x_1^+ = -m \Big/ \tau_1 \frac{d\theta_0}{d\tau}$$

$$x_-(t) = [\tau(t) - 1] + x_1^- t + x_2^- \frac{t^2}{2} + \ldots \; ; \; x_1^- = m \Big/ \tau_1 \frac{d\theta_0}{d\tau}$$

Note that at the principal branch $(x_+^r)^0 = x_+^0 = x_-^0 = \tau - 1$. Continuing with the $0(\tau^2)$ term in the rate equilibrium equation (3.9) one obtains (by using also (3.4), (3.8), (3.10)-(3.12)):

$$\left[ 3\left( \frac{d\lambda_0}{d\tau} + K\frac{d\theta_0}{d\tau} \right)L - 4\frac{m^2}{1+h} \right] \frac{d\theta_0}{d\tau} (\tau_1)^2 = \frac{m}{1+h}\left[ m + \frac{d\theta_0}{d\tau}\,\tau_1\, x_1^r \right]^2 \tag{3.13}$$

which finally leads to the following expression for $\tau_1$:

$$\tau_1 = \left( m \Big/ \frac{d\theta_0}{d\tau} \right)^{1/2} \left\{ 1 \pm \left[ 3\left( \frac{1+h}{m^2} \right)\left( \frac{d\lambda_0}{d\tau} + K\frac{d\theta_0}{d\tau} \right)L - 4 \right]^{1/2} \right\}^{-1} \tag{3.14}$$

where it is understood that all derivatives of the principal solution are evaluated at $\tau = \tau_c$.

At this point it should also be noted that for adequately large K, (3.11) will admit a solution (with $\tau_c \to 0$ as $K \to \infty$) and the expression for $\tau_1$ in (3.14) will also make sense. Only $\tau_1 > 0$ solutions are acceptable (in view of the initial assumption that the controlled parameter in the experiment is the external load $\lambda$ which is considered to increase monotonically). It appears from (3.14) that for certain values of $\tau_c$ instead of one two different secondary ($\psi$) bifurcation branches are possible for $\psi > 0$ (and also another two are possible for $\psi < 0$).

The same results for $\tau_1$ could have been obtained by a direct application of (2.37) to the present model. Noting that:

$$3\lambda_1 \int_\Omega \frac{d}{d\lambda}\left\{ \underset{\sim}{\varepsilon}_{,u}(\underset{\sim}{u}_2) \bullet \left[ \frac{\partial^2\phi}{\partial\underset{\sim}{\varepsilon}\partial\underset{\sim}{\varepsilon}} - \chi_{I_2^+}\left( \frac{\partial^2\phi}{\partial\underset{\sim}{\varepsilon}\partial\underset{\sim}{\alpha}} \bullet \frac{\partial f}{\partial\underset{\sim}{A}} \right)\left( \frac{\partial f}{\partial\underset{\sim}{A}} \bullet \frac{\partial^2\phi}{\partial\underset{\sim}{\alpha}\partial\underset{\sim}{\alpha}} \bullet \frac{\partial f}{\partial\underset{\sim}{A}} \right)^{-1}\left( \frac{\partial f}{\partial\underset{\sim}{A}} \bullet \frac{\partial^2\phi}{\partial\underset{\sim}{\alpha}\partial\underset{\sim}{\varepsilon}} \right) \right] \bullet \underset{\sim}{\varepsilon}_{,u}(\underset{\sim}{u}_2) \right.$$

$$\left. + \frac{\partial\phi}{\partial\underset{\sim}{\varepsilon}} \bullet \underset{\sim}{\varepsilon}_{,uu}(\underset{\sim}{u}_2, \underset{\sim}{u}_2) \right\} dV = 3\tau_1\left( \frac{d\lambda_0}{d\tau} + K\frac{d\theta_0}{d\tau} \right)_{\tau=\tau_c} L \tag{3.15}$$

$$\Delta\left[ \int_{\partial I_2^+} \underset{\sim}{\varepsilon}_{,u}(\underset{\sim}{u}_2) \bullet \frac{\partial^2\phi}{\partial\underset{\sim}{\varepsilon}\partial\underset{\sim}{\alpha}} [[\mu]]\, v_n\, dS \right]^{\cdot} = \left[ -\frac{m}{1+h}(\dot{u} + x_+^r\,\dot{\theta} + m\dot{\psi})\,\dot{x}_+^r \right]^{\cdot}_{x=\tau_c-1,\ t=0}$$

$$\Delta\left[ \int_{\partial I_2^+} \underset{\sim}{\varepsilon}_{,u}(\underset{\sim}{u}_2) \bullet \frac{\partial^2\phi}{\partial\underset{\sim}{\varepsilon}\partial\underset{\sim}{\alpha}} [[\dot{\mu}]]\, v_n\, dS \right] = \left\{ -\frac{m}{1+h}\left[ -1\ddot{u} + x\ddot{\theta} - m\ddot{\psi})\,\dot{x}_- + (\ddot{u} + x\ddot{\theta} + m\ddot{\psi})\,\dot{x}_+^r \right]_{x=\tau_c-1,\ t=0}\right.$$

one recovers (3.14). It should be noted however that the relation between $x_1^r$ and the principal solution depends on the details of the propagating discontinuity of $\mu$ along $\partial I_2^+$ and requires the exploitation of (3.8). The derivation of a general expression in terms of the principal solution and the eigenmode for the discontinuity terms in (2.37) is not at all easy as this relatively simple example indicates. The independent derivation of (3.14) using the general approach developed in the previous section provides an additional check for the results obtained in this example.

## 4. CONCLUSIONS

In the present work, the general bifurcation and postbifurcation analysis for generalized standard continua has been presented.

By adopting the genaralized standard material formalism, one can have certain advantages (in terms of clarity and generality) over Hill's classical formalism of the problem. More specifically, the issue of higher than rate one bifurcations can be addressed in a straightforward manner. Hill's bifurcation exclusion criterion is found to be more general than what proved thus far, since it is valid for bifurcations of any rate. In addition, a consistent methodology that takes properly into account all the field quantities discontinuities, is given for the postbifurcation expansions in boundary value problems, without the so far needed resrictions of total loading in the principal branch of the solution.

The theory, in its present form, is applicable to small strain - moderate rotations rate independent elastoplastic solids obeying the maximum dissipation principle. The reason for this limitation is the lack (at least up to the present) of a satisfactory generalization of the generalized standard material formulation for the case of finite strains. For this case, the Hill - Hutchinson approach still remains the only method for analysing the angular bifurcation and post bifurcation problem. Moreover, the smooth bifurcation and postbifurcation issue (always in the large strain context) has only been addressed via the use of a Christoffersen - Hutchinson type model as discussed by TRIANTAFYLLIDIS (13), an approach that is limited to a particular class of constitutive laws that are very difficult to verify experimentally. Once an acceptable generalized standard material formalism can be found for rate independent materials in the large strain regime, the present bifurcation and postbifurcation analysis can easily be extended using essentially the same approach.

## REFERENCES

( 1) CONSIDERE, A. Resistance des Pieces Comprimees. *Congr. Intl. Proc. Const.* (1891), p. 371, Paris

( 2) VON KARMAN, T. Untensuchungen Uder Knickfestigkeit, Mitteilungen Uder Forchungsarbeiten. *Ver. Deut. Ing.*, (1910), Vol. **81**

( 3) SHANLEY, F.R. Inelastic Column Theory. *J. Aeronaut. Sci.*, (1947), Vol. **14**, p. 261

( 4) HILL, R. On the Problem of Uniquness in the Theory of a Rigid - Plastic Solid. *J. Mech. Phys. Solids*, (1956), Vol. **4**, p. 247

( 5) HILL, R. A General Theory for Uniquness and Stability in Elastic - Plastic Solids. *J. Mech. Phys. Solids*, (1958), Vol. **6**, p. 236

( 6) HUTCHINSON, J.W. Post - Bifurcation Behavior in the Plastic Range. *J. Mech. Phys. Solids*, (1973), Vol. **21**, p. 163

( 7) HUTCHINSON, J.W. Plastic Buckling. *Advances Appl. Mech.*, (1974), Vol. **14**, p. 67

( 8) KOITER, W.T. On the stability of Elastic Equilibrium. *Doctoral Thesis* (1945), Delft.

( 9) BUDIANSKY, B. Theory of Buckling and Post - Buckling Behavior of Elastic Structures. *Advances Appl. Mech.* (1974), Vol. **14**, p.1

(10) NGUYEN, Q.S. Bifurcation and Post - Bifurcation Analysis in Plasticity and Brittle Fracture. *J. Mech. Phys. Solids*, (1987), Vol. **35**, p.123

(11) NGUYEN, Q.S. Contribution a la Theorie Macroscopique de l' Elastoplasticite avec Ecrouissage. *Doctoral Thesis*, (1973) Paris

(12) HALPHEN, B.and NGUYEN, Q.S. Sur les Materiaux Standards Generalises. *J. Mecanique*, (1975), Vol. **14**, p. 39

(13) TRIANTAFYLLIDIS, N. Bifurcation and Postbifurcation Analysis of Elastic - Plastic Solids Under General Prebifurcation Conditions. *J. Mech. Phys. Solids*, (1983), Vol. **31**, p. 499

(14) CHRISTOFFERSEN, J. and HUTCHINSON, J.W. A Class of Phenomenological Corner Theories of Plasticity. *J. Mech. Phys. Solids*, (1979), Vol. **27**, p. 465

(15) NGUYEN,Q.S.and TRIANTAFYLLIDIS, N. Plastic Bifurcation and Post - Bifurcation Analysis for Generalized Standard Continua. *J. Mech. Phys. Solids*, (1989), Vol. **27**, p. 465

(16) NGUYEN, Q.S. and STOLZ, C. Sur la Methode du Developpement Asymptotique en Flambage Plastique.*C.R. Acad. Sci. Paris*, (1985), V. **300**, Ser. II, No.7, p. 235

(17) NGUYEN, Q.S. Problemes de Plasticite et de Rupture, *Course Notes*, (1980), Univ. of Orsay.

(18) PETRYK, H. and THERMANN, K. Second Order Bifurcations in Elastic - Plastic Solids. *J. Mech. Phys. Solids*, (1985), Vol. **33**, p. 577

# TENSILE INSTABILITIES AT LARGE STRAINS

**V. Tvergaard**
**The Technical University of Denmark, Lyngby, Denmark**

## ABSTRACT

After a short introduction to basic formulations for elastic–plastic and elastic–viscoplastic material behaviour at finite strains the conditions for uniqueness and bifurcation of the incremental solution are discussed. The strong dependence of bifurcation predictions on the constitutive model is emphasized, and specific analyses of tensile instabilities are mentioned. The effect of viscoplastic material behaviour on stability analyses is emphasized. Furthermore, studies of localization in shear bands are discussed, and some recent results for cavitation instabilities in highly constrained plastic flow are presented.

## 1. INTRODUCTION

Instabilities under tensile loading occur in elastic materials as well as inelastic materials, with the common feature that usually such instabilities take place at finite strains. Localized necking in elastic–plastic bodies is a well–known consequence of some types of tensile instabilities that is frequently observed, for example in various metal forming processes. The most widely known example of necking is that of a uniaxial tensile bar made of a ductile material, for which the localization of plastic deformation becomes clearly visible shortly after that the maximum load has been passed. Localization of plastic flow in a shear band is another type of instability, which is associated with loss of ellipticity of the governing differential equations.

Much basic understanding of elastic–plastic stability problems has been developed in connection with studies of plastic buckling (see Hutchinson [1], Needleman and Tvergaard [2]). For columns Shanley [3] explained the significance of the so–called tangent modulus load as the critical bifurcation point, at which the straight configuration looses its uniqueness, but not its stability. A firm theoretical basis was provided by Hill's [4,5,6] general theory of uniqueness and bifurcation in elastic–plastic solids, which applies to bifurcation under large strain conditions as well as to the more classical area of structural buckling.

Localization of plastic flow by either necking or shear band formation are important failure modes. Typically, in uniformly stressed specimens or structural elements plastic flow is rather homogeneous until localization occurs at some stage, and subsequently the strains grow very large in the neck region or the shear band, without much further increase of the overall average strains. Usually this failure mechanism competes with failure by the development of damage on the micro–level, and in many cases final failure involves the interaction of instabilities and damage, with damage developing rapidly in the localization region as the strains grow large.

Hill's [4,5] theory of uniqueness and bifurcation applies to time–independent plasticity. For elastic–viscoplastic material behaviour the same type of approach can be used, and it is found that only elastic bifurcations occur. Thus, only the effect of imperfections can explain the instabilities observed for elastic–viscoplastic solids (e.g. see Tvergaard [7]). Therefore, some recent attempts to analyse the buckling of elastic–viscoplastic structures in terms of bifurcation theory must be characterized as *ad hoc* approximations.

Unstable void growth is an example of a tensile instability that has received some attention recently, both in the context of nonlinear elasticity (Ball [8], Horgan and Abeyaratne [9]) and in the context of elastic–plastic material response (Huang *et al.* [10], Tvergaard *et al.* [11]). The analysis of such cavitation instabilities will be discussed here, and examples of results will be presented.

## 2. GOVERNING EQUATIONS

The finite strain formulation to be used here is a Lagrangean, convected coordinate formulation of the field equations (e.g. see Green and Zerna [12], Budiansky [13]). Relative to a fixed Cartesian frame, the position of a material point in the reference configuration is denoted by the vector $\underset{\sim}{r}$ , and the position of the same point in the current configuration is $\underset{\sim}{\bar{r}}$ . The displacement vector $\underset{\sim}{u}$ and the deformation gradient $\underset{\sim}{F}$ are given by

$$\underset{\sim}{u} = \underset{\sim}{\bar{r}} - \underset{\sim}{r} \ , \ \underset{\sim}{F} = \frac{\partial \underset{\sim}{\bar{r}}}{\partial \underset{\sim}{r}} \tag{2.1}$$

In many cases the reference configuration is identified with the initial undeformed configuration, but there are also cases where another choice of reference is more convenient.

Convected coordinates $\xi^i$ are introduced, which serve as particle labels. The convected coordinate net can be visualized as being inscribed on the body in the reference state and deforming with the material. The displacement vector $\underset{\sim}{u}$ is considered as a function of the coordinates $\xi^i$ and a monotonically increasing time–like parameter $t$ . Covariant base vectors $\underset{\sim}{e}_i$ and $\underset{\sim}{\bar{e}}_i$ of the material net in the reference configuration and the current configuration, respectively, are given by

$$\underset{\sim}{e}_i = \frac{\partial \underset{\sim}{r}}{\partial \xi^i} \ , \ \underset{\sim}{\bar{e}}_i = \frac{\partial \underset{\sim}{\bar{r}}}{\partial \xi^i} \tag{2.2}$$

The metric tensors in the reference and current configurations are given by the dot products of the base vectors

$$g_{ij} = \underset{\sim}{e}_i \cdot \underset{\sim}{e}_j \ , \ G_{ij} = \underset{\sim}{\bar{e}}_i \cdot \underset{\sim}{\bar{e}}_j \tag{2.3}$$

and their determinants are denoted g and G , respectively, while the inverse of the two metric tensors are denoted by $g^{ij}$ and $G^{ij}$ . Latin indices range from 1 to 3 , and the summation convention is adopted for repeated indices.

Components of vectors and tensors on the embedded coordinates are obtained by dot products with the appropriate base vectors. Thus, the displacement components on the reference base vectors satisfy

$$u_i = \underset{\sim}{e}_i \cdot \underset{\sim}{u} \ , \ u^i = \underset{\sim}{e}^i \cdot \underset{\sim}{u} \ , \ \underset{\sim}{u} = u^i \underset{\sim}{e}_i \tag{2.4}$$

Substituting (2.1a) in (2.2b) and using (2.4c) gives

$$\underset{\sim}{\bar{e}}_i = \underset{\sim}{e}_i + u^k_{,i} \underset{\sim}{e}_k \tag{2.5}$$

where $(\ )_{,i}$ denotes the covariant derivative in the reference frame. The Lagrangean strain tensor $\eta_{ij} = \frac{1}{2}(G_{ij} - g_{ij})$ , expressed in terms of displacement components, is then found using (2.3) and (2.5)

$$\eta_{ij} = \frac{1}{2}\left[u_{i,j} + u_{j,i} + u^k_{,i} u_{k,j}\right] \tag{2.6}$$

The true stress tensor $\underset{\sim}{\sigma}$ in the current configuration (the Cauchy stress tensor) has the contravariant components $\sigma^{ij}$ on the current base vectors, where

$$\underset{\sim}{\sigma} = \sigma^{ij} \underset{\sim}{\bar{e}}_i \underset{\sim}{\bar{e}}_j \ , \ \sigma^{ij} = \underset{\sim}{\bar{e}}^i \cdot \underset{\sim}{\sigma} \cdot \underset{\sim}{\bar{e}}^j \tag{2.7}$$

The contravariant components $\tau^{ij}$ of the Kirchhoff stress tensor on the current base vectors are defined by

$$\tau^{ij} = \sqrt{G/g}\ \sigma^{ij} \tag{2.8}$$

where $\sqrt{G/g} = d\bar{V}/dV = \rho/\bar{\rho}$ , expressed in terms of the volume element dV and the density $\rho$ .

The requirement of equilibrium can be specified in terms of the principle of virtual work

$$\int_V \tau^{ij}\delta\eta_{ij}\, dV = \int_S T^i\delta u_i\, dS + \int_V \rho f^i\delta u_i\, dV \tag{2.9}$$

where V and S are the volume and surface, respectively, of the body in the reference configuration, and $\underset{\sim}{f} = f^i \underset{\sim}{e}_i$ and $\underset{\sim}{T} = T^i \underset{\sim}{e}_i$ are the specified body force per unit mass and surface tractions per unit area in the reference frame. The Euler equations of (2.9) express the same requirement directly in terms of the equilibrium equations and the corresponding boundary conditions

$$-\left[\tau^{ij} + \tau^{kj}u^i_{;k}\right]_{,j} = \rho f^i \tag{2.10}$$

$$u_i = 0 \text{ on } S_U \,, \quad \left[\tau^{ij} + \tau^{kj}u^i_{;k}\right]n_j = T^i \text{ on } S_T \tag{2.11}$$

where displacements and tractions are specified on the surface parts $S_U$ and $S_T$, respectively, and $\underset{\sim}{n} = n_j \underset{\sim}{e}^j$ is the surface normal in the reference state.

For the solution of boundary value problems incremental equilibrium equations are needed, due to the material path dependence. When the current values of all field quantities, e.g. stresses $\tau^{ij}$, displacements $u_i$ and tractions $T^i$, are assumed known, an expansion of (2.9) about the known state gives to lowest order

$$\begin{aligned}\Delta t \int_V \left\{\dot{\tau}^{ij}\delta\eta_{ij} + \tau^{ij}\dot{u}^k_{;i}\delta u_{k,j}\right\}dV &= \Delta t \int_S \dot{T}^i\delta u_i\, dS + \Delta t \int_V \rho\dot{f}^i\delta u_i\, dV \\ &- \left[\int_V \tau^{ij}\delta\eta_{ij}\, dV - \int_S T^i\delta u_i\, dS - \int_V \rho f^i\delta u_i\, dV\right]\end{aligned} \tag{2.12}$$

Here, $(\dot{\ }) \equiv \partial(\ )/\partial t$ at fixed $\xi^i$, and $\Delta t$ is the prescribed increment of the "time" t, so that $\Delta T^i = \Delta t\, \dot{T}^i$ are the components of the prescribed traction increments. The terms bracketed in (2.12) vanish according to (2.9), if the current state satisfies equilibrium. However, in linear incremental analyses the solution tends to drift away from the true equilibrium path, due to incrementation errors, and the bracketed terms in (2.12) can be included to avoid such drifting.

The constitutive relations considered here are based on the assumption that the total strain–rate is the sum of the elastic and plastic parts, $\dot{\eta}_{ij} = \dot{\eta}^E_{ij} + \dot{\eta}^P_{ij}$. Thus, with an elastic relationship of the form $\overset{\nabla}{\sigma}{}^{ij} = R^{ijkl}\dot{\eta}^E_{kl}$, as is often assumed (Hutchinson [14]), the constitutive relations can be written as

$$\overset{\nabla}{\sigma}{}^{ij} = R^{ijkl}\left(\dot{\eta}_{kl} - \dot{\eta}^P_{kl}\right) \tag{2.13}$$

Here, the Jaumann (co–rotational) rate of the Cauchy stress tensor $\overset{\nabla}{\sigma}{}^{ij}$ is related to the convected rate by

$$\overset{\nabla}{\sigma}{}^{ij} = \dot{\sigma}^{ij} + \left[G^{ik}\sigma^{jl} + G^{jk}\sigma^{il}\right]\dot{\eta}_{kl} \tag{2.14}$$

An incremental stress–strain relationship in terms of $\dot{\tau}^{ij}$, needed in (2.12), is obtained from (2.13) by substituting (2.14) and the incremental form of (2.8)

$$\dot{\tau}^{ij} = \sqrt{G/g}\,\dot{\sigma}^{ij} + \tau^{ij}G^{kl}\dot{\eta}_{kl} \tag{2.15}$$

The type of elastic relationship considered here is actually hypo–elastic, since it cannot be derived from a work potential [14,15]. However, in the limit of small stresses relative to Young's modulus it reduces to a standard small–strain elastic stress–strain relation. In elastic–plastic analyses the elastic contribution to the total straining is usually very small so that the use of this hypo–elastic relationship rather than a truly elastic one is a reasonable approximation.

For time–independent plasticity the plastic part of the strain increment $\dot{\eta}^P_{ij}$ is homogeneous of degree one in $\overset{\nabla}{\sigma}{}^{kl}$. Using this in (2.14) and solving with respect to the stress increments leads to constitutive relations of the form

$$\dot{\tau}^{ij} = L^{ijkl}\dot{\eta}_{kl} \tag{2.16}$$

where $L^{ijkl}$ is the tensor of instantaneous moduli.

Viscoplastic material models are time–dependent, so that here the parameter t denotes time. In this type of material model the plastic part of the strain rate is a function of the current stresses and strains, but not of the stress–rate. The relationship may be of the form

$$\dot{\eta}^{P}_{ij} = F(\sigma_e, \epsilon_e)\, \frac{\partial \Phi}{\partial \sigma^{ij}} \tag{2.17}$$

where $\Phi$ is a plastic potential function, while $\sigma_e$ and $\epsilon_e$ are the effective stress and strain. Using this in (2.13) leads to constitutive relations of the form

$$\dot{\tau}^{ij} = L_*^{ijkl}\dot{\eta}_{kl} + \dot{\tau}_*^{ij} \tag{2.18}$$

where $\dot{\tau}_*^{ij}$ acts as an initial stress increment that represents the viscous terms.

## 3. DIFFUSE BIFURCATION MODES

For time–independent plasticity, with a constitutive law of the form (2.16), boundary value problems are solved incrementally, using the incremental form of the equilibrium equations (2.9) or (2.10) and (2.11). A theory for the uniqueness and bifurcation in such solids has been developed by Hill [4,5,6]. At the current point of the loading history it is assumed that there are at least two distinct solution increments $\dot{u}_i^a$ and $\dot{u}_i^b$ corresponding to a given increment of the prescribed load (or displacement). The difference between these two solutions is denoted by $(\tilde{\ }) = (\dot{\ })^a - (\dot{\ })^b$, so that subtraction of the incremental principle of virtual work for the two solutions gives

$$\int_V \left\{ \tilde{\tau}^{ij}\delta\eta_{ij} + \tau^{ij}\tilde{u}^k_{,i}\delta u_{k,j} \right\} dV - \int_S \tilde{T}^i \delta u_i \, dS - \int_V \rho \tilde{f}^i \delta u_i \, dV = 0 \tag{3.1}$$

where $\tau^{ij}$ are the current stresses. Thus, if the solution of the incremental boundary value problem is non–unique, (3.1) has a non–zero solution $(\tilde{\ })$.

For elastic–plastic solids obeying normality Hill [4,5] has made use of the expression

$$I = \int_V \left\{ \tilde{\tau}^{ij}\tilde{\eta}_{ij} + \tau^{ij}\tilde{u}^k_{;i}\tilde{u}_{k,j} \right\} dV - \int_S \tilde{T}^i\tilde{u}_i \, dS - \int_V \rho\tilde{f}^i\tilde{u}_i \, dV \tag{3.2}$$

to prove uniqueness. A comparison solid is defined by choosing fixed instantaneous moduli $L_c^{ijkl}$, which are equal to the current plastic moduli for every material point currently on the yield surface, and the elastic moduli elsewhere. For this comparison solid with fixed moduli (3.2) reduces to the quadratic functional

$$F = \int_V \left\{ L_c^{ijkl}\tilde{\eta}_{kl}\tilde{\eta}_{ij} + \tau^{ij}\tilde{u}^k_{;i}\tilde{u}_{k,j} \right\} dV - \int_S \tilde{T}^i\tilde{u}_i \, dS - \int_V \rho\tilde{f}^i\tilde{u}_i \, dV \tag{3.3}$$

A smooth yield surface and normality of the plastic flow rule are often used to model metal plasticity. Then the instantaneous moduli in (2.16) are of the form

$$L^{ijkl} = \mathscr{L}^{ijkl} - \mu M^{ij} M^{kl} \tag{3.4}$$

where $\mu$ is zero or positive for elastic unloading or plastic loading, respectively, while $M^{ij}$ is normal to the yield surface. For this type of material model it can be proved that the relation

$$\tilde{\tau}^{ij}\tilde{\eta}_{ij} \geq L_c^{ijkl}\tilde{\eta}_{kl}\tilde{\eta}_{ij} \tag{3.5}$$

is satisfied at every material point, and thus $F \leq I$. Therefore, since any non–trivial solution of (3.1) gives $I = 0$, the requirement $F > 0$ is a sufficient condition for uniqueness.

Equality in (3.5) and thus $I = 0$ for $F = 0$ requires that both solution increments, $\dot{u}_i^a$ and $\dot{u}_i^b$, give plastic loading at all material points currently on the yield surface. Then, if $\dot{u}_i^a$ is identified with the prebifurcation solution and $\tilde{u}_i$ with the bifurcation mode, the variation of the prescribed load (or deformation) parameter $\lambda$ with the bifurcation mode amplitude $\xi$ initially after bifurcation can be written on the form

$$\lambda = \lambda_c + \lambda_1 \xi + \dots \;, \quad \xi \geq 0 \tag{3.6}$$

with $\lambda_1$ chosen sufficiently large. Generally the minimum value of $\lambda_1$ is positive. The relation (3.6) has been extended into an actual post–bifurcation expansion by Hutchinson [16], accounting for elastic unloading zones that spread in the material.

Non–normality of the plastic flow rule is an important feature of elastic–plastic models describing the frictional dilatant behaviour of rocks or soils [17] and is also important during void nucleation in ductile metals. Here, the instantaneous moduli are of the form [18]

$$L^{ijkl} = \mathscr{L}^{ijkl} - \mu M_G^{ij} M_F^{kl} \tag{3.7}$$

where $M_F^{ij}$ represents the yield surface normal, and $M_G^{ij} \neq M_F^{ij}$. In this case the relation (3.5) cannot be proved for the usual comparison solid. Raniecki and Bruhns [19] have proposed an alternative comparison solid, in which both $M_G^{ij}$ and $M_F^{ij}$ are replaced by $\frac{1}{2}(M_G^{ij} + rM_F^{ij})/\sqrt{r}$, for $r > 0$. These alternative comparison moduli satisfy (3.5), so that a lower bound to the first critical bifurcation point is obtained, while the usual comparison solid gives an upper bound. Uniqueness is only guaranteed up to the highest lower bound, but various solutions indicate that the actual bifurcation occurs much closer to the upper bound, which tends to be well above the lower bound [19,18].

The formation of a vertex on the yield surface is implied by physical models of polycrystalline metal plasticity, based on the concept of single crystal slip [20,21]. At a vertex the instantaneous moduli are functions of the stress–rate direction, e.g. of the form

$$L^{ijkl} = L^{ijkl}(\overset{\nabla}{s}{}^{mn}) \tag{3.8}$$

where $s^{ij}$ is the stress deviator. For a pyramidal vertex Sewell [22] has shown that the total loading moduli (all slip systems active) satisfy the inequality (3.5). For a phenomenological corner theory of plasticity, $J_2$ corner theory [23], similar conditions apply [2]. If the prebifurcation solution satisfies total loading everywhere, so that this condition can also apply to the bifurcated solution, the initial post–bifurcation behaviour is of the form (3.6), with the minimum value of $\lambda_1$ determined by the requirement of total loading. If total loading is not satisfied everywhere, application of the total loading moduli in (3.3) will only give a lower bound, while application of the moduli corresponding to the prebifurcation solution gives an upper bound [24].

A smooth bifurcation, i.e. $\lambda_1 \to \infty$ in (3.6), occurs in the case where the total loading region at the vertex shrinks to a single stress rate direction, as has been shown by Needleman and Tvergaard [2]. Such smooth bifurcations can also occur for the classical elastic–plastic solid (3.4). The comparison solid does show a normal bifurcation, but due to the extra requirement of plastic loading (or total loading) at all material points currently at the yield surface, the elastic–plastic solid may not allow for a non–zero $(\tilde{\ })$ solution. At smooth bifurcations the linear lowest order $\xi$ contribution in (3.6) will be replaced by a term with exponent smaller than unity [2]. A treatment of smooth bifurcation based on a special continuum formulation for small strains has been given by Nguyen and Triantafyllidis [25].

Necking in a uniaxial tensile test specimen is the most well known example of a tensile instability. Some insight in this problem is gained by considering the limit of a long, thin tensile bar, in which all other stresses than that in axial direction can be neglected. Assuming incompressibility, using the current configuration as reference, and replacing stresses, strains and displacements by the cross–sectional averages, the bifurcation functional (3.3) for the comparison solid reduces to

$$F = \int_V (\tilde{\sigma}\tilde{e} - \sigma\tilde{e}^2)dV - \tilde{P}\tilde{u}_p = (E_t - \sigma)A\int_0^{\ell} \tilde{e}^2 dx \tag{3.9}$$

Here, $\sigma$ is the true stress, $E_t$ is the slope of the uniaxial true stress–natural strain curve at stress level $\sigma$, the linear strain is $\tilde{e} = \partial\tilde{u}/\partial x$, the cross–sectional area is $A$, and the bar has zero displacement at $x = 0$, with axial load $P$ and displacement $u_p$ at $x = \ell$. If load is the prescribed quantity, uniqueness of the uniform state of deformation (constant $\tilde{e}$) is lost at $\sigma = E_t$, which is the point of maximum load. If $u_p$ is the prescribed quantity, the mode of uniform deformation

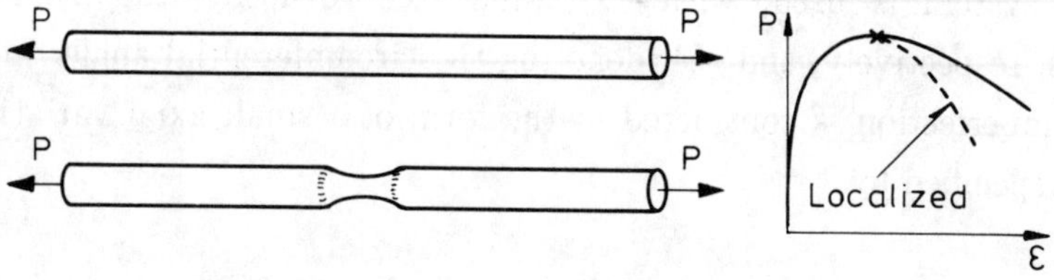


Fig. 1. Localized necking in a uniaxial tensile test specimen.

remains unique; but at the same point $\sigma = E_t$ bifurcation still occurs into all modes of the form

$$\tilde{u} = \sin \frac{m\pi x}{\ell} \; , \quad m = 1,2,3... \tag{3.10}$$

or in any other mode with zero end displacements (Fig. 1). These results represent the classical necking criterion of Considère [26].

In a full axisymmetric analysis for the onset of necking in a round tensile test specimen Hutchinson and Miles [27] have used the general theory based on eqs. (3.1)–(3.4) to show that 3D–effects delay bifurcation beyond the maximum load point. When the values at the maximum load point are $\sigma_{max} = (E_t)_{max}$ , R and $\ell$ are the current radius and length, respectively, and $\gamma = m\pi R/\ell$ , they find the following expression for the critical stress $\sigma_c$ at bifurcation

$$\frac{\sigma_c}{\sigma_{max}} = 1 + \frac{\gamma^2}{8} + \frac{E}{3E_t}\frac{\gamma^4}{192} + ... \tag{3.11}$$

Thus, $\sigma_c$ exceeds $\sigma_{max}$ so that bifurcation is delayed, but it is seen that Considère's one dimensional result corresponds to the limit of a long thin bar $(\gamma \to 0)$ .

As an example of the application of Hill's bifurcation theory an elastic–plastic tube under internal pressure will be considered (Tvergaard [28]). The tube is taken to have the initial mean radius $R_0$ , the initial wall thickness $t_0$ and the length $\ell$ . It is assumed that the length remains constant throughout the deformation, in agreement with experimental observations of Larsson *et al.* [29], and idealized symmetry boundary conditions are assumed at the ends. A cylindrical reference coordinate system is used, with $x^1$ and $x^2$ denoting the axial and radial coordinates, respectively, and $x^3$ denoting the circumferential angle. In some cases an initial imperfection is considered in the form of a small axial variation $\Delta R$ of the radius, specified by

$$\Delta R = \bar{\xi}\left[R_0 - \frac{t_0}{2}\right]\cos\frac{\pi x^1}{\ell} \tag{3.12}$$

where $\bar{\xi}$ is the imperfection amplitude.

The idealized symmetry boundary conditions at the ends of the tube are expressed by

$$\left.\begin{array}{l} u^1 = 0 \ , \ T^2 = T^3 = 0 \ \text{for} \ x^1 = 0 \\ u^1 = 0 \ , \ T^2 = T^3 = 0 \ \text{for} \ x^1 = \ell \end{array}\right\} \tag{3.13}$$

On the outside of the tube zero tractions are specified

$$T^i = 0 \ \text{for} \ x^2 = R_0 + \Delta R + \frac{t_0}{2} \tag{3.14}$$

while a hydrostatic pressure $p$ is applied on the inside

$$T^i = -p\alpha^{ir}n_r \ \text{for} \ x^2 = R_0 + \Delta R - \frac{t_0}{2} \tag{3.15}$$

Here, $n_r$ is the outward reference normal and the tensor $\alpha^{ir}$ gives the configuration dependence of hydrostatic loading

$$\alpha^{ir} = \frac{1}{2}\,\epsilon^{ijk}\epsilon^{lmr}(g_{jl} + u_{j,l})(g_{km} + u_{k,m}) \tag{3.16}$$

where $\epsilon^{ijk}$ is the alternating tensor (see Sewell [30]).

Now, the expression (3.15) is used in the bifurcation condition (3.2), and the quadratic functional (3.3) for the comparison solid. If either the pressure or the enclosed volume are the prescribed quantity, the bifurcation condition (3.3) is expressed by the requirement of a non–zero solution to the eigenvalue problem given by the variational equation

$$\delta F = 0 \tag{3.17}$$

$$F = \int_V \left\{ L_c^{ijkl} \tilde{\eta}_{kl} \tilde{\eta}_{ij} + \tau^{ij} \tilde{u}^k_{,i} \tilde{u}_{k,j} \right\} dV + \int_{S^*} p \tilde{\alpha}^{ir} n_r \tilde{u}_i dS \tag{3.18}$$

The inner surface is denoted by $S^*$ and

$$\tilde{\alpha}^{ir} = \frac{1}{2} \epsilon^{ijk} \epsilon^{lmr} \left\{ \tilde{u}_{j,l} (g_{km} + u_{k,m}) + (g_{jl} + u_{j,l}) \tilde{u}_{k,m} \right\} \tag{3.19}$$

If pressure is the prescribed quantity, uniqueness is lost at the maximum pressure point, with a corresponding circular cylindrical mode of instability. However, this mode of instability is excluded if the enclosed volume is the prescribed quantity, as will be the case when a given volume of an incompressible fluid is pumped into the tube. The analyses to be presented here focus on cases where the enclosed volume is prescribed, so that the bifurcation modes must satisfy the requirement of zero volume change.

For tubes made of a strain hardening rigid–plastic material Storåkers [31] and Strifors and Storåkers [32] have found that bifurcation away from the cylindrically symmetric state of deformation may occur beyond the point of maximum pressure. The problem has been reinvestigated by Chu [33] for elastic–plastic material behaviour, both for very thin–walled tubes and for more thick–walled tubes, and this study has shown that all bifurcation points occur well after the maximum pressure point. Larsson, Needleman, Tvergaard and Storåkers [29] have investigated the instabilities and the final failure modes in pressurized tubes, both experimentally and theoretically. These experiments were carried out for closed–end aluminium or copper cylinders with the internal pressure applied by means of a hydraulic loading device. The failure mode observed in the experiments corresponds to a bifurcation mode with a single circumferential wave, leading to necking at one side of the tube and finally shear fracture in the neck region. In the analyses it is assumed that the material develops a vertex on the yield surface, represented in terms of $J_2$ corner theory (Christoffersen and Hutchinson [23]), and the analyses predict the neck development as well as the onset of localized shearing in narrow bands of material inside the neck, as observed experimentally.

All these analyses were based on the assumption of plane strain conditions; but the final failure mode observed experimentally by Larsson *et al.* [29] involves the development of a localized bulge on one side of the tube where final shear fracture occurs. The agreement between computed and experimentally observed failure modes found by Larsson *et al.* [29] refers to a cross–section through the centre of the bulge. To investigate the possibility of non–uniformities in the axial direction, non–planar bifurcation modes are included in a subsequent analysis (Tvergaard [28]), and also localized post–bifurcation solutions leading to secondary bifurcations are considered. These analyses are based on classical plasticity theory with isotropic hardening, $J_2$ flow theory; but also bifurcation predictions based on a finite strain generalization of deformation theory (Støren and Rice [34]) are included, to approximately model the effect of a vertex on the yield surface.

For a thin–walled tube the analysis can be carried out within the framework of a plane stress approximation, in which the in–plane stress components $\sigma^{11}$, $\sigma^{33}$ and $\sigma^{13}$ are considered as thickness averages. The current radius and thickness of the tube are denoted $R$ and $t$, and the current state is used as reference. Then, assuming elastic incompressibility, the in–plane constitutive relations are of the form

$$\dot{\sigma}^{\alpha\beta} = \hat{L}^{\alpha\beta\gamma\delta}\dot{\eta}_{\gamma\delta} \tag{3.20}$$

where Greek indices range over 1 and 3 (not 2). In the pre–bifurcation solution the only non–zero stresses are the axial true stress $\sigma_1 = \sigma^{11}$ and the circumferential true stress $\sigma_3 = R^2\sigma^{33}$, and the corresponding plane–stress instantaneous moduli are (see Needleman and Tvergaard [35]).

$$\hat{L}^{1111} = \frac{4}{3}E - (E - E_t)\left[\frac{\sigma_1}{\sigma_e}\right]^2 - 2\sigma_1 \tag{3.21}$$

$$R^4\hat{L}^{3333} = \frac{4}{3}E - (E - E_t)\left[\frac{\sigma_3}{\sigma_e}\right]^2 - 2\sigma_3 \tag{3.22}$$

$$R^2\hat{L}^{1133} = \frac{2}{3}E - (E - E_t)\frac{\sigma_1\sigma_3}{\sigma_e^2} \tag{3.23}$$

$$R^2\hat{L}^{1313} = \frac{1}{3}E - \frac{1}{2}(\sigma_1 + \sigma_3) \tag{3.24}$$

These plane–stress moduli correspond to $J_2$ flow theory, but the Støren–Rice moduli are directly obtained from (3.21)–(3.24) if Young's modulus $E$ is replaced by the secant modulus $E_s$. Furthermore, when the length of the tube remains constant and the internal pressure is $p$, the stress values to be substituted into (3.21)–(3.24) are

$$\sigma_3 = \frac{R}{t}p\,, \quad \sigma_1 = \frac{1}{2}\sigma_3\,, \quad \sigma_e^2 = \frac{3}{4}\sigma_3^2 \tag{3.25}$$

For the membrane the bifurcation functional (3.18) reduces to

$$F = t\int_A \left\{\tilde{\sigma}^{\alpha\beta}\tilde{\eta}_{\alpha\beta} + \sigma^{\alpha\beta}\tilde{u}^k{}_{,\alpha}\tilde{u}_{k,\beta}\right\}dA - \int_A p\tilde{a}^{i2}\tilde{u}_i dA \tag{3.26}$$

where $A$ is the current middle surface area. Here, $\tilde{\sigma}^{\alpha\beta}$ is expressed in terms of the plane stress constitutive relations (3.20), with the current stress values (3.25) substituted into the expressions (3.21)–(3.24) for the instantaneous moduli. Bifurcation modes are considered of the form

$$\tilde{u}_1 = \tilde{U}\sin\frac{\pi x^1}{\ell}\cos(mx^3)$$

$$\tilde{u}_2 = \tilde{W}\cos\frac{\pi x^1}{\ell}\cos(mx^3) \tag{3.27}$$

$$\tilde{u}_3 = R\tilde{V}\cos\frac{\pi x^1}{\ell}\sin(mx^3)$$

where $m$ is the circumferential wave number and the half wave length in the axial direction is equal to the length $\ell$ of the tube analysed. Substitution of (3.27) into (3.26) leads to three homogeneous algebraic equations for the amplitudes $\tilde{U}$, $\tilde{W}$ and $\tilde{V}$, and the critical circumferential stress for bifurcation is determined by the condition that the determinant of these equations vanishes.

In the limit of a very long axial wavelength, $\pi R/\ell \to 0$, these equations for $m \geqq 1$ reduce to those obtained by Chu [33] for plane strain conditions. The corresponding critical circumferential stresses are readily found

$$\sigma_3 = \frac{4}{3} E_t \,, \quad \text{for } m = 1,2,\ldots \text{ and } \frac{\pi R}{\ell} \to 0 \tag{3.28}$$

and it is noted that no axisymmetric bifurcation solution $(m = 0)$ is allowed for in the context of plane strain when the enclosed volume is prescribed. The bifurcations (3.28) occur well beyond the maximum pressure point

$$\sigma_3 = \frac{2}{3} E_t \text{ at } p_{max} \tag{3.29}$$

In the case of axisymmetric modes the critical hoop stress at bifurcation is determined by the following quadratic equation

$$\sigma_3 = \frac{2}{3} E_t \left[ 1 - \frac{1}{4} \left[ \frac{\pi R}{\ell} \right]^2 \left\{ 1 + \frac{1}{2E} \left[ \frac{2}{3} E_t - \sigma_3 \right] \right\} \right]^{-1} \tag{3.30}$$

Thus, for a very long axial wavelength, $\pi R/\ell \to 0$, bifurcation occurs at the maximum pressure point for $\sigma_3 = 2/3\, E_t$, and even for a relatively short tube with a length $\ell$ of the order of the diameter a simple explicit expression for $\sigma_3$ is obtained from (3.30) by noting that $[2/3\, E_t - \sigma_3]/(2E)$ is negligible compared to unity in the range of interest.

In general the bifurcation points corresponding to modes of the form (3.27) may be obtained by a simple numerical determination of the roots for a 3×3 determinant. Figure 2 shows the dependence of the critical hoop stress $\sigma_3$ on the axial half wave length $\ell$ for various values of the circumferential wave number $m$. Here, the material is taken to be power hardening with the initial yield stress $\sigma_y = 0.004\, E$ and the strain hardening exponent $n = 10$. For all values of the circumferential wave number $m$ the value of the critical stress $\sigma_3$ for bifurcation tends towards infinity as $\ell/(\pi R) \to 0$. However, for $\ell/(\pi R) > 6$ all bifurcations corresponding to circumferential wave numbers $m \geqq 1$ occur essentially at the

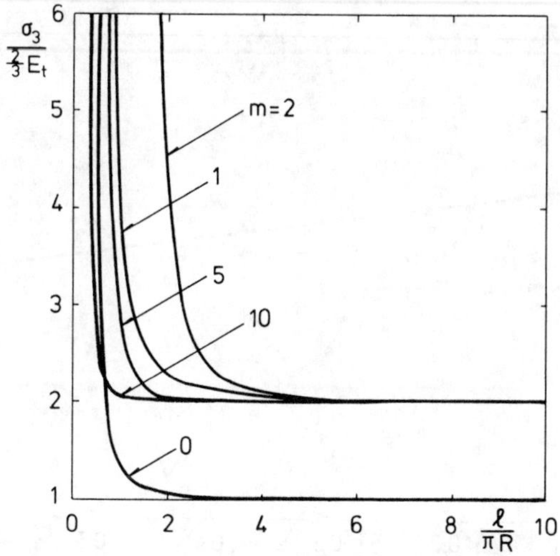


Fig. 2. Critical hoop stress $\sigma_3$ for bifurcation vs. length $\ell$. Membrane theory predictions for $\sigma_y/E = 0.004$ and $n = 10$ (from [28]).

stress level (3.28). For the case of axisymmetric bifurcation modes $(m = 0)$ the delay beyond the maximum pressure point, $\sigma_3 = 2/3\ E_t$, is negligible when $\ell/(\pi R) > 4$.

It is noted from Figure 2 that even for rather short tubes the bifurcation into an axisymmetric mode $(m = 0)$ is first critical. Thus the plane strain bifurcation modes (3.28) for $m \geq 1$ are hardly ever relevant as the first critical modes for pressurized tubes. Following the axisymmetric bifurcation just after the maximum pressure point a localized post–bifurcation deflection pattern develops, as is typical of many structural problems where bifurcation occurs at the maximum load (Tvergaard and Needleman [36]). A secondary bifurcation into a mode with $m \geq 1$ may occur at some point of the post–bifurcation solution, but this bifurcation is not governed by the above equations, since here the pre–bifurcation shape of the tube is not cylindrical.

The possibility of secondary bifurcations that occur after the development of a localized axisymmetric bulge on the tube can be analysed numerically by considering a tube with a small initial imperfection of the form (3.12), Tvergaard [28]. Then, during the development of the axisymmetric deformations the possibility of bifurcation into a non–axisymmetric mode of the form

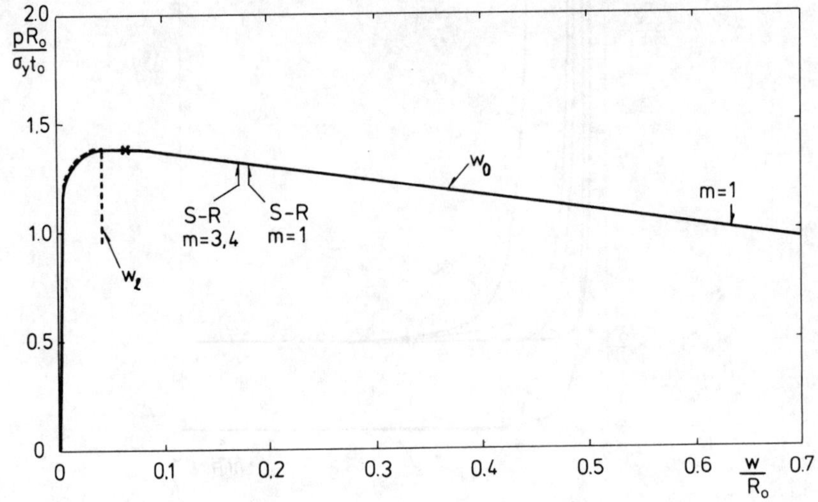


Fig. 3. Pressure vs. radial expansion for tube with $t_0/R_0 = 0.4$ , $\ell/R_0 = 8$ and $\bar{\xi} = 0.005$ . Bifurcation points are indicated by arrows (from [28]).

$$\tilde{u}_1 = \tilde{U}_1(x^1,x^2)\cos(mx^3)$$

$$\tilde{u}_2 = \tilde{U}_2(x^1,x^2)\cos(mx^3) \tag{3.31}$$

$$\tilde{u}_3 = \tilde{U}_3(x^1,x^2)\sin(mx^3)$$

is checked by using (3.31) in (3.18). Results for a thickwalled tube with $t_0/R_0 = 0.4$ , $\ell/R_0 = 8$ and $\bar{\xi} = 0.005$ are shown in Figs. 3 and 4. It is seen in Fig. 3 that elastic unloading near $x^1 = \ell$ occurs just after the maximum pressure point (instead of the primary bifurcation), so that the corresponding radial displacement $w_l$ stops growing. Subsequently an axisymmetric bulge develops on the tube as shown by the deformed meshes in Fig. 4. The first bifurcation into a non–axisymmetric mode predicted by $J_2$ flow theory is that for $m = 1$ , which occurs at a high radial displacement in the centre of the bulge, $w_0/R_0 = 0.635$ , much later than the corresponding bifurcation predicted for plane strain conditions. Much earlier bifurcations are predicted by deformation theory modelling a vertex (see

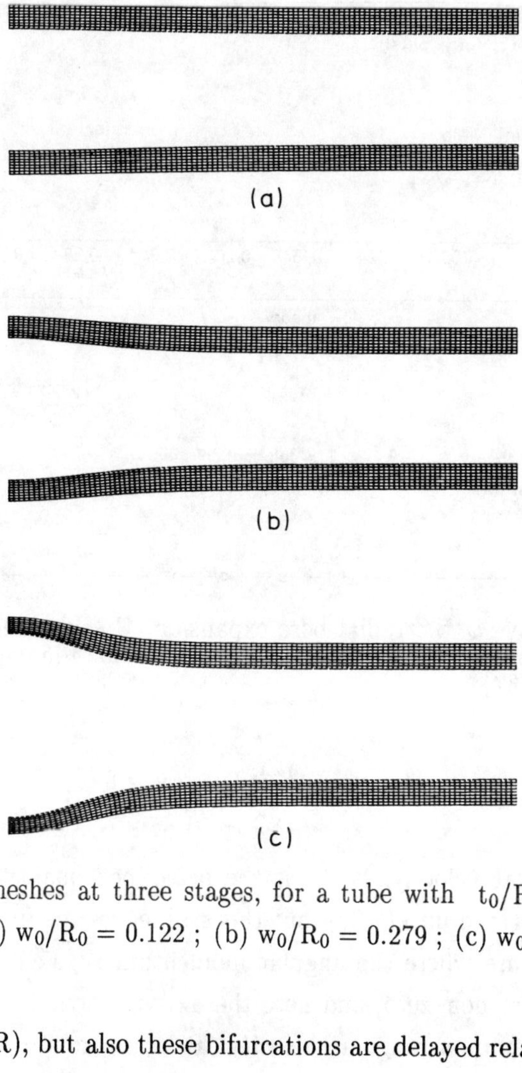


Fig. 4. Deformed meshes at three stages, for a tube with $t_0/R_0 = 0.4$ , $\ell/R_0 = 8$ and $\bar{\xi} = 0.005$ . (a) $w_0/R_0 = 0.122$ ; (b) $w_0/R_0 = 0.279$ ; (c) $w_0/R_0 = 0.588$ (from [28]).

arrows denoted S–R), but also these bifurcations are delayed relative to plane strain predictions.

Another example, in which configuration dependent loading plays a role in a bifurcation analysis, is that of a rotating turbine disk analysed by Tvergaard [37]. In a co–rotating cylindrical coordinate system with $\xi^1$ , $\xi^2$ and $\xi^3$ denoting the initial radial, angular and axial coordinates, respectively, the external loading is the centrifugal force $\rho f^i$ . Here, the body force term in (3.3) takes the form

$$\tilde{f}^i = \omega^2 \tilde{\phi}^i + 2\omega\tilde{\omega}\phi^i \tag{3.32}$$

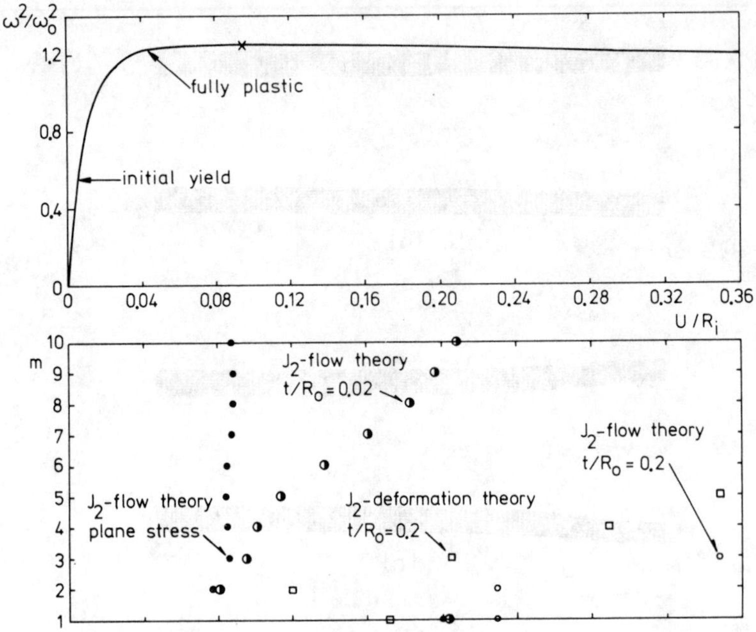


Fig. 5. Angular velocity $\omega$ vs. radial bore expansion U for a ductile turbine disk with $R_i/R_0 = 0.1$ . Lower diagram shows bifurcation mode number m vs. critical bore expansion (from [37]).

$$\phi^1 = \xi^1 + u^1 \ , \ \phi^2 = u^2 \ , \ \phi^3 = 0 \ , \ \tilde{\phi}^\alpha = \tilde{u}^\alpha \ , \ \tilde{\phi}^3 = 0 \qquad (3.33)$$

where $\omega$ is the angular velocity. If $\omega$ is the prescribed quantity, the constraint $\tilde{\omega} = 0$ has to be substituted into (3.32); but this still leaves the first term non–zero. In the more realistic case where the angular momentum is taken to be prescribed, both terms in (3.32) are non–zero, and here the axisymmetric solution can remain unique beyond the maximum angular velocity. Bifurcation into sinusoidal modes with wave number m around the circumference has been investigated [37] for a disk with a central bore, having the initial inner and outer radii $R_i$ and $R_0$ and the thickness h . Bifurcation predictions in Fig. 5, for different cases and material models, show that the $m = 2$ mode is first critical in all cases, which agrees very nicely with experimental observations by Percy *et al.* [38] who found necking at two opposite sides of the bore before bursting. This necking mode is illustrated by results of a plane stress post–bifurcation analysis shown in Fig. 6.

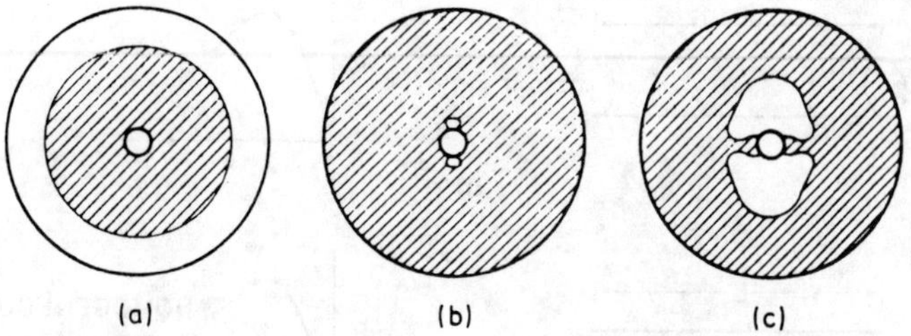


Fig. 6. Current plastic zone (hatched area) for a ductile turbine disk with $R_i/R_0 = 0.1$ (plane stress result). (a) $U/R_i = 0.024$, (b) $U/R_i = 0.078$, (c) $U/R_i = 0.110$ (from [37]).

In connection with the tensile instabilities discussed here it is of interest to note that completely analogous behaviour has been found for a wide class of structural components under compressive loading (Tvergaard and Needleman [36,2]), where the overall behaviour can be represented as that of a softening spring, e.g. for elastic columns on a softening elastic foundation, or long elastic–plastic plates simply supported along the edges. In such structural components the load maximum occurs as a result of buckling into a periodic short–wave pattern, and localization means in reality that one buckle starts to grow more rapidly than the others. The overall behaviour can be represented by a simple analysis for a homogeneous axially compressed bar (Fig. 7) constrained to remain straight but free to slide in the axial direction. The axial force $N$ is taken to be a nonlinear function of the strain $\epsilon$, as represented by the incremental relation

$$\dot{N} = C\dot{\epsilon} \quad , \quad \dot{\epsilon} = \dot{u}_{,x} \tag{3.34}$$

Here, $u$ is the axial displacement, and $C(\epsilon)$ is the instantaneous modulus, which incorporates the combined effects of material nonlinearities and geometric nonlinearities. Then, the bifurcation functional (3.3) takes the form

$$F = \int_0^L C\, \tilde{\dot{\epsilon}}\tilde{\dot{\epsilon}} dx \tag{3.35}$$

when $L$ is the length of the bar, and small strains are assumed. For prescribed end

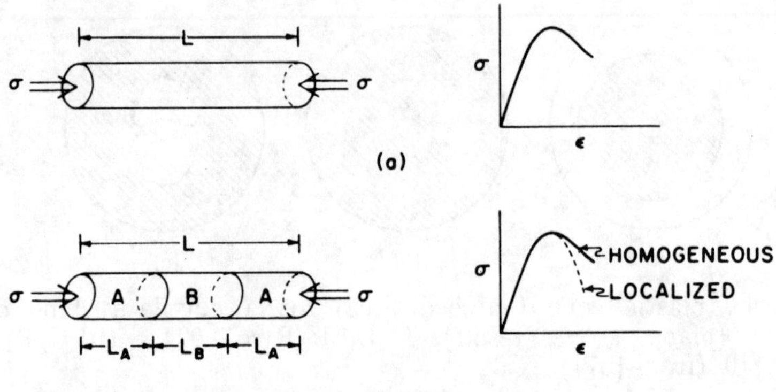


Fig. 7. Axially compressed bar model, representing the overall effect of the localization of buckling patterns (from [36]).

displacements all bifurcation modes of the form

$$u = \sin \frac{m\pi x}{L} \ , \quad m = 1,2,3... \tag{3.36}$$

are critical simultaneously for $C = 0$, i.e. when a load maximum is reached. This simple analysis for buckling localization is closely related to the simple one dimensional analysis (3.9)–(3.10) for the onset of necking in a tensile test specimen.

## 4. EFFECT OF VISCO–PLASTIC BEHAVIOUR

For elastic–viscoplastic materials conditions for uniqueness and bifurcation can be derived in a manner analogous to that employed in Section 3 for elastic–plastic materials. Then, the constitutive relations are of the form (2.18), with the inelastic part of the strain–rate given by a purely viscous expression such as (2.17). It is again assumed that there are two distinct solutions $\dot{u}_i^a$ and $\dot{u}_i^b$ (displacement rates) corresponding to a given rate of change of the prescribed load (or displacement), and the difference between the two velocity fields is denoted $(\tilde{\ }) = (\dot{\ })^a - (\dot{\ })^b$. Now, since the current stress and strain fields are known, the inelastic part of the current strain rate is uniquely determined by (2.17) so that the only possible difference between the two total strain rates $\dot{\eta}^a_{ij}$ and $\dot{\eta}^b_{ij}$ is an elastic contribution. Thus, $\tilde{\eta}_{ij}$ is an elastic strain rate field, $\tilde{u}_i$ is compatible with this field, and $\tilde{\tau}^{ij}$ is the corresponding elastic stress rate field. Consequently, (3.1) shows directly that uniqueness and bifurcation is entirely governed by the elastic part of the material response.

Often the function $F$ in (2.17) is taken to be proportional with $[\sigma_e/g(\epsilon_e)]^{1/m}$, where the rate–hardening exponent $m$ is a measure of the strain–rate sensitivity, while the function $g(\epsilon_e)$ incorporates the strain hardening. Thus, for a rate–sensitive version of $J_2$ flow theory (2.17) takes the form

$$\dot{\eta}^P_{ij} = \dot{\epsilon}^P_e \frac{3}{2}\frac{s_{ij}}{\sigma_e} \quad , \quad \dot{\epsilon}^P_e = \dot{\epsilon}_0 \left[\frac{\sigma_e}{g(\epsilon_e)}\right]^{1/m} \tag{4.1}$$

where $s_{ij}$ is the stress deviator and $\dot{\epsilon}_0$ is a reference strain–rate. In the limit $m = 0$ (4.1) reduces to the corresponding expression for time independent plasticity, and bifurcation is then governed by the instantaneous elastic–plastic moduli as discussed in connection with (3.4); but for any positive value of $m$ (even very small) bifurcation is governed by the elastic moduli. Usually, the elastic bifurcation occurs much later than the plastic bifurcation, so that only imperfections can explain the instabilities observed in the case of elastic–viscoplastic solids. However, for small rate–sensitivities, such as $m = 0.001$, it is found that even extremely small imperfections give essentially identical response of the rate–sensitive and the time independent solids in the vicinity of the elastic–plastic bifurcation point.

The influence of strain–rate sensitivity on tensile instabilities has been investigated by Hutchinson and Neale [39,40] for necking in uniaxial tension and in biaxially stretched sheets, respectively. Elastic strains are neglected in these analyses, so that no bifurcation at all corresponds to the perfect case, and accordingly it is found that the critical strain for localization tends towards infinity for vanishing initial thickness inhomogeneity.

Both for uniaxial tension and for plane strain sheet necking an asymptotic expression for the critical logarithmic, axial strain, $\epsilon_c$, is found of the form

$$\frac{\epsilon_c - \epsilon_c^o}{\sqrt{N}} \simeq \frac{m}{2\sqrt{2\bar{\xi}}} \ln\left[\frac{4\pi\bar{\xi}}{m}\right] \qquad (4.2)$$

where N is the strain hardening exponent, $\epsilon_c^o = N$ is the critical strain for necking according to time–independent plasticity, and $\bar{\xi}$ is the initial thickness reduction [39,40]. This relationship, valid for $\bar{\xi} << 1$, $m < 2\bar{\xi}$ and small $m/N$, shows how $\epsilon_c \to \epsilon_c^o$ in the time–independent limit, with the difference $\epsilon_c - \epsilon_c^o$ proportional to $\sqrt{N}$ for small m. An interesting feature of these analyses is that the necking delay is independent of the rate of deformation, but depends strongly on the degree of rate–sensitivity as measured by m. It is noted that $\epsilon_c - \epsilon_c^o \to \infty$ for $\bar{\xi} \to 0$, because there is no elastic bifurcation.

For a planar double slip model of a single crystal the influence of rate–sensitivity in the slip systems has been analysed by Peirce *et al.* [41]. The focus of this investigation is on necking, the formation of macroscopic shear bands and the occurrence of patchy slip. It is found that rate sensitivity of the slip mechanism does result in a significant delay of flow localization.

A column mode analysis for an eccentrically stiffened panel gives some insight in the influence of rate–sensitivity on plastic buckling [42]. Corresponding to a fixed speed of average axial shortening, $\dot{\epsilon}_a = -\dot{\epsilon}_o$, Fig. 8 gives computed load maxima versus imperfection amplitude $\bar{\xi}$. The load is normalized by the bifurcation load corresponding to time–independent plasticity and $\bar{\xi}$ denotes the imperfection amplitude relative to the flange thickness. Naturally, the maximum load attained is quite sensitive to the rate of axial compression (in contrast to the critical strain for sheet necking [40]). Therefore it should be noted that the uniaxial stress–strain curve for the time–independent plasticity model in Fig. 8 is chosen such that the three materials compared in the figure have identical response when strained uniformly at the strain–rate $\dot{\epsilon}_o$.

For the rate–sensitive materials in Fig. 8 the first critical bifurcation point corresponds to the elastic value $\lambda_E$, which is quite high here, $\lambda_E = 2.18\lambda_c$. Both an asymptotic analysis and numerical analyses in [42] show that even for

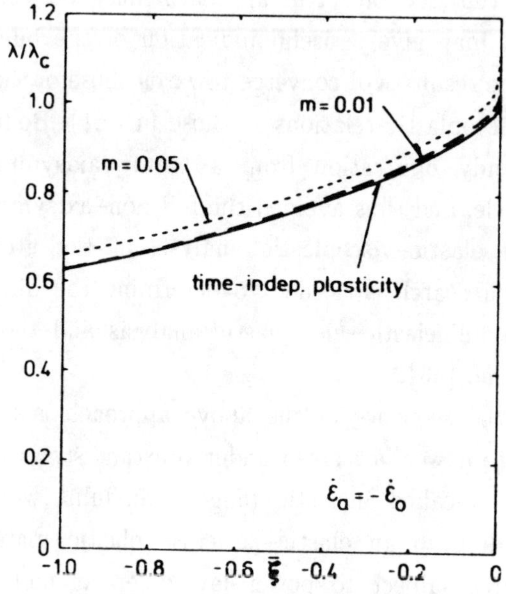


Fig. 8. Maximum load vs. imperfection amplitude for a column with an asymmetric cross–section, made of a rate–sensitive material (from [42]).

imperfections $\bar{\xi}$ as small as $10^{-7}$ or $10^{-12}$ the maximum load is at most $1.2\lambda_c$ (for $m = 0.05$) and thus far below the bifurcation load. The results in Fig. 6 show that although the role of bifurcation is strongly changed by accounting for strain–rate sensitivity, the classical elastic–plastic buckling predictions remain good approximations for imperfection amplitudes that are not extremely small, as long as the rate hardening exponent m does not exceed values of the order of $0.01 - 0.05$ .

Quite a different approach to the analysis of elastic–viscoplastic instability problems has been used by Bodner *et al.* [43] in a study of the inelastic buckling of axisymmetric shells. Here, in the pre–buckling solution a so–called inelastic tangent modulus $E_t^*$ is defined, for a given plastic strain rate, as $E_t^* = d\sigma_e/d\epsilon_e^P$ , for each material point. Then these inelastic tangent moduli are substituted in expressions for the instantaneous moduli of a time–independent elastic–plastic material, and these moduli are used in Hill's bifurcation criterion. By this approach the viscous nature of the elastic–viscoplastic material model is forgotten during the bifurcation analysis, which does not agree with elastic–viscoplastic theory, as has been explained in the beginning of this section. The approach may be considered an *ad hoc* approximation, which may give a useful indication of the buckling load in some cases, and certainly the results will converge towards those of the time–independent limit for $m \to 0$ in viscoplastic relations as those in (4.1). Bodner *et al.* [43] used their approach to study bifurcation from a purely axisymmetric state into a non–axisymmetric mode, and thus avoided the full non–axisymmetric shell analysis needed to do a real elastic–viscoplastic analysis of the growth from a small imperfection. Further research is needed to determine the difference between the results obtained by a full elastic–viscoplastic analysis and those obtained by the approach of Bodner *et al.* [43].

An interesting consequence of the above approach is found for a material subject to steady–state power law creep under constant stress (i.e. constant $g(\epsilon_e)$ in (4.1)). Then the so–called inelastic tangent modulus would be zero, giving predictions corresponding to an elastic–perfectly plastic material, which is not realistic. For structures subject to power–law creep at high temperatures it is well–known that the only bifurcations found are elastic and that creep buckling initiates from small imperfections (e.g. see Obrecht [44], Hoff [45]).

## 5. LOCALIZATION IN SHEAR BANDS

The discussion in Sections 3 and 4 has focussed on bifurcation into a diffuse mode, which will often occur while the governing differential equations are elliptic. Localization of plastic flow in a narrow shear band is a different type of instability, which is observed in a wide variety of materials as a rather sudden change from a smooth deformation pattern.

The basic phenomenon of localization can be studied by a relatively simple model problem for solids subject to uniform straining, as illustrated in Fig. 9. The localized shearing is assumed to occur in a thin slice of material with reference normal $n_j$, while the strain fields outside this band are assumed to remain uniform throughout the deformation history. The quantities inside and outside the band are denoted by $(\ )^b$ and $(\ )^o$, respectively, and a Cartesian reference coordinate system is used. Since uniform deformation fields are assumed both inside and outside the band, equilibrium and compatibility inside the solid are automatically satisfied,

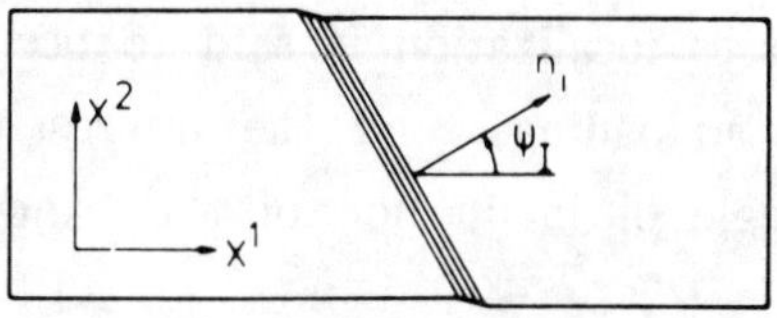


Fig. 9. Plastic flow localization in a uniformly strained solid.

apart from the necessary conditions at the band interface. These conditions are

$$u^b_{i,j} = u^o_{i,j} + c_i n_j \tag{5.1}$$

$$(T^i)^b = (T^i)^o \tag{5.2}$$

where $c_i$ are parameters to be determined, while $T^i$ are the nominal tractions on the interface.

The incremental form of (5.2) can be written as (see (2.11))

$$\left[(\dot{\tau}^{ij} + \dot{\tau}^{kj}u^i_{;k} + \tau^{kj}\dot{u}^i_{;k})n_j\right]^b = \left[(\dot{\tau}^{ij} + \dot{\tau}^{kj}u^i_{;k} + \tau^{kj}\dot{u}^i_{;k})n_j\right]^o \tag{5.3}$$

When the constitutive relations (2.16) are substituted herein, using the incremental form of (5.1), the result is the following set of incremental algebraic equations for the unknown parameters $c_\ell$

$$(A^{iq\ell})_b\, n_q \dot{c}_\ell = \left[(A^{iq\ell})_o - (A^{iq\ell})_b\right] \dot{u}^o_{\ell,q} \tag{5.4}$$

where $A^{iq\ell}$ denotes the expression

$$A^{iq\ell} = \left[L^{kjpq}(g^i_k + u^i_{,k})(g^\ell_p + u^\ell_{,p}) + \tau^{qj} g^{\ell i}\right] n_j \tag{5.5}$$

If there is an initial inhomogeneity inside the band, such as a lower yield strength or a higher degree of damage, the incremental equations (5.4) are inhomogeneous, since

$(A^{iq\ell})_b$ differs from $(A^{iq\ell})_o$, and thus the equations describe the gradual evolution of localization. In such cases localization is said to occur when straining stops outside the band (elastic unloading), and the interest is usually focussed on determining the initial angle of inclination of the band, $\psi_I$, that gives first localization.

If there is no initial inhomogeneity, equations (5.4) are homogeneous, so that a non–trivial solution for $\dot{c}_\ell$ can only occur at a bifurcation point. The first such bifurcation into a shear band mode coincides with the loss of ellipticity of the equations governing incremental equilibrium, and the corresponding critical band inclination is that of the characteristics (Hill [46], Rice [47]). Thus, the analysis of this material instability follows the theoretical framework due to Hadamard [48].

The classical elastic–plastic solid with a smooth yield surface and normality of the plastic flow rule is very resistant to localization, unless the strain hardening level is very low. However, several investigations of shear localization have shown that localization at more realistic strain levels can be found, when deviations from the classical material model are accounted for, such as plastic dilatation, non–normality of the plastic flow rule, or the formation of a vertex on subsequent yield surfaces. Fig. 10 shows an example of localization predictions for a material that forms a vertex on the yield surface, with an inhomogeneity in the form of a lower initial yield stress inside the band [49]. The initial post–bifurcation behaviour is stable, analogous to $\lambda_1 > 0$ in (3.6). For small imperfections it is found that the localized flow saturates in this case, while larger imperfections give failure by the development of large strains inside the band.

It is noted that in a study as that illustrated in Fig. 8 the assumption is made that the state of deformation outside the band remains uniform, even though the critical strain $\epsilon_c$ for bifurcation into a shear band is exceeded. In reality, shear bands would also start to form in the outer field when the critical strain $\epsilon_c$ is exceeded, but based on solutions like that of Fig. 10 it is expected that such shear bands would also saturate. However, a full understanding of this behaviour would require a complete solution of the nonlinear field equations in the range where ellipticity is lost, and such solutions are not yet available.

The model problem described here is an important tool for the understanding of finite strain behaviour. The incremental solution of the three simultaneous

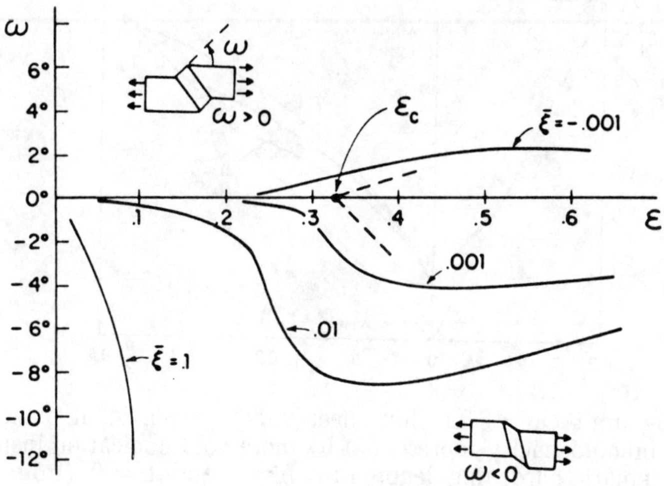


Fig. 10. Growth of a shear band in a solid that develops a vertex on the yield surface. Dashed lines show initial post–bifurcation slopes for a homogeneous solid, and the imperfection amplitude $\bar{\xi}$ represents a material inhomogeneity (from [49]).

algebraic equations (5.4) for $c_\ell$ is so simple that parameter studies are feasible, e.g. regarding the effect of stress state on loss of ellipticity, the degree of imperfection sensitivity, and the effect of the choice of constitutive equations. In fact, such model studies should be recommended prior to any elaborate numerical study for non–uniformly strained solids, in which plastic flow localization could become an issue.

In metal forming problems involving deep drawing the onset of localized necking in thin sheets is an important failure mode that limits the sheet metal formability. In terms of three dimensional theory bifurcation occurs into a diffuse mode that should be analysed based on eqs. (3.1)–(3.6); but often a plane stress formulation is used as a simplified model. When a plane state is assumed both inside and outside the band, sheet necking is the plane stress analog of the shear localization problem described above. Here, equations (5.1) and (5.2) are replaced by

$$u^b_{\alpha,\beta} = u^o_{\alpha,\beta} + c_\alpha n_\beta \tag{5.6}$$

$$h^b (T^\alpha)^b = h^o (T^\alpha)^o \tag{5.7}$$

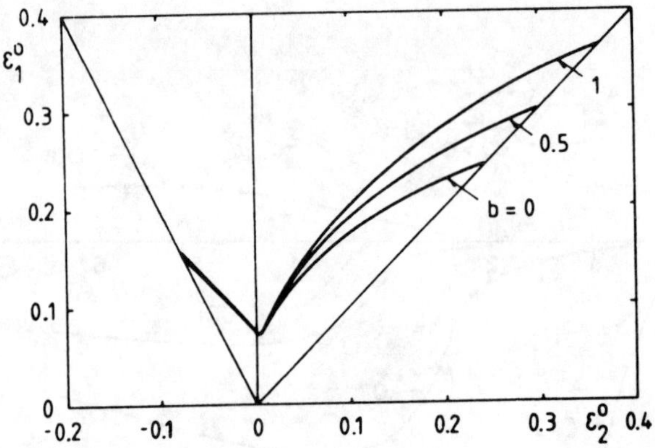


Fig. 11. Forming limit curves for thin sheet under proportional biaxial straining, with a material inhomogeneity represented by more void nucleation inside the band. Isotropic and kinematic hardening denoted by $b = 1$ and $b = 0$ (from [51]).

where Greek indices range from 1 to 2 , and the initial sheet thicknesses inside and outside the band are denoted by $h^b$ and $h^o$ , respectively. The incremental algebraic equations for the two unknown parameters $c_\alpha$ resulting from (5.6) and (5.7) are analogous to (5.4).

The simple plane stress (M–K) analysis for the effect of a thickness inhomogeneity on the onset of localized necking in biaxially stretched sheets was first introduced by Marciniak and Kuczynski [50], while the strong effect of the constitutive model was shown by Støren and Rice [34]. In the sheet metal forming literature results are presented in terms of a forming limit diagram, which shows the principal logarithmic strains $\epsilon_1^o$ and $\epsilon_2^o$ outside the band at localization failure. Fig. 11 shows such a forming limit diagram, which illustrates the effect of the yield surface curvature ($b = 0$ and $b = 1$ denote kinematic and isotropic hardening, respectively) for a power hardening material with $N = 0.1$ [51]. A porous ductile material model has been used in Fig. 11, with no initial inhomogeneity, but more void nucleation inside the band. The increased yield surface curvature ($b = 0$) has the effect of a rounded vertex, which tends to reduce the strains at failure. As in most forming limit diagrams the curves in Fig. 11 correspond to proportional straining; but it should be emphasized that deviations from this assumption have a strong effect on localization.

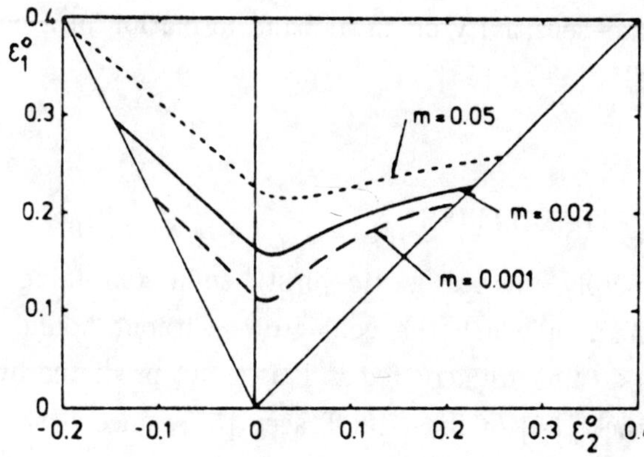


Fig. 12. Forming limit curves for thin sheet under proportional biaxial straining, for various rate hardening exponents. The same material inhomogeneity is used in all cases, represented by more void nucleation inside the band than outside (from [52]).

In the case of an elastic–viscoplastic material the equations (5.1) and (5.2) for localization in a shear band and the equations (5.6) and (5.7) for sheet necking are still valid. Also the incremental algebraic equation (5.4) for $c_\ell$ retains its form, apart from two additional terms on the right–hand side that result from the viscous terms $\dot{\tau}_*^{ij}$ in (2.18). Since the elastic moduli $L_*^{ijkl}$ replace the instantaneous moduli in (5.5), it is clear that bifurcation into a localized mode is entirely governed by elasticity. Thus, bifurcation is not predicted at realistic stress levels, and therefore localization in visco–plastic solids relies on the gradual amplification of initial inhomogeneities, analogous to the behaviour discussed in Section 4 for diffuse bifurcation modes.

When the stress dependence of the viscous part of the strain rate follows a power law analogous to (4.1), with a rate hardening exponent $m$, the predictions for the corresponding time–independent plastic solid appear in the limit $m \to 0$. Relative to these predictions, viscosity tends to delay the onset of localization, both in the case of shear bands and sheet necking. This delay is illustrated by the forming limit diagram in Fig. 12, for a power hardening material with $N = 0.2$ [52]. As in Fig. 11, a porous ductile material model has been used, with more void nucleation inside the band than outside. It is interesting that the delay found for a given value of the rate hardening exponent $m$ is essentially independent of the strain–rate applied, as was also found by Hutchinson and Neale [40]. Similar results for the effect of material rate–sensitivity on shear band formation in a porous ductile solid have been found by Pan *et al.* [53].

## 6. CAVITATION INSTABILITIES

For an infinite, remotely stressed elastic–plastic solid containing an isolated void a state may be reached, in which the void grows without bound, even though the remote stresses and strains are kept fixed. This is not predicted by the rigid–plastic analyses of McClintock [54] or Rice and Tracey [55], since elasticity is neglected in these investigations, and it is the release of elastic stored energy in the surrounding material that drives the void expansion at a cavitation instability.

For an elastic–plastic material under pure hydrostatic tension the existence of cavitation instabilities was early recognized by Bishop, Hill and Mott [56]. Recently, much interest has been devoted to such instabilities in the context of nonlinear elasticity (Ball [8], Horgan and Abeyaratne [9]), still with focus on the spherically symmetric problem, or on the analogous problem of a cylindrical void under plane strain axisymmetric conditions. In these nonlinear elastic investigations the occurrence of a cavitation instability has been interpreted either as a bifurcation from a homogeneously stressed solid to a solid containing a void, or as the growth of a preexisting void.

For a spherical void in an incompressible elastic–plastic solid under hydrostatic tension spherical symmetries apply, and the analysis given by Huang, Hutchinson and Tvergaard [10] is relatively simple. A brief version of this analysis will be given here, to introduce the basic effect of strain hardening on cavitation instabilities.

The uniaxial true stress vs. logarithmic strain relationship is taken to be of the form $\sigma/\sigma_y = f(\epsilon)$, where $\sigma_y$ is the initial yield stress. When $R_i$ and $R_0$ denote the initial and current void radii, respectively, the logarithmic radial and hoop strains of the material element at the current radius $R$ are found by incompressibility as

$$\epsilon_R = -2\epsilon_\theta = \frac{2}{3}\ln\left[1 - \left(R_0^3 - R_i^3\right)/R^3\right] \tag{6.1}$$

Expressed in terms of the true stress components $\sigma_R$ and $\sigma_\theta$ in the current configuration at radius R the equation of equilibrium is

$$\frac{d\sigma_R}{dR} + \frac{2}{R}(\sigma_R - \sigma_\theta) = 0 \tag{6.2}$$

with the boundary conditions

$$\sigma_R = 0 \ , \ R = R_0 \ \text{ and } \ \sigma_R \to \sigma^\infty \ \text{ as } \ R \to \infty \tag{6.3}$$

Integration of (6.2), using (6.3), gives an expression for the remote stress

$$\sigma^\infty = -2\int_{R_0}^{\infty} (\sigma_R - \sigma_\theta)\frac{dR}{R} \tag{6.4}$$

Due to incompressibility the strains are not affected by superposing a hydrostatic pressure $\sigma_\theta$ on the stress state $(\sigma_R, \sigma_\theta, \sigma_\theta)$, leading to a uniaxial stress $\sigma_R - \sigma_\theta$, and therefore $(\sigma_R - \sigma_\theta)/\sigma_y = f(\epsilon_R)$. Substituting this into (6.4) and using (6.1) gives the expression

$$\frac{\sigma^\infty}{\sigma_y} = -2\int_1^\infty f\left[\frac{2}{3}\ln\left\{1 - \frac{1-(R_i/R_0)^3}{\eta^3}\right\}\right]\frac{d\eta}{\eta} \tag{6.5}$$

The cavitation limit stress $\sigma_c^\infty$ is obtained from (6.5) by letting $R_0/R_i \to \infty$

$$\frac{\sigma_c^\infty}{\sigma_y} = -2\int_1^\infty f\left[\frac{2}{3}\ln\left\{1 - \eta^{-3}\right\}\right]\frac{d\eta}{\eta} = -\int_0^\infty \left[e^{\frac{3}{2}\xi} - 1\right]^{-1} f(-\xi)d\xi \tag{6.6}$$

For a power hardening solid the uniaxial stress–strain law is

$$\frac{\sigma}{\sigma_y} = f(\epsilon) = \begin{cases} \epsilon/\epsilon_y & , \ |\epsilon| \le \epsilon_y \\ \text{sign}(\epsilon)\left\{|\epsilon|/\epsilon_y\right\}^N & , \ |\epsilon| > \epsilon_y \end{cases} \tag{6.7}$$

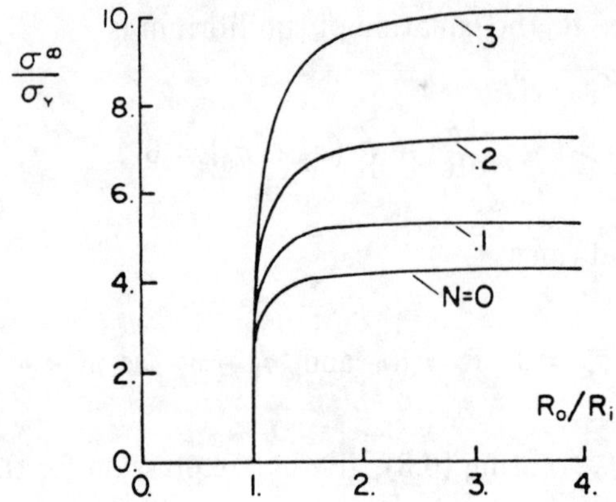


Fig. 13. Remote stress vs. cavity radius for power hardening incompressible solids with $\sigma_Y/E = 0.003$, subject to spherically symmetric loading (from [10]).

where $\epsilon_Y = \sigma_Y/E$. For this stress–strain behaviour, with $N = 0$, $0.1$, $0.2$ or $0.3$, Fig. 13 shows the development of $\sigma^\infty/\sigma_Y$ vs. $R_0/R_i$, corresponding to $\epsilon_Y = 0.003$. The cavitation instability stresses are approached asymptotically by these curves, and it is seen that the critical stress increases significantly with strain hardening. Values of $\sigma_c^\infty$ computed by Huang *et al.* [10] for different values of $\epsilon_Y$ show that $\sigma_c^\infty$ decreases with increasing $\epsilon_Y$.

A cavitation instability type of failure mode has been observed in experiments of Ashby, Blunt and Bannister [57] for a lead wire bonded to a glass matrix. These experiments were designed to obtain insight into the role of ductile reinforcements in toughening brittle matrix materials by a crack bridging mechanism. In the ductile wires highly constrained plastic flow leads to high stress levels at failure, and the specimens failed by an enormous enlargement of a single void in the wire near the crack plane. Since the stress state is not spherically symmetric in the wire, these experimental results motivated an interest in the possibility of cavitation instabilities under other remote stress states.

Huang, Hutchinson and Tvergaard [10] have analysed an elastic–perfectly plastic material under a remote axisymmetric stress–state, with axial tensile stress S and transverse stresses T, and with an initially spherical void. The analyses were carried out by coupling a closed form approximate solution for an outer region with

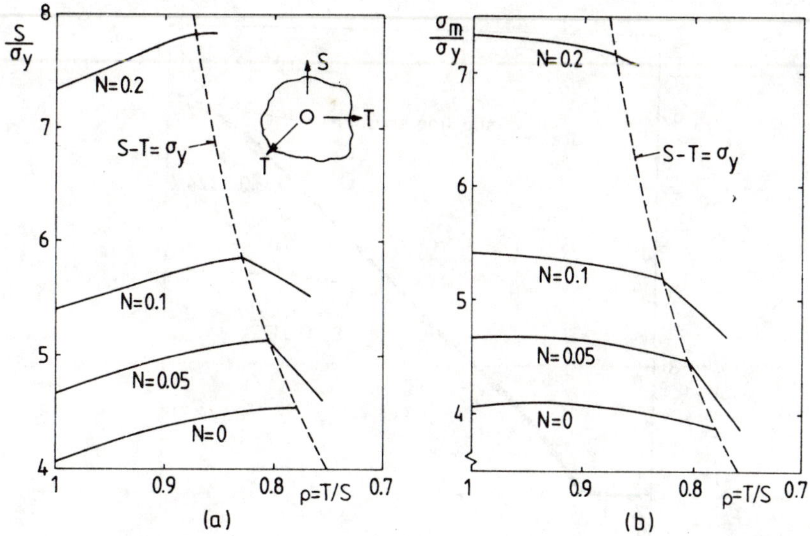


Fig. 14. Cavitation limits for a spherical void in elastic–plastic power hardening solid subject to remote axisymmetric stressing ($\sigma_y/E = 0.003$ and $\nu = 0.3$). (a) Axial tensile stress. (b) Mean stress. (from [11]).

a full finite strain finite element solution for an inner region, using a Lagrange multiplier method. It is found that cavitation instabilities occur at high axial tensile stresses S, or mean stresses $\sigma_m$, of the order of $4\sigma_y$ to $4.5\sigma_y$. The instabilities are found for a range of stress ratios T/S, ending at the value where remote plasticity sets in, i.e. where $S - T = \sigma_y$.

The investigation of Huang *et al.* [10] has been extended by Tvergaard, Huang and Hutchinson [11] to consider axisymmetric cavitation states for a power hardening elastic–plastic material. The ductile reinforcing wire used in the experiments by Ashby *et al.* [57] was made of lead, which did show a significant amount of deformation hardening, and the measured peak stresses were significantly above the critical levels corresponding to elastic–perfectly plastic solids. Fig. 14 shows the cavitation limit $S/\sigma_y$ and $\sigma_m/\sigma_y$ as a function of the remote stress ratio, $\rho = T/S$, for the elastic–perfectly plastic solid, $N = 0$, and for three different values of the power hardening exponent, $N = 0.05$, $N = 0.1$ and $N = 0.2$. The other material parameters are $\sigma_y/E = 0.003$ and $\nu = 0.3$. For spherically symmetric conditions ($\rho = 1$) the critical values of $S/\sigma_y$ found in Fig. 14a are slightly lower than those found in Fig. 13 (of the order of 5 pct.), because

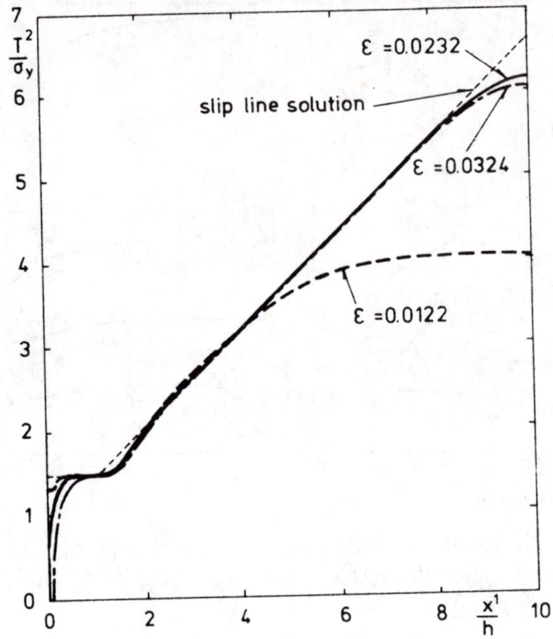


Fig. 15. Distribution of normal traction $T^2$ at metal/ceramic interface for a metal bond layer with aspect ratio $w/h = 10$ and $\sigma_y/E = 0.003$ (from [59]).

elastic compressibility is accounted for in Fig. 14a. The dashed curves in Fig. 14 indicate the onset of remote yielding.

It is noted that the high stress levels needed for the onset of cavitation instabilities are not reached in usual structural components. Even in front of a blunting crack tip in an elastic–perfectly plastic material (McMeeking [58]) the peak stress levels reached are well below the levels required according to Fig. 14. However, bonding with a material that remains elastic, as in the experiments of Ashby, Blunt and Bannister [57], gives rise to constraints that significantly increase the stress levels. Recently, the plastic deformations have been analysed for a metal bond layer between two ceramics, with the metal represented as elastic–perfectly plastic (Tvergaard [59]). For a material with the same initial yield strain, $\sigma_y/E = 0.003$, as that in Fig. 14 the distribution of normal tractions $T^2$ on the metal/ceramic interface are shown in Fig. 15 for various levels of the average strain

$\epsilon$ across the bond layer. It is seen that here plastic flow is also so constrained that the stress levels required for the onset of a cavitation instability, according to Fig. 14, are reached. Also the growth of a void at the metal/ceramic interface has been analysed by Tvergaard [59], and it has been found that unstable growth of such cavities requires higher stresses that than for a cavitation instability in an infinite solid.

It is noted that the analyses of cavitation instabilities have not here been interpreted as growth from an infinitely small void (bifurcation from a uniformly stressed state with no void), but rather as the growth of an initial hole to an unstable state where growth will continue without bound, driven by the elastic energy stored in the infinite medium. The latter interpretation is preferred because an infinitely small void in a metal is hard to visualize. But the unstable growth driven by stored elastic energy is quite analogous to other tensile instabilities. Thus, a long thin tensile test specimen (a piano string), will show unstable neck development, even in a rigid testing machine, due to the relatively large amount of stored elastic energy in the string outside the neck region.

## REFERENCES


1. Hutchinson, J.W.: Plastic buckling, Advan. Appl. Mech., 14 (1974), 67–144.
2. Needleman, A. and Tvergaard, V.: Aspects of plastic post–buckling behaviour, Mechanics of Solids, The Rodney Hill 60th Anniversary Volume (eds. H.G. Hopkins and M.J. Sewell), Pergamon Press, Oxford, (1982), 453–498.
3. Shanley, F.R.: Inelastic column theory, J. Aeronaut. Sci., 14 (1947), 261–267.
4. Hill, R.: A general theory of uniqueness and stability in elastic–plastic solids, J. Mech. Phys. Solids, 6 (1958), 236–249.
5. Hill, R.: Bifurcation and uniqueness in nonlinear mechanics of continua, Problems of Continuum Mechanics, Soc. Ind. Appl. Math., Philadelphia, Pennsylvania, (1961), 155–164.
6. Hill, R.: Aspects of invariance in solid mechanics, Advan. Appl. Mech., 18 (1978), 1–75.
7. Tvergaard, V.: Plasticity and creep at finite strains, in Theoretical and Applied Mechanics (eds. P. Germain *et al.*), (1989), 349–368.

8. Ball, J.M.: Discontinuous equilibrium solutions and cavitation in nonlinear elasticity, Phil. Trans. R. Soc. Lond., A306 (1982), 557–610.

9. Horgan, C.O. and Abeyaratne, R.: A bifurcation problem for a compressible nonlinearly elastic medium: growth of a microvoid, J. of Elasticity, 16 (1986), 189–200.

10. Huang, Y., Hutchinson, J.W. and Tvergaard, V.: Cavitation instabilities in elastic–plastic solids, J. Mech. Phys. Solids, 39 (1991), 223–241.

11. Tvergaard, V., Huang, Y. and Hutchinson, J.W.: Cavitation instabilities in a power hardening elastic–plastic solid, Danish Center for Appl. Math. and Mech., Report No. 415 (1990).

12. Green, A.E. and Zerna, W.: Theoretical elasticity, Oxford University Press, Oxford (1968).

13. Budiansky, B.: Remarks on theories of solid and structural mechanics, in Problems of Hydrodynamics and Continuum Mechanics (eds. M.A. Lavrent'ev *et al.*), SIAM, Philadelphia (1969), 77–83.

14. Hutchinson, J.W.: Finite strain analysis of elastic–plastic solids and structures, in Numerical Solution of Nonlinear Structural Problems (ed. R.F. Hartung), ASME, New York (1973), 17.

15. McMeeking, R.M. and Rice, J.R.: Finite–element formulations for problems of large elastic–plastic deformation, Int. J. Solids Structures, 11 (1975), 601–616.

16. Hutchinson, J.W.: Postbifurcation behavior in the plastic range, J. Mech. Phys. Solids, 21 (1973), 163–190.

17. Rudnicki, J.W. and Rice, J.R.: Conditions for localization of deformation in pressure–sensitive dilatant materials, J. Mech. Phys. Solids, 23 (1975), 371–394.

18. Tvergaard, V.: Influence of void nucleation on ductile shear fracture at a free surface, J. Mech. Phys. Solids, 30 (1982), 399–425.

19. Raniecki, B. and Bruhns, O.T.: Bounds to bifurcation stresses in solids with non–associated plastic flow law at finite strain, J. Mech. Phys. Solids, 29 (1981), 153–172.

20. Hill, R.: Generalized constitutive relations for incremental deformation of metal crystals by multislip, J. Mech. Phys. Solids, 14 (1966), 95–102.

21. Hutchinson, J.W.: Elastic–plastic behavior of polycrystalline metals and composites, Proc. Roy. Soc. London, A318 (1970), 247–272.

22. Sewell, M.J.: A survey of plastic buckling, in Stability (ed. H. Leipholz), University of Waterloo Press (1972), 85.

23. Christoffersen, J. and Hutchinson, J.W.: A class of phenomenological corner theories of plasticity, J. Mech. Phys. Solids, 27 (1979), 465–487.

24. Tvergaard, V.: Plastic buckling of axially compressed circular cylindrical shells, Int. J. Thin–Walled Struct., 1 (1983), 139–163.

25. Nguyen, S.Q. and Triantafyllidis, N.: Plastic bifurcation and postbifurcation analysis for generalized standard continua, Ecole Polytechnique, Palaiseau, France (1988).

26. Considere, M.: Annales des Ponts et Chaussees, 9 (1885), 574.

27. Hutchinson, J.W. and Miles, J.P.: Bifurcation analysis of the onset of necking in an elastic–plastic cylinder under uniaxial tension, J. Mech. Phys. Solids, 22 (1974), 61.

28. Tvergaard, V.: Bifurcation in elastic–plastic tubes under internal pressure, European Journal of Mechanics, A/Solids, 9 (1990), 21–35.

29. Larsson, M., Needleman, A., Tvergaard, V. and Storåkers, B.: Instability and failure of internally pressurized ductile metal cylinders, J. Mech. Phys. Solids, 30 (1982), 121–154.

30. Sewell, M.J.: On the calculation of potential functions defined on curved boundaries, Proc. R. Soc. London, A286 (1965), 402–411.

31. Storåkers, B.: Bifurcation and instability modes in thick–walled rigid–plastic cylinders under pressure, J. Mech. Phys. Solids, 19 (1971), 339–351.

32. Strifors, H. and Storåkers, B.: Uniqueness and stability at finite–deformation of rigid plastic thick–walled cylinders under hydrostatic pressure, in Foundations of Plasticity (ed. A. Sawczuk), Nordhoff, Leiden, 1 (1973), 327.

33. Chu, C.–C.: Bifurcation of elastic–plastic circular cylindrical shells under internal pressure, J. Appl. Mech., 46 (1979), 889–894.

34. Støren, S. and Rice, J.R.: Localized necking in thin sheets, J. Mech. Phys. Solids, 23 (1975), 421–441.

35. Needleman, A. and Tvergaard, V.: Necking of biaxially stretched elastic–plastic circular plates, J. Mech. Phys. Solids, 25 (1977), 159–183.

36. Tvergaard, V. and Needleman, A.: On the localization of buckling patterns, J. Appl. Mech., 47 (1980), 613–619.

37. Tvergaard, V.: On the burst strength and necking behaviour of rotating disks, Int. J. Mech. Sci., 20 (1978), 109–120.

38. Percy, M.J., Ball, K. and Mellor, P.B.: An experimental study of the burst strength of rotating disks, Int. J. Mech. Sci., 16 (1974), 809–817.

39. Hutchinson, J.W. and Neale, K.W.: Influence of strain–rate sensitivity on necking under uniaxial tension, Acta Metallurgica, 25 (1977), 839–846.

40. Hutchinson, J.W. and Neale, K.W.: Sheet necking – III. Strain rate effects, in Mechanics of Sheet Metal Forming (eds. D.P. Koistinen and N.–M. Wang), Plenum Publ. Corp., New York (1978), 111–126.

41. Peirce, D., Asaro, R.J. and Needleman, A.: Material rate dependence and localized deformation in crystalline solids, Acta Metallurgica, 31 (1983), 1951–1976.

42. Tvergaard, V.: Rate–sensitivity in elastic–plastic panel buckling, in Aspects of the Analysis of Plate Structures, A volume in honour of W.H. Wittrick (eds. D.J. Dawe *et al.*), Clarendon Press, Oxford (1985), 293–308.

43. Bodner, S.R., Naveh, M. and Merzer, A.M.: Deformation and buckling of axisymmetric viscoplastic shells under thermomechanical loading, Int. J. Solids Structures, 27 (1991), 1915–1924.

44. Obrecht, H.: Creep buckling and postbuckling of circular cylindrical shells under axial compression, Int. J. Solids Structures, 13 (1977), 337–355.

45. Hoff, N.J.: Creep buckling of plates and shells, in Proc. 13th Int. Congr. Appl. Mech. (eds. E. Becker and G.K. Mikhailov), Springer–Verlag (1973), 124–140.

46. Hill, R.: Acceleration waves in solids, J. Mech. Phys. Solids, 10 (1962), 1–16.

47. Rice, J.R.: The localization of plastic deformation, in Theoretical and Appl. Mech. (ed. W.T. Koiter), North–Holland (1977), 207–220.

48. Hadamard, J.: Lecons sur la propagation des ondes et les équations de l'hydrodynamique, Libraire Scientifique A, Hermann, Paris (1903).

49. Hutchinson, J.W. and Tvergaard, V.: Shear band formation in plane strain, Int. J. Solids Structures, 17 (1981), 451–470.

50. Marciniak, K. and Kuczynski, K.: Limit strains in the process of stretch forming sheet metal, Int. J. Mech. Sci., 9 (1967), 609–620.

51. Tvergaard, V.: Effect of yield surface curvature and void nucleation on plastic flow localization, J. Mech. Phys. Solids, 35 (1987), 43–60.

52. Needleman, A. and Tvergaard. V.: Limits to formability in rate–sensitive metal sheets, in Mechanical Behavior of Materials – IV (eds. J. Carlsson and N.G. Ohlson), Pergamon Press, Oxford (1984), 51–65.

53. Pan, J., Saje, M. and Needleman, A.: Localization of deformation in rate sensitive porous solids, Int. J. Fracture, 21 (1983), 261–278.